ED LABINOWICZ

LEARNING FROM CHILDREN

NEW BEGINNINGS FOR TEACHING NUMERICAL THINKING

A Piagetian Approach

Illustrations by
TERENCE D. WILSON and MARY JORGENSEN

ADDISON-WESLEY PUBLISHING COMPANY

Menlo Park, California • Reading, Massachusetts • London • Amsterdam • Don Mills, Ontario • Sydney

To teachers willing to learn from children;

To children, who have so much to teach us about thinking and learning.

This book is published by the Addison-Wesley Innovative Division.

Copyediting: **Lorraine Anderson**
Design and production: **Marvin R. Warshaw**
Composition: **Etc. Graphics**

ISBN-0-201-20321-9
BCDEFG-ML-89876

CONTENTS

PREFACE

THE NEED FOR THIS BOOK

Mathematics is a remarkable human achievement. Yet in recent years there has been a trend toward first reducing it to computation and then presenting it as isolated facts and procedures, in an attempt to simplify it. Rather than deal with the complexity of mathematics, teachers in the back-to-basics movement are expected to bypass it. At the same time, the advent of calculators and computers is bringing into focus the computational needs of a rapidly changing society. Although computational abilities are still important, it is now obvious that obtaining right answers by single standard paper-and-pencil methods is not sufficient. In today's society we need the flexibility of choosing among available resources—calculator or computer, paper and pencil, or mental calculation methods—and the power of judging the reasonableness of answers through estimation or cross-checking with alternative methods. Furthermore, decision-making ability is required to apply correct computational procedures to problems not previously experienced. Since these abilities require a flexible understanding of the numeration system and operations, the current emphasis on teaching computation as isolated facts and procedures leads nowhere. Limiting children's mathematical knowledge to memory alone is, according to O'Brien (1977b), "to cheat them and mislead them, if not cripple them."

In recent years there also has been slow but continuous growth in research exploring the mathematical thinking process of children up to age eight. This research has culminated in a virtual explosion of knowledge about children's numerical thinking. Interview methods that explore children's reasoning regardless of the correctness of their answers have gleaned a wealth of knowledge concerning both the capacities of children for numerical thinking and their constraints. Many of these capacities go unnoticed and unappreciated in the classroom. For example, children come to school with surprising counting abilities developed largely through their own activity. Similarly, constraints in children's thinking in relation to the demands of the school curriculum largely go unnoticed. These research findings from listening to children can help teachers redefine what is basic in children's education and teach congruently with children's natural learning processes.

We currently know enough about children's numerical thinking to drastically revise the arithmetic program for primary-grade children. Yet there is a glaring disparity between what we know about children together with the level of computational ability required for survival in the real world, and what takes place in the classroom. Although recent research on children's numerical thinking has much to say to teachers, little attempt has been made to communicate it directly to them. This book attempts to build a bridge between theory, research, and classroom practice with the aim of supporting informed change in the schools. The ultimate goal is to promote children's numerical thinking and to ensure their survival in a rapidly changing world.

FEATURES OF THE BOOK

In recent years research in mathematics education has shifted its focus from the superficiality of large group comparison to both the complexities of children's thinking and learning and the complexities of the teaching-learning process. This research is characterized by clinical interviews of individual children and microscopic observation of the teaching-learning process through experiments with small groups of children. Such is the focus of the research described in this book.

The research reported here consisted of listening to and videotaping primary-grade children in interactions on number tasks during one full year. The following table summarizes the year's research activities.

SCHEDULE	TYPE OF STUDY	GRADE	SOCIOECONOMIC STATUS	AGE	NUMBER OF CHILDREN	FOCUS OF STUDY	LOCATION IN BOOK
Sept.	pilot interviews	2 & 3	upper middle	7–8	12	assorted tasks	—
Oct. to Jan.	series of 3 interviews	2	upper middle	7	30	counting methods non-counting methods equality sentences number story problems	Ch. 4 5 7 8
Oct. to Feb.	series of 3 interviews	3	upper middle	8	29	meaning of place value models counting by tens place value concepts addition w regrouping subtraction w regrouping	1, 10 10 11 12 13
Mar. to April	teaching-learning experiment	2	upper middle	7½	4	equality sentences	7, 9
May	single interview	1	lower middle	6½	30	equality sentences	6

Rather than merely replicating Piaget's classical tasks, the research adapted his open method of exploration—the clinical interview method—to the study of school-related tasks. The videotapes contained such a wealth of information on children's thinking, the interview process itself, and the teaching-learning process, that it has taken nearly three additional years to replay and analyze the tapes, receive feedback from others, attempt to communicate the results in early drafts of the manuscript, and revise it based on critical reviews by students, teachers, researchers, and mathematics educators. My interaction with the various sources of feedback has provoked rethinking that has ultimately strengthened this work. At no time in my adult life have I learned so much as in my nearly four years of intensive study, research, and writing about thinking, learning, and teaching. The children have been exceptional teachers. This resulting book has a number of features that I hope will make it uniquely useful to the reader.

Involves The Reader as Researcher

Rather than presenting only the results of the research, this book involves the reader in the process of doing research and in considering its implications for the classroom. Stop-action graphics from videotapes of my interviews illustrate numerous interactions with children in subtle detail. The diversity of responses for the same tasks are placed side by side, illustrating not only the products of children's thinking but also the processes involved. Numerous episodes of children's thinking in process serve as springboards for productive discussion. By attempting to follow children's thinking, the reader gains a better understanding of the numerical concepts underlying primary school arithmetic, and of the intellectual demands placed on children. Since the graphics also illustrate ways of interacting with children that facilitate their thinking and probe the depth of their understanding, numerous opportunities are provided for the reader to reexamine existing notions of teaching and learning. Many behind-the-scenes perspectives are provided—for example, the interviewer's reasons for asking a certain question at a particular point in the interaction. Some of the explorations raise more questions than they answer. Such unanswered questions are highlighted. Similarly, limitations in the interviewer's abilities in specific situations are discussed openly. The teacher as researcher is a theme running through the book as the reader is invited to conduct her or his own explorations of children's thinking, while improving on the tasks presented. Although the book's approach is informal, its intent is serious—to provoke the reader to interact with numerical ideas, children's thinking, and the complexities of the teaching-learning process.

Provides an Integrated Summary of Contemporary Research on Children's Numerical Thinking.

Although the book focuses primarily on my videotaped research of children's number ideas, it also serves as a vehicle for integrating the findings of diverse research projects. This integration is facilitated by the replication of other studies or the adaptation of tasks from these sources in my own work. Also, in areas I did not study personally, summaries of research bridge the gap to present a more complete picture of children's numerical thinking.

Develops a Theoretical Framework for the Interpretation of Children's Responses.

Selected interview episodes and personal experiences of the reader are used as springboards for an introduction to Piaget's theory of intellectual development in the first chapter. This theoretical framework forms the basis of interpretations of children's numerical behaviors throughout the book. From this framework the child is viewed as an active learner—not merely absorbing information passively, but interpreting selectively and constructing his own meaning. Furthermore, the child is viewed as initiating interactions and generating his own knowledge. At the same time that numerous examples of children's numerical behaviors are explained and predicted by Piaget's theory, they in turn serve to clarify the theory. No prior understanding of Piaget's theory is assumed.

Weaves Theory and Research About Children's Thinking Into Classroom Practice.

Discussions of children's thinking are liberally sprinkled with issues concerning the teaching-learning process to provoke a rethinking of existing notions. Guidelines for teaching are abstracted directly from knowledge of children's numerical thinking and are discussed at the ends of chapters under the heading "Reflections and Directions." These guidelines for teaching include extensions of the interviewer behaviors in Piaget's clinical method, already supported by classroom research (Rowe, 1973). Some of these guidelines for teaching are demonstrated and analyzed in explicit detail in the report of my teaching-learning experiment.

Throughout the book, two viable alternatives to current methods of teaching computation are presented. One alternative involves the careful design of developmental learning sequences for standard methods of computation, based on observed levels of children's numerical thinking. Another alternative facilitates children's invention of nonstandard methods. Although the latter alternative requires more creativity and spontaneity on the part of the teacher, both demand considerable understanding of children's thinking and how it develops, together with an ability to interact with children and provoke rethinking at higher levels of numerical relations. Frameworks for comparing alternative teaching methods are found throughout the book. These comparisons of applications of theory, in turn, serve to clarify the theory itself.

Encourages Teachers to Develop a Coherent View of Education to Support Undertaking the New Role of Change Agent.

The book views not only children but also teachers as active learners. Teachers are seen as capable of independent thinking and innovation, like children. Although this theme is communicated throughout the book, it is clearly brought into focus in the last chapter. Teachers require a coherent view of education, not only to guide their day-to-day functioning with children, but to critically examine educational materials, to resist pressures to conform to questionable practices, and to initiate change in their classrooms based on knowledge of children's numerical thinking. The last chapter suggests a variety of possibilities for undertaking professional development and initiating informed change in the schools. The potential for positive change through collaborative efforts of teachers is highlighted in this chapter. A primary aim of this book is to contribute to the development of teachers as self-sustaining professionals.

AUDIENCE AND USE OF THIS BOOK

Although this book was written primarily with teachers in mind, I believe that it can be a valuable handbook for anyone interested in children's numerical thinking. This multiple audience includes developmental psychologists, researchers, mathematics educators, supervisors and curriculum developers, legislators and parents. The book contains a wealth of explorations, information, and ideas that can stimulate interaction and invite replication, adaptation, and elaboration, and ultimately initiation of informed change.

Although written for teachers, the book has not been designed with any particular course in mind. Rather than fitting into any existing molds, the book has been designed to break them. Its use need not be restricted to the structure of a university course, which comes to an end after forty-five semester hours and exists only when course credit is given. The book can be used for group study by teachers with a common interest in children's thinking that brings them together for regular meetings. It's a book that can be returned to at many different levels of understanding—when things previously unnoticed come into focus. The development of teacher autonomy is not a short-range concern. It cannot be scheduled within a fifteen-week period or accomplished by means of a single reading of a book.

Drafts of the manuscript have been pilot-tested with in-service teachers in both a graduate-level math methods course and a math education seminar. Selected chapters, or portions thereof, have been used successfully with pre-service teachers in an undergraduate methods class. Following the thinking of children in intimate detail, grappling with the complexities of teaching and learning, and initiating informed change are neither easy tasks nor short-term concerns. Whereas some students were ready to undertake the endeavor, others resisted. Direct involvement in interview explorations of their own increased the level of interest and commitment, as well as enlivened the class discussions. For some, these experiences prompted a reexamination of their role as teacher. The activities involved in using this book are geared toward helping the teaching profession in its quest for autonomy, but they do demand some personal struggle in stretching one's thinking and in rethinking.

A NOTE ON STYLE AND METHOD

Throughout the book the pronoun *he* is used in the traditional manner when discussing a hypothetical child, or a child in a general sense (as opposed to a particular child from the illustrated episodes). This form is used for convenience of communication and is not meant to suggest that boys are somehow more important than girls. Selection of children for illustration and discussion was based on an attempt to present examples of the range of responses observed, to present the clearest example in making a point, and to provide the most interesting example for discussion. When similar responses were available from different children, consideration was given to a balanced presentation of sexes and races.

ACKNOWLEDGEMENTS

Although only one person is solely responsible for the quality of the research reported here and the content of this book, much of the content has been derived from the work of others and considerable assistance has been received along the way.

First and foremost, I owe a large debt of gratitude to the late Jean Piaget, whose clinical method of inquiry and theory of intellectual development formed the basis for my own interview explorations and interpretations.

In planning the interview explorations I selected and adapted tasks from existing research, in particular the videotapes of Leslie Steffe and Larry Hatfield of the University of Georgia, and the research reports of Constance Kamii, University of Illinois, Chicago Circle; Tom Carpenter and Jim Moser, University of Wisconsin; Merlyn Behr, Northern Illinois University; and Herb Ginsburg, Rochester University.

In the schools, I am grateful to the fourteen cooperating teachers who not only collected parental permission slips, but also tolerated numerous interruptions of their classroom routines as children went in and out.

In the videotaping of the interactions with children, I acknowledge the technical assistance of the Media Center at California State University, Northridge, which included providing a video camera operator when available.

During the videotape analysis and interpretation phase of this research I received considerable feedback from numerous people as I used videotaped episodes in my teaching and carried my videotape recorder around the country to conferences and informal meetings. I am grateful to these people.

Throughout the research and writing I corresponded with other researchers currently exploring aspects of children's numerical thinking, and received prepublication reports of their work. Many of these research findings were integrated with my own in presenting a more complete picture of children's numerical thinking. In rewriting the first draft of the manuscript I made extensive use of Karen Fuson's research reviews of children's counting (Northwestern University). Similarly, Pat Thompson's dissertation (University of Georgia) on young children's concepts of whole number numeration filled in gaps and stimulated a deeper analysis of my own explorations in that area.

Many people reacted to drafts of the entire manuscript, or parts of it, enduring my rough sketches in the process. In particular I would like to thank the following people: the students at California State University, Northridge, for discussions that provided regular feedback on the clarity of communication and for reflections on their own interview experiences that enlivened the manuscript; Dean Hendrickson of the University of Minnesota, Duluth, for invaluable feedback on the first draft of the manuscript and on selection and adaptation of the interview tasks and systematic review of the videotapes; Constance Kamii of the University of Illinois, Chicago Circle, for her detailed critical analysis of early chapters and hours of related discussion that provoked much rethinking; Jane Beasley Raph, professor emeritus, Rutgers University, for her support throughout the project and her editorial assistance ranging from improvement of communication to reorganization of the manuscript; the people at the Teachers' Center Project at Southern Illinois University, Edwardsville, including directors Tom O'Brien and Shirley Casey, and teachers who provided detailed, practical written feedback on various drafts of the manuscript, sometimes after meeting to discuss their reactions; and Tom O'Brien for his continuing encouragement and support in terms of critical feedback, editorial assistance, and direct contribution of ideas through a stream of letters, phone conversations, and face-to-face meetings. A final note of appreciation goes to Shirley Marini Labinowicz, my wife, for her professional and emotional support throughout the project.

The publication of this book also represents hundreds of hours of commercial production, with a multitude of difficult decisions made at various stages. In particular, I would like to thank the following people: Lorraine Anderson, for the superb copyediting that smoothed out wrinkles in style and sharpened the clarity of communication; Terry Wilson and Mary Jorgensen, for bringing my rough sketches to life with their sensitive drawings; Marv Warshaw, for masterfully pulling the diverse features of the book together into a coherent design and keeping production on schedule; Ginger Johnson, for patiently anchoring the whole project at Addison-Wesley; and Stuart Brewster, director of the Innovative Division, for caring about the quality of public education and believing in the viability of the project.

Most of all, I owe much to the children who shared their thinking so openly in order that I might learn from them. Without their cooperation this book would not have been possible.

Los Angeles
March, 1984

ABOUT THE AUTHOR

Ed Labinowicz (LabiNOwich) began his teaching career in Canada, where he taught in public schools for several years. He studied both science and education at the University of Manitoba, did graduate work in chemistry at the University of Hawaii, and earned his Ph.D. degree in science education at Florida State University (1970). Since that time he has been teaching science and mathematics education courses at California State University, Northridge.

Intrigued by his first encounter with child watching during his doctoral research, the author has continued to study Piaget's theory of intellectual development and clinical interview method, together with their classroom applications. He is also the author of THE PIAGET PRIMER: THINKING, LEARNING, TEACHING (Addison-Wesley). As this book goes to press he is exploring the microcomputer Logo environment—both as a means of enhancing children's problem-solving capacities and as another window to children's thinking.

CREDITS

17, 18, 431

1. From *The Mechanisms of the Mind* by Edward de Bono. © Edward de Bono 1969. By permission of A. P. Watt Ltd., London, England.

31, 224, 237

2. From *Teaching Science as Continuous Inquiry* by Mary Budd Rowe. New York: McGraw-Hill Company, 1973. Reproduced by permission.

45, 150, 217, 325

3. Adapted from *Children's Arithmetic: The Learning Process* by Herbert Ginsburg, pp. 143, 113, 114. Copyright © 1977 Van Nostrand Reinhold Co., Inc.

112, 441

4. Illustration and adaptation of questions on pages 21, 23, and table on page 30 of *Thursday Math Sampler* by Robert Wirtz. Copyright © 1982 Curriculum Development Associates, Inc.

144, 145

5. "Making Sense Out of Word Problems" by Marilyn Burns and Kathy Richardson. Reprinted by special permission of LEARNING, The Magazine for Creative Teaching, © 1981 by Pitman Learning, Inc., 19 Davis Drive, Belmont, CA 94002.

226–227

6. "Piaget Takes a Teacher's Look," Reprinted by special permission of LEARNING, The Magazine for Creative Teaching, © 1973 by Pitman Learning, Inc., 19 Davis Drive, Belmont, CA 94002.

231

7. "Groups of Four" in "Organizing the Learners" from *Meeting the Challenge* by Marilyn Burns. The Learning Institute, Palo Alto, 1980.

232, 372, 374, 375, 376

8. From *Notes on Mathematics for Children* by D. W. Wheeler (ed.), 1977, Cambridge University Press. Reprinted by permission.

268

9. Adapted from the chapter "A Developmental Theory of Number Understanding," in *The Development of Mathematical Thinking* by H. Ginsburg (ed.). Copyright 1982, Academic Press.

271, 272

10. From David J. Fuys and Rosamond Welchman Tischler, *Teaching Mathematics in the Elementary School.* Copyright © 1979 David J. Fuys and Rosamond Welchman Tischler. Reprinted by permission of Little, Brown and Company.

385, 388

11. From "Making Problem Solving Come Alive in the Intermediate Grades" by Marilyn Jacobson, Frank Lester, and Arthur Stengel in *Problem Solving in Mathematics*, S. Krulick, R. Reys (eds.). Reston, VA: National Council of Teachers of Mathematics, 1980, pp. 127–135.

394

12. From *Guiding Children's Arithmetic Experiences: The Experience-Language Approach to Numbers*, by J. Allen Hickerson © 1952, renewed 1980, p. 67. Reprinted by permission of Prentice-Hall, Inc., Englewood Cliffs, N.J.

399

13. From "Forward to Basics" from *Genetic Epistemologist*, July 1977, p. 1-2, by Thomas O'Brien (1977). Reprinted by permission of the Jean Piaget Society.

405, 408

14. From *Developing Computational Skills* by Marilyn Suydam and Robert Reys (eds.). Reston, VA: National Council of Teachers of Mathematics, 1980, 1978, pp. 204, 207-08.

409, 410

15. From "'Buggy'—Outfitting for the Great Error Hunt" by Tim Barclay, *Classroom Computer News*, March/April 1982, pp. 25-27.

421, 422

16. "A Declaration of Professional Conscience for Teachers" by Kenneth Goodman. Copyright © 1980 by Kenneth Godman. Reprinted by permission.

442, 443

17. From the chapter "Development of Children's Problem Solving Ability in Arithmetic," in *The Development of Mathematical Thinking*, by H. Ginsburg (ed.). © 1982, Academic Press.

PART ONE

INTRODUCTION

1

A PIAGETIAN FRAMEWORK FOR OBSERVING CHILDREN'S NUMERICAL THINKING

PREVIEW

Many aspects of children's numerical thinking are not anticipated by teachers. Difficulties that children experience with school arithmetic often are unnoticed, unexplained, or misinterpreted. Similarly, children's successful methods of computation that do not coincide with the singular methods taught in schools often go unnoticed or unappreciated in the classroom. Both the limitations in children's numerical thinking and their surprising capacities are highlighted in this book. The explorations of children's numerical thinking to be described are guided in method and theory by the work of Jean Piaget, a Swiss psychologist and multidisciplinary scholar. Becoming acquainted with his method of listening to children will allow you to observe what previously went unnoticed in your classroom. Developing an understanding of his theory of how children learn will allow you to anticipate their behaviors on numerical tasks and to act upon them in ways likely to expand their learning.

Piaget's monumental work consisted of elaborating a theory of how we come to know our world—how we develop knowledge. This theory was based on numerous studies of children's thinking conducted over a span of more than half a century, and evolved from attempts to explain the patterns in his findings for children from infancy through adolescence. This lifelong endeavor had an unusual beginning. Early in his

career, while working as a psychologist scoring children's correct responses on an IQ test and developing test norms, Piaget became intrigued by the pattern of similar incorrect responses given by children of like ages. From a need to understand the thinking underlying children's responses, Piaget gradually developed a flexible interview method. He devised ingenious tasks involving common objects from the child's environment together with provoking questions that engaged the child in thought and conversation about his notions of space, number, time, causality, and so on. In this *clinical interview method*, as it came to be known, questions were not only phrased in the child's language, but also restated in different ways when needed to clarify the child's thinking processes. Thus rather than asking standard questions of each child in a predetermined sequence, as on an IQ test, interviewers using Piaget's clinical method aimed to interact with the child's thinking. The surprising observations reported by Piaget and his colleagues have provoked many psychologists and teachers to appreciate the importance of listening to children.

In this chapter, basic notions of Piaget's theory of intellectual development will be interwoven with references to both your personal learning experiences and to those of children as they respond to interview tasks dealing with space and number. This discussion is designed to help you open a small window to children's minds in preparation for interpreting explorations found throughout the book and for reexamining your role as teacher.

As a teacher, your view of learning is reflected directly in how you respond to errors in children's answers. To become conscious of your reactions, examine the following example of a child's computational errors and locate the specific teacher reactions with which you can identify.

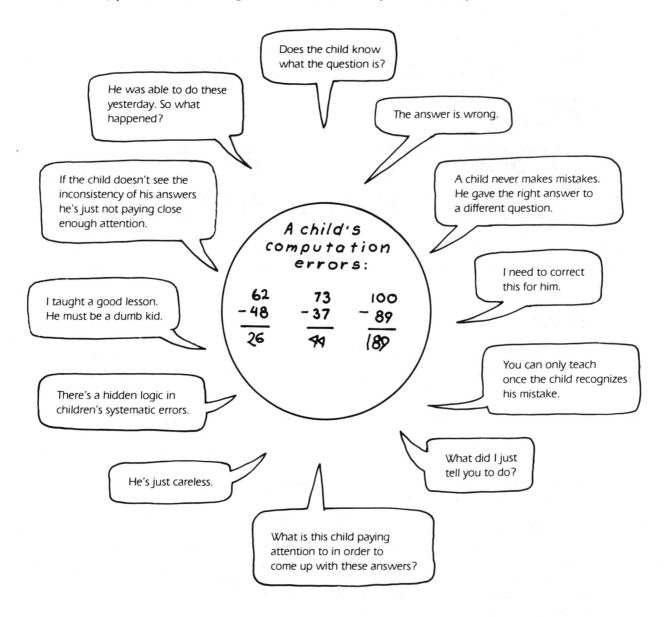

As a springboard to a discussion of Piaget's theory of intellectual development, examine the following aspects of your personal learning process (O'Brien, 1982b):

---

1. Think of a time when you mastered something without direct aid of a teacher or a textbook. Reflect on the quality of that learning.

2. Think of a time when you read a novel or saw a movie and then disagreed with a friend or a critic on your interpretation of the plot, the characters, or the quality of the production.

3. Think of a time when you read a novel or saw a movie and then, years later, you reread the novel or saw the movie again and had a different interpretation of it the second time.

4. Compare your recollections with those of at least one other person.

PIAGET'S CONSTRUCTIVIST VIEWPOINT

Rather than view knowledge as absorbed passively from the environment, or preformed in the child's mind and ready to emerge as he matures, Piaget considers it to be constructed actively by the child in the process of adapting to his environment. This view of the development of knowledge is described as *constructivist*. An interesting outcome of this active internal construction is that children and adults have different perspectives of the world around them.

The Learner Is an Active Agent in His Own Learning

We need only to observe a young child's exploration of his immediate environment to see self-determined learning in action. Rather than passively waiting for things to happen, he begins to relate to the world by experimenting and testing his ideas about how the world works. Piaget writes that to know something one must act upon it and transform it. The child's activity is not restricted only to actions or objects; the activity may later involve reflections on these actions to construct ideas. Eventually these ideas are coordinated with other existing ideas at a higher level of understanding (Piaget and Duckworth, 1973). Despite the fact that the majority of classrooms offer only limited opportunity for such do-it-yourself learning, we can each recall examples of children's self-initiated exploration resulting in significant learning.

Rather than passively copying knowledge that exists "out there," we actively construct our knowledge of the world internally through continuous interaction with the environment. Each person's understanding is like a personal painting resulting from one's own interpretation and synthesis of reality. It is not a photograph of reality. It is a person's internal network of ideas that interacts with reality in a mutual transformation. Although you and your friends may see the same movie or read the same book, your personal network of ideas about the world brings different meaning to the "common" experience. Without bringing your own meaning to a printed page, it remains a mass of black squiggles framed by orderly margins. In the process of your interactions with movies and books, and with your friends, your network of ideas may undergo some alteration.

Children View the World Differently From Adults

As adults we make personal constructions of shared experiences, which may contrast with those of our friends. At the same time, there is sufficient commonality in our network of ideas that we agree on many aspects of reality. For example, although our drawings of three-dimensional objects might be styl-ized differently, we would agree in terms of perspective. Children's drawings of familiar objects created from memory illustrate a common view of the world, which in turn differs drastically in terms of perspective from the drawings of adults (and older children). Our current understandings of spatial relations interact with sensory input from the environment. In our interpretations of sensory data, we bend reality to the way we are organized to receive it.

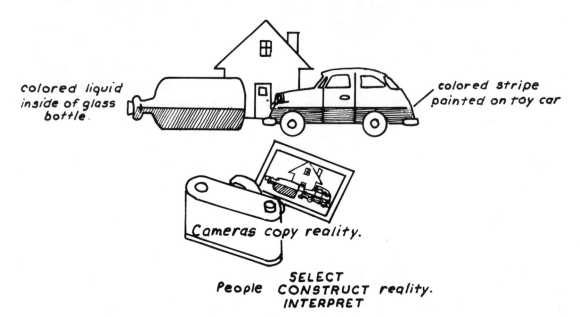

colored liquid inside of glass bottle.

colored stripe painted on toy car

Cameras copy reality.

People SELECT CONSTRUCT reality. INTERPRET

PERSPECTIVE OF AN ADULT

PERSPECTIVE OF A FIVE-YEAR-OLD CHILD

For example, look at the preceding figure. Memory drawings made several minutes following visual presentation of the objects reflect the contrasting mental images of the adult and young child despite common sensory input. When presented with feedback from the actual objects for comparison with their drawings, most five-year-olds would be unaware of anything "wrong" with their drawings. They would still see the slantiness of the roof in the doll house, which for them includes the chimney as well. They would also see the water level as consistent within the bottle, seemingly unaware that such a situation could not exist in reality. An older child, having more structured ideas of space, might recognize discrepancies in his drawings based on the feedback from the original objects. However, it might take a few years before the child could draw all aspects of the models in an adult three-dimensional perspective (Duckworth, 1979; Sinclair, 1973).

Continuous interaction between our expanding network of ideas and the environment accounts for our changing perspectives of the world from early childhood through adulthood. Thus a book can trigger different meanings for the same individual at different periods of time. *Huckleberry Finn* may have been assimilated as an adventure story when read by a child or as a social commentary or a metaphoric statement of personal identity when read again by a teenager or an adult (Ziajka, 1978). Similarly, classics such as *Alice in Wonderland, Gulliver's Travels,* and *The Hobbitt* may be enjoyed at different levels of sophistication at different times in one's life.

We See What We Understand

From the preceding examples of our interpretations of reality we can conclude that *we see what we understand* rather than understand what we see. Our drawings of reality and interpretations of films and books reflect the internal organization of our idea networks. In the classroom this means that all the children may not grasp what you are attempting to explain. What is learned by children is not a photocopy of what is being taught. The children may be able to focus on only part of what is being taught or they may actively deform it to fit their perspective of the world. Before we can undertake teaching children in any new content area, we need an understanding of their existing views.

With these fundamental notions in mind about the constructivists' view of how we relate to the world, let us now turn to an example of children's responses to a specific numerical task. Then we will resume our discussion of Piaget's theory with this example in mind.

AN EXAMPLE OF CHILDREN'S NUMERICAL THINKING

I conducted a series of three individual interviews with a group of children to explore their understanding of the whole-number numeration system. One of the tasks I presented to them involved a three-dimensional model—Dienes blocks—of our numeration system. Although these materials are found in many classrooms, the third graders interviewed had not experienced them previously. An open-ended question introduced the task, with subsequent questions being more specific as needed. The questions were as follows.

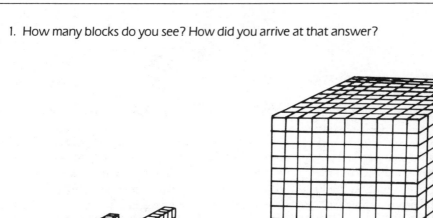

1. How many blocks do you see? How did you arrive at that answer?

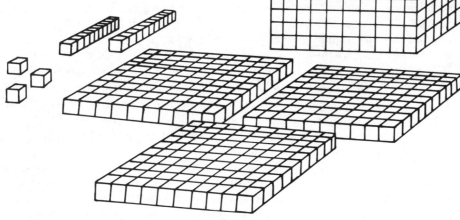

2. These blocks stand for a certain number.
   What number do you think they stand for?

3. How many of these small cubes would you need to build all of these blocks (while showing the child blocks of increasing size)?

Number of Blocks Seen by the Children

Piaget became intrigued by children's errors because of their similarity of perspective. By the same token, in my interview, although the third graders' responses to the initial, open-ended question fell within a wide range, there was some similarity among them. All but the last response illustrated below were given by more than one child. The "correct" response was 1,323.

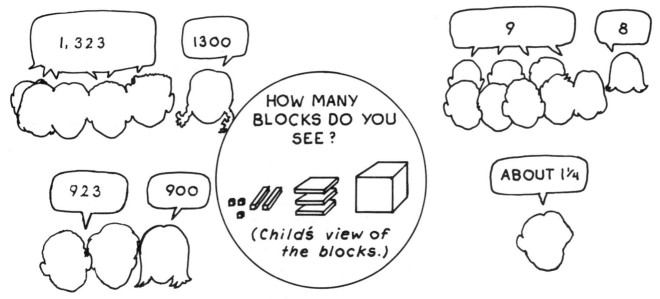

How did these children see the blocks?

In studying children's responses, Piaget was interested in the reasoning underlying both the expected and the unexpected ones. Let us now look at the reasoning behind each of the answers in the preceding task. Assume that each child did not guess and believed he was right. Try to take a child's-eye view of the blocks to determine the reasoning behind each response.

The perspective taken by nine of the twenty-nine third graders disregarded the differences in size and the markings on the blocks and counted the number of discrete objects. One of these children miscounted by one. Perhaps these children would have seen the blocks differently if asked a more specific question.

Three of the children took a surface-area perspective. These children were observed counting each of the six surfaces of the large cube as 100 and attributing a representation of 600 to it. Hence they arrived at responses of 900 and 923.

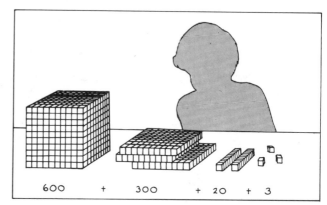

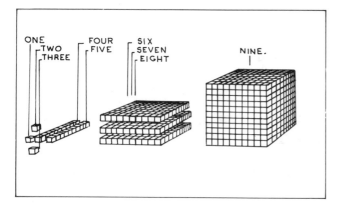

In a unique perspective of the blocks, Joey chose the larger cube as his reference unit and tried to relate the other blocks to it. Later, these children were asked more specific questions to clarify my perspective. However, here I regret that I was unable to adapt the next question to Joey's perspective so that he could develop his own line of thinking further.

ONE, BECAUSE THIS IS THE ONLY BLOCK... AND A HALF, A QUARTER, OR SOMETHING.

Joey R.

A sophisticated view of the blocks was taken by 5 of the 29 children, who gave a response of 1,323 that they were able to justify. A sixth child identified the larger cube as 1,000 but had difficulty counting beyond 1,300.[1] These children had no prior experience with physical models of the numeration system that included a representation of 1,000. However one child did indicate seeing illustrations of the model in the textbook. These six children were able to see the blocks at another level of complexity and were able to abstract complex relationships from them. Counting by units of different magnitude was the predominant strategy, although some children used multiplication, either to obtain or to justify their total. Some of the specific strategies are illustrated in the following stop-action sequence.

Before the question can be completed.

OH, I KNOW, IT'S 1,323 BECAUSE THERE'S TEN OF THESE (FLATS). THIS FLAT IS ONE HUNDRED AND TEN TIMES A HUNDRED IS A THOUSAND.

Frank M.

After a false start...

A HUNDRED?(Pointing to large cube)... THESE ARE ONES, TENS, HUNDREDS, AND THIS IS A THOUSAND...1,323.

She spontaneously stacks the three flats and counts by hundreds to a thousand.

Cindy G.

After a careful analysis...

ARE THERE TEN?... (He starts counting as indicated.)...1,323 BECAUSE TEN TIMES TEN IS A THOUSAND.

He gave his answer following his fifth count of the block dimensions.

Steven R.

Despite a very open-ended question, these six children were able to integrate information from the context of the task to read more into the question. These children were not questioned any further on this task. I wondered if more specific questions would allow other children to see the blocks at a higher level of complexity.

How the Children Interpreted the Number Represented by the Blocks

In response to the second question, "These blocks stand for a certain number. What number do you think they stand for?" all of the children previously giving answers of nine or less increased their estimates. The question helped one child to refocus his thinking to obtain a total of 1,323. The other responses ranged up to "thousands." If the children were unable to read the

context of the interview into this question, they could only guess the number that was in my head. Rather than showing a lack of understanding by the children, this may illustrate my inability to coax out their understanding with the right question.

How the Children Measured the Blocks

I thought that perhaps the third question would be more accessible to the twenty-two children remaining. "How many small cubes would you need to build all of these blocks?" was usually restated for blocks of

---

1. Some children experienced difficulty in counting by ones, tens, and hundreds at various stages in the task. Although these counting difficulties will not be described in this chapter, a detailed discussion of similar counting difficulties experienced by the same children may be found in Chapter 10.

increasing size as needed, starting with the rod. The question of the grand total was saved until the end of the task. With this opportunity to examine the blocks in more detail, would the children be able to determine that 1,000 small cubes are needed to construct the larger cube? Would the children then be able to determine the number of small cubes for the original arrangement of blocks?

Rather than prompting a range of responses, the third question brought an almost uniform response from the children. All but 2 of the 22 children indicated that 600 small cubes would be needed to construct the large cube. An interview episode with one of these children, Teddy, illustrates his responses to this question and to a guided demonstration that followed. Here, Teddy shows other surprising behaviors.

WHAT NUMBER DO YOU THINK ALL OF THESE BLOCKS STAND FOR?

423.

Teddy D.

How does the child see the largest block?

A closer examination of the blocks

NOW LET'S CHECK YOUR IDEA. WHAT DO THESE (Rods) STAND FOR?

SO TEN LITTLE BLOCKS MAKE UP ONE OF THESE LONGS.

TEN.

AND WHAT DO EACH OF THESE FLATS STAND FOR?

A HUNDRED. A HUNDRED. A HUNDRED.

A rod is moved across one flat. He identifies each of the flats as a hundred.

SO HOW MANY HUNDREDS DO WE HAVE?

100, 200, 300.

500
100
300
400   200
?

He starts counting the flats and continues counting to include the surfaces of the large block in the indicated sequences.

IS THERE ANYTHING ON THE BOTTOM?

WELL,... IT'S GOT TO BE... (He starts counting again)... 923.

LETS TAKE ANOTHER LOOK AT THIS BIG ONE. HOW MANY FLATS DO YOU NEED TO STACK UP TO BUILD A BLOCK LIKE THIS ONE?

TEN.

AND WHAT DO EACH OF THESE FLATS STAND FOR?

ONE HUNDRED.

SO HOW MANY LITTLE CUBES WOULD THERE BE IN THE LARGE BLOCK?

SIX HUNDRED.

2. Often, the children hesitated in counting the bottom surface but decided to do so after peeking underneath. Occasionally, I lifted the large cube as a suggestion to consider the bottom side.

A resistance to change was displayed by many of these children on closer examination of the materials. Like Teddy, 13 of the 22 children exposed to this guided demonstration persisted in their surface-area focus. Despite identifying the fact that 10 flats of 100 were needed to make up the large cube, they all concluded that 600 small cubes would be needed for its construction. These children identified potential evidence in support of another conclusion based on a volume strategy, but their focus on surface area didn't allow them to integrate the evidence. Actually, they deformed the evidence to fit their existing perspective.

On closer examination of the materials, another 6 of the 22 children became aware of the conflict between their original surface area interpretation and the need for 10 flats of 100 to build a large cube. These children were able to reorganize their idea network to incorporate the evidence. The following interview episode shows Harold in the process of reorganizing his thinking.

Harold's original estimate of the total

NINE HUNDRED.

Harold S.

The original estimate for the total suggests a surface area strategy.

IF WE HAD ENOUGH OF THESE WHITE CUBES, HOW MANY DO YOU THINK WE'D NEED TO BUILD THIS LARGE BLOCK?

I'M NOT SURE,... A THOUSAND. THERE'S TEN HERE AND HERE (Pointing at the large cube) AND THERE'S SOME INSIDE.

DO YOU MEAN TEN OF THESE?

NOT TEN,... A HUNDRED. THERE'S A HUNDRED ON EACH SIDE.

SO THERE'S A HUNDRED ON EACH SIDE AND YOU SAID THERE'S ALSO SOME INSIDE.

SO HOW MANY SIDES ARE THERE?

YEAH.

FOUR,... NO, SIX.

SO THAT'S AT LEAST HOW MANY HUNDRED?

AT LEAST SIX HUNDRED.

Harold looks puzzled.

BUT YOU SAID THERE'S MORE INSIDE. MAYBE THIS WILL HELP,... SUPPOSE WE STACKED THESE FLATS, HOW MANY WOULD WE NEED TO BUILD THE LARGE BLOCKS?

HOW MANY ARE THERE IN EACH FLAT?

SO HOW MANY ARE IN THE LARGE BLOCK?

TEN.

A HUNDRED.

A THOUSAND.

A look of satisfaction.

Although Harold's initial estimate suggests a surface-area focus, his concern for the inside of the cube also indicates awareness of other possibilities. Most of the other children making progress during the guided demonstration made no specific reference to the inside of the cube. Instead they reacted directly to the evidence of 10 flats of 100 to determine the value of the block. These children demonstrated the ability to respond to a tight sequence of questions and to count by tens to one hundred and by hundreds to one thousand (or ten hundred). Are these abilities synonymous with having a concept of a thousand? Since no further questions were asked to pursue the depth of the children's understanding, I continue to wonder how these children would have reacted to the idea of an "inside". However, the responses to the question on the grand total value of the original arrangement gives an indication of the stability and depth of their understanding.[3]

In my delight with the children's new constructions of a thousand I lost sight of the original focus on

3. You will notice that the phrasing of the grand question was varied from child to child.

the value of the entire group of blocks. The following episode jolted me into realizing the importance of pursuing that question to seek more information on the children's understanding.

Eventually, three of the children who had reorganized their thinking about the large cube were asked the value of the entire group of blocks. Since these children had counted successfully by ones, tens, and hundreds, or had counted the group of smaller blocks as 323 and had identified the largest block as a thousand, one might think that obtaining the total would present no problem. Having this orientation, I was unprepared for what was to transpire. Although no further questions were asked to clarify the children's unusual responses, observation of their behaviors does provide us with some clues as to their thinking.

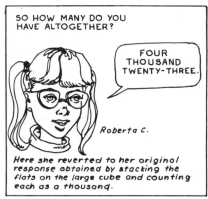

How can children understand something at one minute and not the next?

Here's another interesting episode from the interviews with third graders. It may provide us with some insight into the thinking of the children illustrated above.

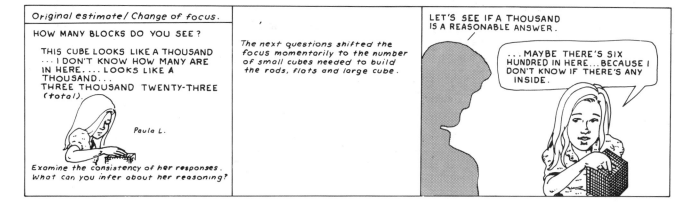

Notice the change of expression during the long pause prior to responding.

Whereas identifying the need for 10 flats of 100 to construct the large cube was sufficient evidence to take care of Harold's concerns for " inside," it wasn't enough to convince Paula. What additional experiences might have provoked her to construct a stable understanding of the thousand unit?

Looking Back at the Exploration

Although the open-ended question "How many blocks do you see?" prompted an interesting range of children's perspectives, and even revealed Joey's unique perspective (1¼), it did not encourage all of the children to view the blocks from a sophisticated perspective. The third question, focusing on the small cube as a building block or unit of measure, seemed to have the greatest potential for engaging children in higher-level thinking.

Illustrations of episodes from my interviews underscore the complexity of Piaget's clinical interview technique. In addition to asking questions that can be understood by children, the interviewer must possess the flexibility of adapting the task to the child's perspective and spontaneity in probing the depth of children's understanding. Besides telling us something about the interview method, the interview episodes have revealed a pattern of children's responses together with unusual behaviors that demand explanation. Chapter 10 will discuss children's miscounting when exposed to mixed units of ones, tens, hundreds, and thousands. Here, we will consider Piaget's theory of intellectual development in search of explanations for the following:

1. The children's contrasting perspectives of the blocks even following closer examination
   a. The ability of some children to benefit from the guided demonstration
   b. The resistance of other children to the evidence at hand
2. The apparent understanding by children at one moment and not the next
3. The ability of some children to give the sophisticated response of 1,323 even without prior experience with this specific numeration model

Not only is Piaget's theory useful in explaining many of the children's unusual responses described earlier, it also serves as a basis for predicting and explaining children's responses to a variety of number tasks described throughout this book and observed in your classroom. The knowledge that you acquire in this study of children's numerical thinking can serve as a framework for making decisions in teaching. The ensuing discussion may help you to make decisions about the following teaching concerns.

1. Should you postpone using the block model for one thousand until the perspective of more children is open to the possibility of an "inside"?
2. What additional experiences could you provide that might provoke Paula to construct a stable understanding of the thousand unit?

DEVELOPMENTAL ASPECTS OF PIAGET'S VIEWPOINT

Rather than knowledge existing preformed in the child's mind—ready to emerge fully formed at certain ages—Piaget theorizes its gradual construction and transformation by the child—resulting in distinctive capacities at different periods or *stages*.

Stages of Intellectual Development

Piaget noticed that at certain ages, patterns in children's responses to intellectual tasks were remarkably different from adult responses. Furthermore, he found differences in response patterns for children of unlike ages. Based on repeated observation of these patterns, Piaget categorized levels of children's thinking into four major stages. At the same time that each stage is marked by the emergence of new capacities for adapting to the environment, it also is characterized by certain constraints. Familiarity with children's behaviors characteristic of each stage alerts teachers to anticipate views of the world that are different from their own.

Children's intellectual development begins in the coordination of body-centered activity as well as activity with physical objects. A major breakthrough in children's expanding abilities occurs when they can mentally re-present these actions for themselves through thought and can represent them to others through language and drawings. Later breakthroughs include developing the capacity for logical thinking based on reflections about actions on objects, and finally, the capacity for conceptualizing in the absence of objects (for instance, dealing with numbers so large that that they are beyond direct experience). Each breakthrough in the development of children's expanding intellectual capacities marks the beginning of a new stage in which these capacities continue to develop. The general characteristics of each stage are summarized in the following table (Piaget and Inhelder, 1969).

PIAGET'S STAGES OF INTELLECTUAL DEVELOPMENT

	STAGE	AGE RANGE	CHARACTERISTICS
Prelogical stages	Sensori-motor	Birth–2 years	Coordination of physical actions; prerepresentational and preverbal
	Preoperational	2–7 years	Ability to represent action through thought and language; prelogical, intuitive
Logical stages	Concrete Operational	7–11 years	Logical thinking, but limited to physical reality
	Formal Operational	11–15 years	Logical thinking, abstract and unlimited

Sequence and Rate of Development

All children pass through the stages of intellectual development in the same sequence, but at different rates. The beginning of the concrete operational stage is associated with seven-year-olds because the earliest benchmarks of logical thinking were demonstrated by 75 percent of the children in Piaget's sample at that age. However, some children may demonstrate their first logical thinking at age five while others may demonstrate it a later age, or not demonstrate it at all. For example, it has been estimated that only one-half of the American adult population has reached the level

of formal operations (Kohlberg and Mayer, 1972). Although there are differences in rates of development, the order in which we pass through the stages does not vary.

Integration of Successive Stages

Each stage is made possible by those stages preceding it. Earlier understandings are integrated into later stages at higher levels of organization and abstraction. Piaget (1974) posits that children's capacity to deal with abstract algebraic concepts in the formal operational stage is made possible through earlier actions on objects and through subsequent reflections on these actions. Thus, in the classroom, children must be allowed to pass through the stage of developing logical thinking in physical contexts before being exposed to highly abstract ideas.

Transitions to the Next Stage

Children are in constant transition to the next stage, often demonstrating thinking characteristic of more than one stage. Children in elementary school may demonstrate logical thinking in one area of study while having a prelogical perspective in another area. In the upper elementary grades they may begin to understand addition and subtraction of multidigit numbers at a high level of abstraction while still developing an understanding of long multiplication and division through physical and mental action on objects. This phenomenon is also true for adults. We may demonstrate facility with abstract ideas (formal operational thinking) in areas of expertise, while in new areas of experience, like physics, we still need to develop our thinking in physical contexts (concrete operational thinking).

Learning and Development

Learning parallels development—taking place within the time constraints of intellectual development. Piaget views intellectual development as a gradual process of reorganization of mental perspectives of the world that is influenced by factors such as the biologi-

cal maturation of the nervous system. (These factors will be discussed later in the chapter.) Whereas intellectual development is gradual spontaneous reorganization of ideas accounting for Piaget's four major stages, learning may be viewed as provoked and accounting for minor reorganizations within the larger network of ideas. Yet both learning and development involve similar processes of reorganization (to be discussed in the next section).

For the teacher, Piaget's description of children's characteristic behaviors at each stage of intellectual development provides general guidelines for the selection of appropriate classroom activities and explanations of children's interpretations of these activities. For example, according to Piaget, most children construct a concept of area at about age eight and a concept of volume at about age eleven. Knowing this, a teacher would not find surprising the tendency of third graders to focus on a surface-area strategy. The children I observed were attempting to coordinate the six surfaces (including the bottom one) without an awareness of an "inside." If most third graders don't understand volume, then Dienes blocks may not be a useful vehicle for helping them understand large numbers such as 1,000. What is surprising, although possible with the average ages used in the theory, is that some third graders appeared to demonstrate a concept of volume. An awareness of the range of abilities within a classroom provides information for the individualization of learning activities.

MECHANISMS OF REORGANIZATION— PIAGET'S EQUILIBRATION

The children's different interpretations of the same reality of the blocks reflect different internal organizations of ideas. Experiencing the following puzzle may provide you with some insight into the process of reorganization of networks of ideas at a higher level of sophistication.

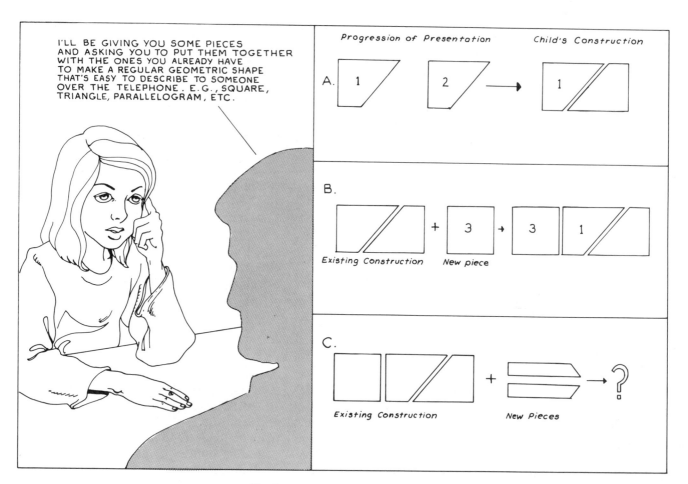

The Progression Puzzle (de Bono, 1969)

How would you solve the puzzle?

Experiencing the solution is critical to understanding the next phase of the discussion.
Cut out the pieces in Appendix A and reproduce the sequence described above.
Now construct a regular geometric shape that incorporates pieces 4 and 5.

Equilibration—Balancing Opposing Tendencies Toward Continuity and Change

When given the first two pieces of the puzzle, most people will construct a rectangle. Upon introduction of piece 3 it is possible to attach it to either end of the original structure. When people are given a new set of pieces, 4 and 5, the comfort of adding new pieces to the existing structure is replaced by a state of discomfort, confusion, or frustration that Piaget calls *disequi-*

librium. A dead end seems to be reached when the new pieces can't be fitted along the edge of the existing rectangle. In a demonstration of resistance to change, there may be a number of returns to unworkable solutions. Despite this resistance to change, the accompanying discomfort provides a motivation to find a solution. Eventually, there is a decision to reorganize the pieces of the original structure in an attempt to incorporate the new pieces. Following the reorganization of pieces 1, 2, and 3 it is possible to maintain the continuity of the square shape by adding pieces 4 and 5. Piaget calls such maintenance of continuity in response to new input *assimilation.* The reorganization of existing structures to incorporate new input is known as *accommodation.* The interaction between these opposing tendencies toward continuity and change results in construction of a shape at a higher level of organization, incorporating all of the pieces. The solution restores the inner balance, or *equilibrium,* and inner contentment.

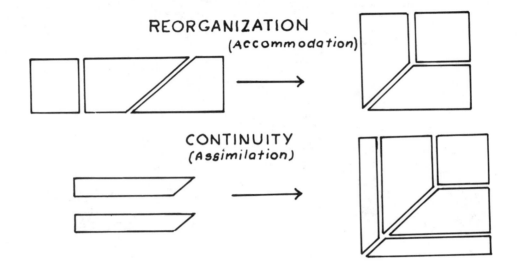

To further clarify this process of restructuring knowledge known as *equilibration,* I will discuss it first in a general context and then in the context of the blocks task. In receiving input from the environment, a learner screens and interprets it through his previously built network of ideas (assimilation). This organization of ideas influences the parts of the environment to which the learner gives attention and the way in which he sees/relates/gives meaning to those parts. The learner's organization of ideas resists change to the extent of distorting the input. Despite the resistance to change, new and compelling environmental input continually stimulates change. Once the limitations of the existing ideas are recognized, new learning is about to begin. The discomfort, confusion, or frustration of the disequilibrium state serves to motivate a search for a better organization of ideas to resolve the discrepancy (accommodation). Intellectual development (or learning) results from the interplay of these complementary processes: assimilation resisting change in exchange for the comfort of the continuity of the familiar, and accommodation providing an impetus for novelty and change.

Equilibration is the interplay between factors inside and outside the child.

The reorganization of knowledge results in a new way of thinking and a new understanding that is accompanied by inner satisfaction. Here, a temporary state of equilibrium is achieved in which there is a balance between interacting factors inside and outside the child. This tentative balance is soon upset, however, as the new organization of ideas stimulates the child to interact further with his environment and to notice different aspects of it, thus leading to additional reorganization of ideas and more effective ways of dealing with the environment. Thus, the child is seen as a mainspring in his own intellectual development, not only uncovering new problems (disequilibrium) but also constructing solutions at higher levels of equilibrium. The direction of this continuous reorganization is towards coherence, stability, economy, and generalizability.

To say this another way, O'Brien (1977a) writes:

The child doesn't merely organize, he quests. He constantly recognizes, engages, resolves—and generates—novelty.

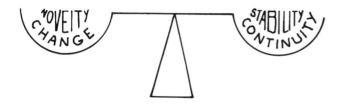

Equilibration and Levels of Thinking Within Stages

Learning and intellectual development involve similar processes of reorganization but on different scales. Major reorganizations of thinking through equilibration account for Piaget's four stages of intellectual development, while smaller reorganizations can account for the different levels of understanding often observed in the learning of concepts and relationships within a stage. Although all of the third-grade children demonstrated thinking characteristic of the concrete operational stage, they understood the blocks at different levels of complexity. At least three levels were apparent from my interviews:

Level 1: The child ignores the relative dimensions of the blocks and their markings to count each block as an individual unit.

Level 2: The child considers the smallest cube as a unit and constructs units of tens and hundreds by reading the ten-to-one relationships into the dimensions of the blocks and coordinating this with his existing network of ideas about tens and hundreds. This understanding is applied to the large cube through a network of ideas that focus only on the surface areas of three-dimensional objects. This network screens out other evidence available.

Level 3: The child considers the smallest cube as a unit and constructs a thousand unit, incorporating units of one, ten, and one hundred together with ideas of "inside" and "outside" into the relationship.

Restating the original question quickly revealed that all the children were capable of viewing the blocks at one of the higher levels. However, it became apparent that to reach level 3 *was not just a matter of looking more closely at the blocks to "see" the evidence, but a matter of looking at them differently* (Forman, 1980). Twelve of the children said that 10 flats were needed

to build the biggest block and that each flat contained 100 small cubes, but then ignored this potential evidence. They were so centered on what they thought was important—surface area—that they seemed unaware of the conflict between that informtion and their answer of 600. Children making level 1 and 2 responses were exposed to physical models reflecting the structure of our numeration system, but did not make a direct mental copy of that structure. Rather, they transformed sensory input to agree with their view of the world. If children perceive (and ignore) certain features of concrete objects according to their own level of understanding, it follows that they are even more likely to misperceive teachers' verbal explanations.

Those children who were able to coordinate all the relationships in constructing the thousand unit were able to do so only after they were aware of the conflict between their existing ideas and the compelling new information from the environment. *Learning begins with an awareness of the problem.* The children were able to decenter from a single surface-area focus and realize there were conflicting ways of viewing the cube. Furthermore, they were able to reorganize their perspective of the cube so that "outside" and "inside" concerns were no longer in conflict. Although children interact with materials on the outside, the major area of interaction is on the inside as they deal with the coordination of conflicting ideas.

The process of equilibration, involving interplay between forces of change and continuity, usually takes place over an extended time period. This period is marked by vacillations in thinking, when the child appears coordinated and advanced at one moment and then reverts to a less mature response at another, before leading to an eventual state of equilibrium. Although Paula gave an initial response of a thousand and had concerns for the inside of the block, she reverted to a surface-area focus having greater familiarity. Despite being aware of the problem, she demonstrated the strong resistance to change characteristic of the equilibration process. Four other children demonstrated the fragility of their newly acquired concept of a thousand, not yet fully integrated, when they experienced difficulty in totaling the blocks. We are reminded of the gradual nature of the equilibration process and the expectation of confusion, by Piaget's colleagues, Inhelder, Sinclair, and Bovet (1974), who write,

Learning takes time. . . . Children who progress the furthest are often those who are most confused initially.

Since not all attempts to coordinate the necessary relationships are successful in terms of adult logic, children's unusual responses can provide invaluable clues to their mental activity.

Equilibration and the Constructive Role of Errors

Another characteristic of the equilibration process is that advanced levels of organization of ideas integrate more primitive ideas from preceding levels. In other words, rather than moving from being totally wrong to being totally right, the less mature ideas are found to contain some logic and may be considered as partial constructions and essential steps in arriving at correct answers (Piaget and Duckworth, 1973). For example, young children recognize that there is something slanty about the roof and chimney of a house, but in their drawings they aren't able to restrict the slantiness of the roof alone. Also, they recognize a perpendicular relationship for the chimney, but to the roof alone. Later in life their drawings reflect a restriction of the slantiness to the roof and a reorganized perpendicular relationship with less obvious indicators of the horizontal plane (Sinclair, 1979, Duckworth, 1979). Similarly, there is some logic in children's responses of 600 for the value of the larger cube. When children add a third dimension to account for the inside of the block in constructing the three-dimensional thousand unit, they incorporate surface-area considerations into the new construction.

 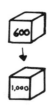

From the preceding discussion of equilibration, we can conclude that *children have a right to be wrong* (except in cases of obvious carelessness). Making errors is a natural part of their intellectual development. Just as babbling is essential before talking and crawling before walking, children's errors can be viewed as constructive—as essential for integration into more elaborate constructions. Yet a popular belief held by the majority of teachers—and supported by the theory of behaviorism—is that errors and confusion have no educational value and, therefore, must be eliminated from the classroom. Hence the teacher's role is interpreted as one of breaking learning down into such small steps that errors are not experienced, and of intervening quickly in times of confusion to rescue the child by giving the correct answer. This common approach does not respect the child's ability to construct his own understanding in his own time. It also assumes that a child makes an exact mental copy of the teacher's correct answer or explanation.

By contrast, Piaget has provided teachers with a developmental-constructivist framework for looking at children's responses apart from their "correctness." Within this framework it is possible to consider children's wrong answers as being appropriate to a certain level of development, as partial constructions to be integrated into more coherent responses in time, and as small windows to their thinking. Piaget (Piaget and Duckworth, 1973) states

> Yes, I think children learn from trying to work out their own ways of doing things—even if it does not end up as we might expect. But children's errors are so instructive for teachers. Above all, teachers should be able to see the reasons behind errors. Very often a child's errors are valuable clues to his thinking. . . . A child always answers his own question correctly; the cause of an apparent error is that he did not ask himself the same question you asked him (p. 24).

The contrasting behaviorist and constructivist interpretations of errors were highlighted in the introduction to this chapter. You are invited to reexamine the teacher reactions to computational errors on page 4 in the light of your current understanding of how children learn.

INTERACTING FACTORS INFLUENCING INTELLECTUAL DEVELOPMENT AND LEARNING

Learning and intellectual development are influenced by four interacting factors: maturation, physical and mental experiences, social interaction and transmission, and equilibration (Piaget and Inhelder, 1969).

Maturation

The older the child is, the more likely that the biological capacity of his nervous system has developed further, organizing new possibilities for the elaboration and coordination of idea networks. Without prior experience with the blocks,

- most first-grade children will see 9 blocks
- most third-grade children will see the large cube as 600 blocks
- most sixth-grade children will see the large cube as 1,000 blocks

Although it is difficult to isolate the maturational effect of time from the other influencing factors, it is an important consideration here.

Physical and Mental Experiences

The more experience a child has with physical objects in his environment, the more likely he is to develop related understandings. Piaget demonstrated that many mathematical concepts have their roots in children's experiences with materials. By observing the blocks, some information, such as their physical properties and composition, is obtained by the child through *simple abstraction*. By acting on the blocks, comparing them, counting the divisional markings, building larger blocks from smaller ones and reflecting on these actions, an indirect relational knowledge is abstracted from the blocks. The process of mental construction of relations that are not directly observable is called *reflective abstraction*. Mindless manipulation of the blocks will not produce such constructions; accompanying mental action is critical to this higher-level process.[4] Rich mental images of prior actions on objects can also provide the basis for reflective abstraction of mathematical relations. The fact that some children were able to abstract a complex relationship from a minimal exposure to the blocks suggests that the amount of physical experience that is essential may vary from child to child. The other children might benefit from extended experiences with the blocks over time.

Social Interaction and Transmission

Learning and intellectual development are enhanced through opportunities for social interaction with both peers and adults. For some children, the interviewer's questions helped focus their interaction with the blocks and may have provoked the reorganization of their perspective. Other children might benefit from a free exploration of the blocks in small groups to determine how many small cubes are needed to construct the large cube. These interactions with their peers would allow them to incorporate the contrasting viewpoints of other children and to share information, such as "Thousand is another name for ten hundred." Furthermore, these interactions would allow the children to raise their own questions about the blocks.

---

4. Mindless manipulation of blocks, according to the teachers at the Teacher's Center Project at Southern Illinois University, Edwardsville, is evidence that children lack the mental structure to engage the issue or that they are bored.

Equilibration and the Interaction Among These Factors

Equilibration, described earlier, is the most fundamental of the four factors, interacting with and coordinating the other three factors. No single factor can account for intellectual development and learning. Rather, there is an interplay of factors.

Despite countless physical experiences with drinking, pouring, bathing, and swimming, young children are unable to represent water levels in relation to the horizon, independent of the position of the container. Rather, they represent situations that don't exist in the real world, since their frame of reference is limited to the inside of the bottle. It takes a few years before biological maturation of the nervous system unleashes the possibility of reconstruction, and equilibration brings the existing network of ideas on space in interaction with the child's physical experiences in the environment (or vivid images of such experiences). This interplay of factors results in the coordination of ideas of horizontal and vertical, creating an external reference system of imaginary lines.

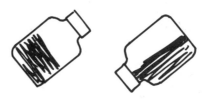

Perspective of a five-year-old

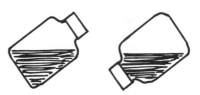

Perspective of a nine-year-old

The children taking a surface-area perspective in the blocks task might be provided with additional opportunities for physical and mental experiences together with social interaction to provoke a reconstruction of their perspective. For example, given more 100-block flats and the opportunity to physically construct a large cube from the flats (rather than imagine it), the notion of an "inside" might occur to the children for the first time as the inside surfaces of the flats got covered. This would make them aware of conflicting perspectives of the large block. A similar experience might provoke Paula, already aware of the possibility of an inside, to reorganize her perspective. Given the opportunity to interact with more blocks, Paula may have chosen to compare the relative weights of the original large cube and her constructed cube to disprove the possibility of the original cube's being hollow. Also, given the opportunity to interact with other children while considering the possibility of an inside for the cube, the children may have reconstructed their perspective in cooperation. These additional experiences would provide three of the necessary factors that might provoke a reorganization of the children's idea networks for a volume perspective. Yet the maturation of a child's nervous system may not have progressed to the point of supporting this high-level organization of ideas. Such a child still may not be aware of any problem with a response of 600. On the other hand, Paula, already aware of conflicting views of the cube, would likely benefit from this additional experience to reorganize her thinking.

In the prevalent view of teaching and learning, task analysis of subject matter is thought to permit teaching in minute steps, with selection of appropriate activities along the way, and to promise a high degree of success in a short period of time. The teacher is seen as being in control of the children's learning through the appropriate manipulation of their environment. By contrast, the preceding discussions have highlighted the active role of the child in screening and interpreting his environment, and the gradual nature of the learning process within the constraints of development. In this constructivist view the teacher is seen as a facilitator of interaction. At the same time that this teacher provides a rich variety of "hands-on/minds-on" activities and opportunities for social interaction and raises potentially thought-provoking questions, he or she also provides the children with time to reflect on their experiences and looks to the children for guidance in the pacing of activities.

In the next section the ideas discussed here will be revisited in the context of a discussion of the different kinds of knowledge children acquire. We will find that the contrasting views of teaching and learning described above place emphasis on the acquisition of different aspects of knowledge.

KINDS OF KNOWLEDGE INTERACTING WITH AND INFLUENCING LEARNING AND DEVELOPMENT

Three distinct kinds of knowledge have been identified by Piaget, varying in source and mode of structuring: social (conventional) knowledge, physical knowledge, and logico-mathematical (relational) knowledge (Kamii, 1982a).

Social (Conventional) Knowledge

Social transmission of cultural conventions and language by modeling and oral communication is a powerful force to ensure continuity of the culture. Although arrived at by social consensus, this specific information from the external world is arbitrary in nature, varying from country to country and even with different cultures within a country. For example, number names can vary with the language spoken (*quatre, cztery,* and *four* represent the same number in different languages), and within a language there can be alternate number names to represent the same quantity (*ten hundred* is the same as *a thousand*). Each culture has its preferred customs. In one culture it is appropriate to sleep on the floor while in another beds are preferred; in one culture it is appropriate to eat with one's fingers while in another this is considered bad manners; in one culture the holy day is scheduled on a Saturday while in another it is scheduled on the day following. It is, therefore, not surprising that different interpretations of books or movies should arise from different social values inculcated within different cultures. The ultimate source of social (conventional) knowledge, consisting of such specific and arbitrary information from the external world, is other people.

Physical Knowledge

Certain knowledge about objects in the environment is directly obtainable from the objects themselves through simple abstraction. In the course of a young child's exploration of blocks, he can determine their color, markings, shape, the sound made when they are dropped on the floor, their bouyancy in water, and so on. Here, the source of knowledge is mainly the objects themselves.

Logico-Mathematical (Relational) Knowledge

The third type of knowledge is not directly available from external sources. In comparing the different properties of the blocks, a young child may identify blocks in different pairings as bigger, the same size, the same color, or heavier. Here the child is constructing relationships that are not inherent in the individual blocks but are rather introduced through his mental activity. In counting nine objects, the "nineness" is created by the child and read into the blocks. To count a single object such as the rod-shaped block as ten, the child must read a ten-to-one relationship into it by comparing it to the smallest block. The relationship does not exist in either or both blocks without the act of mental creation through the process of reflective abstraction. The main source of this relational, or logico-mathematical, knowledge is the child's mind and how he organizes and interprets his reality. Relationships are inventions of the mind.

Through the child's continuous mental activity, not only are relationships being constructed, but they are also being coordinated at higher levels of organization. For example, the expanding concepts of vertical and horizontal are coordinated to create a system for interpreting three-dimensional arrangements in space. Another example is the coordination of ten-to-one relations in the dimensions of the different blocks with "outside" ideas of surface area and "inside" ideas of volume. Such coordination creates a relational framework for interpreting reality.

Another characteristic of logico-mathematical knowledge is that it becomes increasingly detached from physical reality, as it is abstracted and progressively coordinated at a higher mental plane. For example, with time, the child's relationship of "nineness" will be independent of the kinds of objects being counted. Similarly, the thousand unit and its relationship to the hundred, ten, and one units will exist independently of a particular physical model in the child's mind. Although children's first number ideas have their roots in putting objects in the physical world into relationships, these relationships form the basis for further intellectual development and integration into an elaborate network of abstract ideas about numbers in the formal operational stage. This integration of earlier ways of looking at the world can give adults the impression that mathematics exists "out there," independent of the knower. It is this integration phenomenon that makes it difficult both to recognize the concrete roots of abstract mathematical ideas and to appreciate the logical nature of children's errors. Piaget has demonstrated that this difficulty can be overcome by listening to children and by imaginatively taking a child's-eye view of the world.

Interaction of These Three Kinds of Knowledge

Whereas social knowledge consists of arbitrary conventions, varying from culture to culture, logico-mathematical knowledge is universal because of its internal consistency and logical necessity. And whereas social and physical knowledge have external bases, logical mathematical knowledge is internal relational knowledge. Although these three distinct kinds of knowledge have been identified for discussion, in actuality they are thought to interact. Since the three kinds of knowledge develop out of social interaction and transmission, physical and logico-mathematical experiences, equilibration, and their interplay, they cannot exist in total isolation. Logico-mathematical knowledge serves as a relational framework for selecting, interpreting, and organizing both social and physical knowledge. Thus physical knowledge can't be constructed independently of logico-mathematical knowledge. Conversely, logico-mathematical knowledge can't be constructed by the child without physical knowledge gained from objects.

Although the different aspects of knowledge can be mutually supportive, they can also be mutually indifferent, or even destructive (O'Brien, 1982). With the failure of our educational system to make a distinction between the different aspects of knowledge and to provide children with opportunities for developing a relational framework, an inordinate emphasis has been placed on transmitting laundry lists of trivial, unrelated, atomistic information. Children are required to memorize mindlessly addition and multiplication facts in isolation, so when these facts are forgotten the children have no basis for reconstructing them. Also, many elementary teachers view mathematics not as relational knowledge, but as an accumulation of social conventions—that is, arbitrary rules for manipulating symbols that can be transmitted through modeling and by telling. These teachers would, for example, model counting the blocks, identifying the largest one as 1,000, for children's imitation, even though it might contradict the children's view of the block. Piaget warns teachers against being misled by children's ability to parrot "correct" answers at a memorized, verbal level without internalizing that knowledge (in Duckworth, 1964). He refers to this learning as

> . . . false accommodation which satisfies a child because it agrees with a verbal formula he has been given. This is false equilibrium which satisfies a child by accommodating to words—to authority and not to the objects as they present themselves to him (p. 4).

Promoting such superficial learning, which is alien to a child's natural intellectual functioning, may not only contradict the child's view of the world but also result in a loss of confidence in his ability to make sense of the world and in negative attitudes towards mathematics (Piaget, 1973).

To help children achieve their maximum potential, schools must support children's natural intellectual functioning. From a developmental-constructivist perspective, greater emphasis must be placed on supporting children's gradual development of relational knowledge for organizing, accessing, and giving meaning to specific information.

REFLECTIONS AND DIRECTIONS

In this chapter, episodes from interviews with third-grade children on one task have served as a small window to their minds, allowing us to discuss in the context of that task Piaget's theory of children as active learners and to examine some of the implications of this theory for the classroom.

Like the children reorganizing their numerical ideas about the blocks or like your earlier self when reorganizing the puzzle pieces, you may be feeling some disequilibrium at this time. The theory, although having obvious connections with your personal learning, may seem foreign in the prevalent classroom context of teaching and learning. Also, you may be feeling overwhelmed with all the aspects of the theory and how they might affect your teaching. The following suggestions of what to do and to look for may relieve your anxiety and allow you to gain maximum benefits from the discussions ahead about children's thinking on number tasks and alternative teaching methods.

1. Avoid an attempt at mastering the details of the theory immediately. Rather, read the chapter a second time, make a summary of the highlights, and keep these in mind as you continue to read and interact with the book's contents.

2. As you continue to read, use your mental outline of the theory to predict how children will respond to other number tasks and to explain their responses. Additional episodes throughout the book will further illustrate the theory and provide feedback on its usefulness to you.

3. At those times when you doubt seriously that children really do respond in the manner described in this book, try similar tasks with children of the same age in your school or neighborhood. Children have much to teach us. The detailed discussion of the interviewing process in Chapter 2 may help you get started.

4. After you've read half the book, return to Chapter 1 and reread it. You may be astonished at the new meanings that you read into the chapter's contents.

5. As you find yourself reorganizing your views of thinking and learning you will also begin to see children in your classroom from a different perspective.

6. The questions found throughout the book aim to stimulate reflection on learning and teaching, generate productive discussions, and encourage you to raise your own questions and to do your own research on learning and teaching. Some of the questions may not have easy answers, or may have multiple answers, reflecting both the complexities of the learning and teaching processes and our limited

understanding of them. Discuss these questions with at least one other person before reading on.

7. Finally, some teachers find that the theory gives them a basis for expanding what they already know intuitively about children and how children learn. If this is the case for you, the book will serve as an integrator and supporter of your beliefs and hypotheses.

2

LISTENING TO CHILDREN: THE INTERVIEW METHOD

PREVIEW

The most important requirement of assessment of children at school is that it provide information that helps teachers help children. Paper-and-pencil tests that dominate today's schools provide information such as the child's ranking on national norms, based on the number of right or wrong answers. Yet knowing whether a child's response is right or wrong, or knowing his rank is of little value in helping a teacher decide what to do next. Furthermore, having such information for ranking and labeling prevents teachers from really listening to children (O'Brien, 1982a). Throughout this book considerable evidence is given to support a view of the child as an active learner—constructing ideas of number through his own interactions. This alternative view necessitates an alternative method of assessment—one that allows the

teacher/interviewer to follow the child's thinking as he works through tasks presented in the context of materials. In this view the thinking process is of greater interest than the correctness of the response.

The method of studying children's thinking devised by Piaget is known as the *clinical interview method*. This chapter will describe the clinical interview method—both as conducted by Piaget and as adapted for the explorations described in this book. Furthermore, it will identify some general questions about children's numerical thinking that the interview explorations throughout this book attempt to answer. The interview method will be discussed in sufficient detail to allow you to not only understand the method used throughout the book but also to undertake your own interview explorations.[1]

THE CLINICAL INTERVIEW METHOD: THE IDEAL AND THE ADAPTATION

In Piaget's clinical interview method, the adult interviewer involves the child in conversations through verbally-presented problems in the context of physical materials. In order to achieve its goal of exploring children's reasoning on intellectual tasks, the method seeks to encourage each child to verbalize freely and to interact with the objects, thus providing a basis for the interviewer's hypotheses about the underlying thinking. These hypotheses about the child's per-

spective of the problem are tested by spontaneously-invented questions based on earlier responses. Furthermore, the interviewer's role is to encourage the child to

---

1. Although Piaget has used the clinical interview method extensively in his research, he has left the detailed description and analysis of this method to others. A primary source of such information in the preparation of this chapter has been Opper (1977).

consider further, think more specifically, or rethink the process used in arriving at a solution. This is achieved through probing questions or comments that elicit explanations or justifications. Piaget's clinical method is marked by its flexibility in adapting the interview to the individual child in an attempt to follow his thinking.

Although the interview method illustrated throughout the book has features in common with Piaget's clinical method, it may be considered a pale imitation of the idealized method. A standardized introduction of the tasks was used to facilitate communication of the results of a number of interviews. At the same time, some of the flexibility of adapting questions and tasks as needed was retained. The most striking difference came as a result of my lacking considerable prior experience to develop the skill of on-the-spot invention of questions to test my hunches about children's thinking. Nonetheless, despite its limitations relative to the ideals of the clinical interview method, the approach used in this book has uncovered surprising views of number concepts held by children that have gone previously undetected.

QUESTIONS ADDRESSED BY THE CLINICAL INTERVIEW METHOD

The interview explorations conducted with first, second, and third graders have aimed to go beyond the traditional focus on correct answers to answering questions like the following about children's numerical thinking and its facilitation in the classroom.

1. What are children's natural tendencies in approaching numerical tasks? Are there identifiable levels in the ways they approach these tasks?

2. What are the limits of children's numerical abilities and understandings?
    a. What meanings do children assign to their computational procedures?

    b. Are children restricted to a single method of solution or are they capable of multiple solution methods?

    c. What level of conviction do children have in their answers?

    d. Is there any evidence of children's self-construction of strategies/procedures before any formal teaching (or in spite of it)?

3. What numerical meanings do children assign to mathematical models and symbols?
    a. What meanings do children read into materials used in the classroom, such as Dienes blocks or grouped objects?

    b. Do children place more trust in answers obtained through action on objects or in answers obtained by paper-and-pencil methods? Do they see the necessity of answers obtained by both methods coinciding?

    c. Are there identifiable levels of meanings that children assign to mathematical symbols, like equality sentences ($5 + 3 = 8$) or place value notation (235 vs. 198)?

    d. How do these meanings differ from those assigned by mathematicians?

4. Can the study of children's numerical thinking explain perennial stumbling blocks in teaching children mathematics and suggest alternative teaching methods?
    a. How can first graders' difficulty with missing addend tasks ($7 + \underline{\quad} = 9$), or primary graders' difficulties with place value concepts be explained?

b. How can children's understanding at one moment but not the next be explained (also observed as excessive forgetting during a vacation period)?

c. How does the quality of interaction during the interviews compare to the quality of interaction in the classroom?

d. How does the natural inclination of children regarding numerical tasks compare to the expectations of textbooks?

e. How can teaching methods be altered to build on children's natural tendencies?

In attempting to answer such questions, I often began interview tasks as standard school tasks but presented them outside the context of familiar worksheets and pursued them in much greater depth than previously experienced by the children. Totally unfamiliar tasks were also included to gauge the limits of children's abilities. I also pursued in-depth understanding of children's numerical abilities by conducting a series of three interviews with groups of second and third graders. Although much of the discussion of teaching methods will be based on a logical extension of what I learned from children in the interview explorations, the book also includes a detailed study of the teaching-learning process (in Chapter 7).

INITIATING THE INTERVIEW

It takes considerable sensitivity, experience, and skill to become a good interviewer. A raw beginner might not uncover any more information than a standardized achievement test could. This might be due to either a failure to probe or an overbearing manner that inhibits the child. To help you get started with interviewing children, different aspects of the interview method will be discussed and illustrated. Examples will be taken from different interviews described in the book. Excerpts taken from the reports of student interviewers will illustrate both the problems they experienced and the insights they gained in the process of interviewing.

In initiating an interview with a child it is important to establish rapport and to prepare him for the process before beginning. This may be accomplished through a combination of the following procedures.

1. Initiate a conversation by asking personal questions such as the child's name, age, number of brothers and sisters, favorite things to do, and so on.

2. Introduce the materials being used and provide an opportunity for their free exploration. "Have you seen these things before? What can you do with them?"

3. Justify the need for any mechanical recording equipment and restrict the viewing/listening audience.

4. Orient the child to the interview situation. The following is a sample orientation used with second and third graders:

"I'm interested in talking to you to find out how you think about numbers. I've been talking to a lot of children and do you know what I've been finding out? (What?) Everybody doesn't think about numbers in the same way. So, every boy and girl that talks to me teaches me something new. We're going to be doing some things with these materials. Some of the things we'll be doing are like games. The things we'll be talking about may be things you haven't studied in school and others will be things you have studied. I'm not really interested in whether you get the right answer. I'm more interested in how you figure out your answer. So, what we do here has nothing to do with your grade in school."

The interview should begin only after the child appears at ease and has begun to talk freely.

QUESTIONING

Although you may choose to begin a series of interviews with a standard presentation of each task, there will be frequent occasions when you will need to repeat or rephrase the question in the child's language while retaining the essential meaning, or to adapt the task to the level of the child.

When third graders were given the following addition task requiring regrouping, they were asked to demonstrate a solution with interlocking blocks (tens and ones). $\begin{array}{r} 2\overset{8}{8} \\ +37 \\ \hline \end{array}$ Some of the children merely combined the two sets and counted the pieces. Since I was interested in learning whether they could regroup 10 ones as a ten-group, these children received a prompt. If they did not respond to the first question, it was rephrased in one of the following ways.

"Could you show 65 (sum) in another way?"

"Can you show all of these ones (15) in another way?"

"What can you do so that if someone walks in the room he can see there's 65 cubes without doing a lot of counting?"

These questions prompted the children to regroup 10 ones as a ten, which then allowed me to probe their understanding of its connection to the carried "1" in their earlier paper-and-pencil procedure.

When children respond to a task initially, they may do so in superficial terms. A further probing question can encourage the child to consider the task further and allow the interviewer to learn more about the child's numerical ability.

When the children were first asked to explain their procedure for the addition task, many replied in terms of a rule involving carrying the "one." $\begin{array}{r} \overset{1}{2}8 \\ +37 \\ \hline 65 \end{array}$ Their understanding was probed by questions such as

"What does this 1 stand for?"

"How come you don't carry the five instead of the one?"

This probe was extended to the children's demonstration with materials in which 10 ones had been regrouped as a 10-group. To test the limits of each child's understanding, the following question was asked:

"Point to something on the board (of materials) that this 1 stands for."

Here, the child is required to extend his thinking to consider a connection between paper-and-pencil and physical procedures.

Another common response to tasks is in terms of adult vocabulary or with phrasings that suggest that the children are merely parroting what they've been taught. Here, probing would encourage the child to rephrase his response in his own words.

"How would you explain this to a first grader who doesn't understand it?"

The child should be encouraged to elaborate on his responses and to support them with explanations or arguments as far as possible. Each probe has the potential of providing the interviewer with a better understanding of the child's thinking in this problem context.

Additional questions or tasks may need to be introduced to clear up inconsistencies in a child's responses.

When a child did a subtraction reversal, thereby avoiding the need for regrouping, I was able to provide him with feedback on the inconsistency of his method.

$$\begin{array}{r} 43 \\ -26 \\ \hline 23 \end{array}$$

"I noted that the 4 was on top when you took 2 away and that 6 was on the bottom when you took 3 away. Does that make a difference?"

Upon receiving the feedback he became aware of his mistake and initiated steps for regrouping. Other children appeared satisfied with their answer (23). After obtaining a different answer (17) through physical modeling, the children were made aware of the inconsistency in these responses by the following question:

"Is one of these answers better or are they both just as good?"

Such questions may provoke children into becoming aware of inconsistencies and into rethinking their method of arriving at an answer.

Additional questions can be introduced to check the stability of a child's response. By presenting an argument that counters the one given by the child, the interviewer may provoke the child to revert to an earlier level of thinking or to reject the countersuggestion while elaborating on his own argument in the process. This counterargument should be presented as one given by another child of the same age, not as the interviewer's argument, to avoid any interference with the child's autonomy of thought.

Third graders were asked to compare the size of numerals such as 198 and 235 and to justify their choices. Different countersuggestions were given to the children according to their responses.

Children selecting 198 as the largest number were asked to "build the numbers" using Dienes blocks. Children selecting 231 as the largest number were shown chip representations of the two numbers and asked, "Which board stands for the bigger number?" "How do you know?"

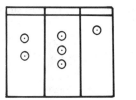 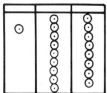

— color coded chips

"The other day a boy/girl said, 'Look at all of these chips here. There's eight green chips here and only one there, nine yellow chips here and only three there. This board has more chips so it stands for the larger number.' What do you think about his/her answer?"

It is important to note that countersuggestions can check the stability of a child's response whether it is "correct" or not.

When an unexpected response is given by a child, it is important to have the child elaborate on or clarify this response. Unless the child's perspective now becomes apparent, the interviewer can construct an appropriate question on the spot to test a hunch about the child's thinking. Because of the sophistication of this technique, I haven't provided many examples of it in the following chapters.

The following excerpts from written reports of student interviewers illustrate both the problems experienced in, and insights gained from, this aspect of interviewing.

. . . I also found that I have very little flexibility as an interviewer. I was always so concerned about asking every question verbatim from my notes that I sometimes lost sight of the purposes of the testing.

M.B.

. . . With regards to changes I would like to make, the first would be to cut down on the amount of talking I do, particularly in rephrasing my own questions and in canceling open-ended questions by following rapidly with closed ones.

B.S.

The major problem with respect to my questioning was excessive interrogation. Instead of asking a single question in response to children's actions, I tended to present a string of questions all at once. . . . I asked fewer questions of the nine-year-olds, thereby providing them with a greater chance for explanation. It seems that with younger children, I tend to overquestion because I want to be sure they are understanding the task. This I feel is a definite area to be changed. Children of all ages require time and patience from the interviewer/teacher.

I would like to change my repetition of questions when met with an inappropriate response. This behavior did not appear to clarify or aid children's understanding and would be beneficially replaced with a rephrased question. Finally, I would like to change the pattern by using the child's answer to go on. In doing this, I believe the interviewer is in a better position to follow the child's line of thinking.

N.L.

. . . I could have focused on more possible tangents to pursue and asked in-depth questions along the way. For example, I could have asked Charles on Task #1 what the zero meant to get a deeper perspective of his thinking.

K.C.

. . . There were times when I became flustered at a child's response, or lack of response, and didn't know what further course to take in questioning. There were also times when a child was confused by my questions on the task and I had a hard time in making my intent clear and understandable.

J.D.

In asking for justifications for their answers, I tried to vary my wording, but relied on just a few phrases. "Why do you think so? What makes you think that? Why is that? How did you know that?"

There were several times when I felt I should have followed up on an answer with some type of response to get more clarification. When listening to the tapes these seemed obvious, but they didn't at that time. . . .

C.B.

I learned that children (at least some third graders) give "reasons" for what they did based on what is often illogical but seems to make sense to them. Unfortunately it occurred to me that some of these "reasons" were given because some children are conditioned to give answers without a chance or appreciation of "thinking them through."

N.L.

The most important point I saw demonstrated was the necessity for having the child justify her answers, whether they were right or wrong. In many instances I heard the child give the correct answer with an incorrect or inconsistent reason. Equally important were those children who gave the incorrect answer and had a valid reason for doing so.

C.B.

Some of the difficulties identified by these student interviewers cannot be avoided consistently even by experienced interviewers because of the sophistication and spontaneity required. Some interviewers have worked in pairs with one person observing behind the camera or behind a one-way mirror and offering occasional suggestions or questions as needed. This may be done unobtrusively through a one-way communication system leading to the interviewer's earplug (O'Brien and Casey, 1982b).

WAITING AND LISTENING

An indirect measure of your listening ability is the amount of wait time or response time that you provide the child. Since reflection on intellectual challenges is to be encouraged, setting a pace that is comfortable for the child is important. By providing an adequate pause following the initial question (called wait time #1) and prior to repeating or rephrasing it, you are allowing the child time to interpret the question and to construct a response. By providing an adequate pause following the child's response (called wait time #2), you are indirectly encouraging the child to elaborate on his response. If you are listening only for a particular response, you will tend to shift abruptly to the next question after hearing what you want to hear. If, however, you are really curious about the child's reasoning, you will want to hear all that the child has to say. You can indicate this by an adequate pause following the child's response.

Although the idea of wait time (Rowe, 1973) is similar in all interactions, the definition of wait time #1 varies with group and individual situations. In group situations the teacher has some control over the length of the pause called wait time #1 before calling on a particular child to respond. In an individual interview, the child has control over the length of the pause before responding, except in the special case when the child has not responded yet and the interviewer interjects a clarification of the initial question. If the original question is followed immediately with a rephrasing, the pause called wait time #1 is nonexistent (0 seconds).[1]

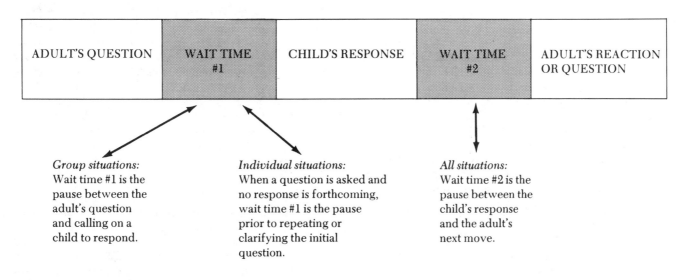

| ADULT'S QUESTION | WAIT TIME #1 | CHILD'S RESPONSE | WAIT TIME #2 | ADULT'S REACTION OR QUESTION |

Group situations:
Wait time #1 is the pause between the adult's question and calling on a child to respond.

Individual situations:
When a question is asked and no response is forthcoming, wait time #1 is the pause prior to repeating or clarifying the initial question.

All situations:
Wait time #2 is the pause between the child's response and the adult's next move.

---

1. Wait time #1 has been redefined for individual situations by Naomi Farina, one of the student interviewers at California State University, Northridge.

Based on small-group and classroom research (Rowe, 1973), an average wait time of at least three seconds seems to provide an adequate pause. However, you may find that the length of your pauses may vary both between and within interviews. Your longest pauses will be given to the child who is hesitant to speak and to the child who is experiencing difficulty with a particular task. As you adjust your pauses according to the natural pacing of the child, be certain that an impulsive responder receives adequate time for reflection. A comfortable pacing of the interview not only provides for the needs of the child, but also provides you with a better opportunity to listen to the child's responses and to reflect on their meaning.

Reactions of student interviewers to this aspect of the interview process include varied concerns for the length of their wait time, suggestions for orienting children to extended wait times and for waiting out "I don't know" responses, awareness of the impact of extended wait times on the quality of children's responses, awareness of personal implications of wait time, and insights into the essence of the interviewing process.

> . . . There were numerous times when my responding too quickly cut the child off. We both began speaking at the same time, and when I stopped to let the child speak, he/she was reluctant to continue or had lost his/her train of thought.
>
> B.S.

> . . . Sometimes I found I was trying so hard to remember what the next task was that I didn't listen to the children very well. When these become familiar I should be able to direct full attention to the child and what she is really saying.
>
> C.B.

> . . . Another point that I noticed was that I frequently presented two related questions to a child without allowing for any response time in between them. My purpose was to rephrase the question so that I would be sure to communicate precisely. The result, however, was that sometimes a child would begin to respond, hear me continuing to talk and therefore cease their response. I think this was probably the most valuable lesson I learned from analyzing my wait time. . . . Incredibly to me, there were a couple of times I actually interrupted a child's response in order to clarify my example. I was so surprised to hear this that I'm sure I will remain conscious of my tendency to be redundant for quite a while to come.
>
> M.S.

> . . . My wait time following children's responses ranged from one-half second to six seconds with an average of just over one second. I feel that this is a weakness in interviewing because at times in listening to the tapes,

I felt as though the children may have expanded their answers when given the opportunity. . . . I found that in instances when more wait time was provided the quality of the children's answers increased. They tended to explain their answers in more than one way.
>
> N.L.

Goals for change include being less of a "talker" and becoming more of a "listener." I think that this goal has import not only in math learning, but for development of self-worth and mutual respect among children. . . . I plan to implement wait time by consciously counting to myself, allowing increases in children's responses. It is hoped that this silent counting can eventually be replaced by a natural pause following questions and responses.
>
> N.L.

As I was interviewing I was very conscious of waiting for the students to think. And I was amazed at how well they did think when I gave them the chance—and how their answers did come in spurts. The students didn't even feel uncomfortable when I just let them think and gave them lots of time. Before each of my first interviews I had told the students that I wanted them to think and would give them lots of time. I did that because I thought they'd be uncomfortable during the pauses.
>
> K.D.

To encourage wait time I often just looked at the child and smiled or said "mmhmm" while still looking at the child. When the pauses were too long, Kenny would repeat the answer again and again until I gave him some kind of confirmation. It would have been better to have asked him if he could tell me anything else rather than making him uncomfortable.
(Average wait time: 2.8 secs.) Z.K.

> . . . I was honestly surprised at the children's unfolding once they had realized that I was willing to wait as long as it took them to answer. I received more responses than I expected to my probing questions. Some strong "I don't knows" (which had obviously worked in the past as a way of dismissing the teacher) yielded answers after the child discovered my habit of waiting forever. Several times the children elaborated upon short answers to make comments that gave me unexpected insights into their ideas about and attitudes towards math.
>
> R.F.

My wait time was variable and changed with the circumstance. . . . I was patient to wait for an answer, even if it took a long, long time. The second wait time, after the child's response, was more telling. At best, as in Hillary's interview, it averaged between four and five seconds. At worst, as in Brian's interview, it averaged less than one second. I was able to sustain a longer wait time 2 if I did not know the child from some other experience. I have known Brian for about five years because his mother works with me and I have listened to him whine at her many times. I have a feeling tone about him, just as I have about other students at school. That feeling tone had an effect on my wait time.[2] When I asked Brian a question, I was patient to wait for an answer, but once the answer was there and right, I was ready to go on. A questionable response would bring more wait time. Had he wanted to, he could have read my wait time as a "correct" or "incorrect" after each response.

I hear myself doing this daily with other teachers also. Sometimes I cut off the last part of their word (How do I know it's the last word?) in order to go on—to move the agenda. Again, the more negative my feeling, the more apt I am to do it.

T.M.

I am feeling more and more the impact of lack of wait time (average of just over one second). In my case, the lack of wait time reflects a hidden agenda for the responder. The implication for the responder is that this is a guessing game in which he might get lucky and give me the "right" answers. My hidden agenda limited my scope of the interview and it was only in rereading the transcripts that I recognized how limited and limit-

ing some interviews were. Examples of this can be seen in Scott's interview. He had one concept of "more" and I had another. In not really probing his concept of the task, I lost the opportunity to really assess his thinking about the number concept. He was answering a different question than I was asking.

Basically, providing wait time requires an inquiring rather than an anticipatory frame of mind. In this set of interviews there were several motivators for me. The largest one was to confirm or deny certain Piagetian observations as they might relate to math concepts. Therefore, I was not keying on the individual child so much as anticipating certain patterns of responses. My frame of mind was out of synch with maximizing wait time.

L.W.

While some of these reflections consider the interview process in isolation, others consider it in relation to teaching. Many of the reflections focus directly on specific interactions and behaviors, or on the general interview process itself. Others extend to concerns for the quality of interaction in the classroom and with other people.

RESPONDING WITH ACCEPTANCE AND ENCOURAGEMENT

As highlighted by the last student interviewer comment, the essence of the interview process is the open spirit of inquiry in which the interviewer is curious to learn everything about the child's perspective on certain problems. By accepting the child's responses with nonjudgment and with encouragement to elaborate further, the interviewer communicates both a respect for the child's thinking and a genuine curiosity to learn more. In this kind of atmosphere the child is likely to feel free not only to do his best thinking, but also to verbalize his thoughts fully.

Each response, whether right or wrong, has the potential of providing information about the child's level of understandng at that moment. Children's responses, therefore, can be acknowledged as having been heard ("Uhuh," "Hmm," nod of the head), but

this must be done in a neutral though friendly manner.[3] The interviewer's posture, facial expression, voice intonation, reaction, or phrasing of questions must communicate no information on the "correctness" of the child's response. Unless the interviewer refrains from direct or indirect evaluative feedback, the child is likely to begin to give responses according to the imagined expectations of the interviewer rather than to share his own spontaneous reasoning. In this way a child's thoughts become "adulterated."

In interviewing second-grade children on their understanding of equality relations and their representation in number sentences, I demonstrated both sensitive acceptance of "incorrect" answers and insensitive biasing of children's responses.

When exploring children's ability to represent

---

2. Another student interviewer, working with her daughter and other children in the neighborhood, reported shock at her impatience with her own child. The shortest average wait time occurred in her daughter's interview. There is risk in interviewing children in your own family or classroom if you are quick to assume responsibility for what they really understand.

---

3. Another form of acceptance is the repetition of a child's response. Not only does it provide an alternative neutral response for the interviewer, it also may encourage the child to clarify or extend his response. However, regardless of the neutral content of an interviewer's response, it is the tone of voice that ultimately determines acceptance. A conversational tone is more effective than a formal "teacher" tone (Scheer, 1980).

two equivalent groups of objects in a number sentence (4 = 4), I was able to accept other responses, such as $4 + 0 = 4$, $4 - 0 = 4$, and maintain the flow of ideas until I learned whether the children would entertain $4 = 4$ spontaneously.

> "That's one way. Is there another way to show they're the same?"

During these interviews my hidden agenda included eliciting only variations of number sentences. Therefore, when some children began to use pictorial representation, I quickly interrupted them and reminded them to use written notation. By not being open to information on children's informal representation of numerical ideas, I drastically limited my study of children's thinking and representation of equality.

The latter example clearly illustrates the importance of accepting children's responses and remaining open to all possibilities.

Most children appear comfortable with nonevaluative acceptance. Yet some children may be so readily attuned to getting the right answer and relying on the teacher for cues that it may take them a while to get accustomed to a different expectation, that the interviewer really is interested in how they figure out the answer, not in the answer per se.

In my interviews, one exception to the children's apparent comfort with nonevaluative acceptance was a boy who was very watchful for any clues about the correctness of his answers. When faced with "uhuh" or silent pauses, he quickly altered his response to see whether this produced a different reaction. His rapid vacillation between responses on one occasion almost resembled a disequilibrium state.

At the same time that the interviewer attempts to maintain a manner of neutral acceptance, he remains friendly and encourages the children's best efforts.

Since second graders found missing addend and comparison number stories to be quite challenging, sometimes their efforts appeared in need of encouragement. In response to a boy's concern about forgetting the details of the story problem, I replied,

> "I can understand. There's a lot to remember."

Then I reread the problem.

In response to children who were extremely hesitant to write a number sentence for comparison number story problems, I replied,

> "I know it's not easy to think about. Would you write down your best idea?"

> "Let's pretend that you knew what to write down. How do you think it would come out?"

The latter forms of encouragement are effective for coaxing out responses about which children are unsure.

The following comments by student interviewers underscore the difficulty of implementing this very sensitive aspect of the interviewing process.

> I used "OK" at least twenty-four times in only ten minutes. I believe half of the time it was used as a gap filler, and the other half as a sanction. I definitely want to reduce the amount of times I say this.
>
> Z.K.

> Numerous sanctions ("OK," "fine," "good," "right") were threaded through my sessions. As I listened to the tape I sounded like a used-car salesperson egging the customer to quickly "close the deal."
>
> M.N.

> I was too quick to evaluate the response as right or wrong. I would then either immediately validate it with "OK" or respond with "but . . . "
>
> L.W.

> I used "OK," "uhuh," and "alright" as accepting responses . . . though at times my "OK" had a note of finality about it . . . .
>
> C.B.

> In my insecurity, I often prodded them gently for the answer, when what I was actually looking for was the *process* of problem solving, regardless of the answer. I had an irresistible compulsion to not leave the child with the wrong answer, and sometimes even supplied it. On the tape, this was usually followed by "—right?" as though that made it a question. . . . My second wait time (following the child's response) is all but nonexistent, replaced by verbal reinforcement in the form of "good," "very good," "right," and "OK"—the latter of which I also overused when moving on from one task to the next. What little instructing experience I've had has involved immediate positive reinforcement for appropriate responses, and that habit is hard to break. I wasn't even aware that I was doing it until I listened to the tape.
>
> K.C.

Interestingly, "OK" was the predominant interviewer response to both correct and incorrect responses by the child. I think this is probably due to the fact that I viewed these interviews as a series of observations without placing an emphasis on factors of right or wrong; therefore every response by the child, right or wrong, was viewed as being "OK. . . . I was surprised to find so many responses with positive connotation ("OK," "good," "oh, very good,"). Despite the apparent judgment of "correctness," the intention of these comments was simply to acknowledge the child's efforts. Nevertheless, since the responses could be misconstrued by the children (as could the absence of such responses) it would probably be a good idea to limit interviewer responses to more neutral utterances.

M.S.

Again, one of the student interviewers captured the essence of acceptance in the interview method of studying children's thinking.

> If I withhold judgment (approval or disapproval) and just act as an active listener, that will give children the impression that what they think is important, rather than what they think I want them to think.
>
> S.B.

RECORDING AND ANALYZING INTERVIEWS

The benefits of recording and analyzing your interviews with children cannot be overemphasized. Since each interview has the potential of uncovering a wealth of information, we cannot depend on our memories to reconstruct each interview. There are a number of ways to record an interview:

1. Jot down notes on what the child says and does during the interview.
2. Have an unobtrusive observer jot down notes during the interview.
3. Audiotape the interview.
4. Videotape the interview.

A major problem in writing notes and interviewing simultaneously is the inability to give full attention to the interview process. This method, therefore, should be avoided whenever possible. An audiotape of the interview allows you to replay it for listening or jotting down notes on what the child and the interviewer said and did. Audiotaping may be supplemented by the interviewer's written notes on the child's inaudible behaviors, like interaction with the materials. Having a third person record observations can provide feed-

In addition to withholding judgment, it's difficult not to get excited about what's happening. Yet letting that excitement show really affects what happens.

At the same time that the interviewer must accept all authentic responses by the child, it is also appropriate not to accept excessively diverting responses and behaviors (Alward and Saxe, 1975). A child who persists in playing with attractive materials may be told, "Right now, I need you to look at this problem. When we're finished you can play with the blocks again." Since you are working within certain time constraints you may decide to assert your needs. You need to be aware, however, that you may be running the risk of closing off behaviors that later prove relevant. If the child persists in diverting behavior you can say, "Since I need to have you look at this problem and you're not ready to do so now, maybe we can try this another time," and terminate the interview. The interview may also be terminated if the child is uncomfortable because of achievement anxiety, illness, missing other activities, and so on. It can be rescheduled at a time that is better for the child. In over two hundred interviews conducted for the explorations reported in this book, only six were terminated.

back on both the child's and interviewer's behaviors. The latter method is used best in conjunction with audiotaping. Although combinations of the first three methods of recording the interview provide adequate documentation of the process, the most complete documentation is offered by videotape recording. Both audible and visible behaviors can be replayed for analysis by the interviewer. The student interviewer reactions described earlier were based on an analysis of audiotapes. The interviews described throughout the book are illustrated from videotapes of the interactions.[4]

Orienting children to videotape equipment can be done best in a group situation prior to individual interviews. In addition to experiencing being videotaped and watching a video playback, the children can gain an understanding of the purpose of the recording during the interview and of its restricted audience.

---

4. The interviews were videotaped usually without the aid of a camera operator. By focusing the camera on the interview area in advance, the interviewer can turn on the videotape equipment quickly as the child enters the room. The availability of a camera operator adds the possibility of close-ups of facial expressions and manipulation of materials.

"So that I can figure out how you think about numbers, I need the camera to help me remember what we did. After talking to you alone, I won't be playing it back for the whole class to see, or on Channel Seven News for the whole city to see. When I play it back it will be only for myself to study."

Each child can be given an opportunity to view a replay of a portion of his own interview at its completion.

No matter what the interview method, having a record of the child's written work is important. In videotaping, the child can be given a felt marking pen and asked to write "large enough so the camera can see it" and given a model of the size required. In the absence of a videotape recorder, you might provide another piece of paper at the point when the child is about to erase or cross out his work and start over. This provides you with a complete record of the child's sequence.

Although this book contains summaries of relevant research on children's numerical thinking, as in the next chapter, most chapters are based primarily on videotaped interviews conducted by me. During the course of the interview explorations, I replayed the videotapes and took written notes on the same day as the interview. Following a series of interviews in the same content area, I summarized and analyzed the data. In the process of writing this book I reexamined the data and replayed many of the interviews. The videotapes contain such a wealth of information on both children's thinking and the interview process itself that each time I viewed the videotapes I observed something that previously went unnoticed. Also, on second viewing of a videotaped interview I often had a different interpretation of a child's response or actions. For this reason it was critical to have feedback from other viewers. Teachers as well as researchers have provided me with such feedback based on the study of both the original videotapes and illustrated interview episodes taken from the videotapes.[5] My interaction with the various sources of feedback has provoked some rethinking that ultimately strengthens this research.

Whereas maximum benefits can be gained from an analysis of your own interview tapes, there is also considerable value in group viewing and discussion of videotaped interviews.

A reaction of one of my students to videotapes of second graders' sophisticated counting strategies was, "Where did these kids come from . . . Mars?" Although she was a second-grade teacher, she hadn't observed such behaviors before. However, once aware of the children's potential, she gradually saw what had previously gone unnoticed and unappreciated in her classroom.

During group viewing of videotapes, teachers become involved in making sense of what the children are doing in group discussion. This involvement is motivated by the realization that the better interpreters they become of how children are seeing a problem, the better their on-the-spot decisions in working with children in the classroom will be (Duckworth, 1981). Getting an intimate look at children in the process of thinking also offers a different perspective of teaching and learning. A teacher from a seminar conducted by Duckworth recorded the following reflections.

> As always the task is to be *really* invested in understanding what a child is thinking. And in the very process of unearthing that, learning, growing, changing is going on. I feel closer to being able to do that after watching than before. I feel closer to changing the vested interest in *my objectives*—or at least believing that the process alone is valuable. I guess for the first time clearly I saw children learning—the process of learning without the answers being fully intact. Ah, so many times around on this issue. (p. 20)

Observing children in the process of learning prompted this teacher to reconsider her role in the classroom.

Throughout this book you will have occasions to interact with illustrated episodes from interviews with children. These will provide opportunities for you to gain a deeper understanding of children's numerical thinking as well as for you to consider alternative ways of interacting with children. The interview tasks themselves may suggest starting points for teaching methods and checkpoints for assessing children's understanding in the classroom. Furthermore, interaction with the interview episodes will offer you opportunities to reexamine the teaching-learning process from different perspectives, and even to restructure your classroom.

---

5. Although considerable feedback provided a check on observations and interpretations of selected task responses, lacking a research grant to support an assistant, I was not able to do a systematic check of all responses. Feedback was sought especially for interpretation of unusual responses.

REFLECTIONS AND DIRECTIONS

Although much can be gained from group viewing and discussion of videotaped and illustrated interview episodes, you can gain maximum benefits from conducting your own interview explorations. Piaget himself has recommended that prospective teachers spend considerable time interviewing children as a preparation for the classroom. He recommends that such a study of children's thinking be conducted with a number of different children on the same tasks. This provides critical experience in rephrasing questions and inventing new ones in your attempt to follow each child's thinking.

You are invited to undertake your own interview explorations and to reflect on the following:

- Children's numerical thinking
- Quality of your interactions with children
- Classroom implications of what you've just learned

This is no simple undertaking. It is neither a one-shot affair nor is there one single way of approaching it. Here are some suggestions to get you started.

1. One way to begin is to engage children in conversation on numerical tasks with the intent of learning from them. If you have a curiosity about learning and a respect for children's thinking, the process of interacting with children will lead you progressively to deeper concerns about thinking, learning, and teaching.

2. At those times when you doubt that children really do respond in the manner described in this book, you are invited to try similar tasks with children of the same age in your school or neighborhood.

3. At those times when you wonder how the children described would have responded if the materials or the questions had been different, you are invited to devise your own tasks and try them out with children.

4. Although giving the same interview tasks to several different children is ideal, it may not always be practical. If the number of available children is limited you may choose to increase the number of tasks given. If the ages of the available children may vary greatly you may choose to vary the tasks for each child.

5. When you are ready to undertake recording and analysis of your own interviews, additional practical suggestions can be found in Appendix B.

PART TWO

EARLY NUMBER CONCEPTS– Thinking, Learning, Teaching

3

COUNTING IN THE EARLY YEARS: CAPACITIES AND CONSTRAINTS

PREVIEW

Most young children enter school with impressive language and counting skills. However, just as children's knowledge of words and subtle meanings in language usage continues to develop throughout their schooling, their capacities to count continue to develop over the next few years. Since counting appears to be a primary avenue to children's acquisition of numeration and number operations, it is important to take a look at young children's counting and how it develops. This chapter will focus on children's capacities and constraints in counting during their early years, from preschool through the first grade. Examples of their counting capacities and constraints will be discussed in the context of related research and current controversies.

To initiate your study of young children's counting, consider the following statements and select those most consistent with your prior observations and understanding.

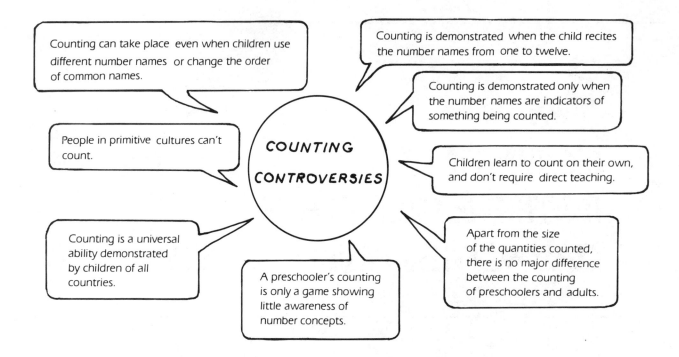

YOUNG CHILDREN AS ACTIVE COUNTERS

Observers of young children marvel at their independent activity in acquiring counting abilities. Evidence of their self-directed activity cited by Ginsburg (1977) includes the following behaviors: spontaneous counting resulting from a willingness to practice and to count almost anything, active experimentation with the decade cycles in the number names in an attempt to expand their number name sequence, and a willingness to ask for more information on number names as needed. In an incident reported by Gelman (1978), one young child indicated his willingness to practice by voluntarily repeating his count twenty times.

Counting appears to be the primary avenue to number ideas for most children. It is a natural response of young children, although not the only response, to the following questions raised by themselves or others:

How many? How many altogether? How many left?

Are both groups the same? Does one group have more? How many more?

Examples of counting and other responses to such questions will be discussed in the sections that follow.

Anthropologists have observed that most cultures use a counting procedure. Since counting procedures may be based on number groupings other than 10, some observers have been misled to conclude that certain primitive tribes have no counting system. On the Ivory Coast of Africa, Ginsburg noticed that all children used counting for solving simple numerical problems regardless of background—unschooled, schooled, not belonging or belonging to a trading society. On the basis of such general observations, psychologists (Ginsburg, 1977; Gelman, 1978) conclude that *counting is a universal ability of normal people.*

Children's counting abilities, although developed largely through their own activity, are also dependent on cultural influences. The counting words used by children depend on the language of the culture as modeled by the adults in their environment. Although initially the counting words are available only from adults, later the same words serve as raw material in which children notice repeating patterns and derive new number names to extend their number-name sequences. An indication of such activity is the invention of original counting-word combinations such as *twenty-ten* and *twenty-eleven* for *thirty* and *thirty-one* prior to hearing and adjusting to the adult conventional names. It is this focus on extension of their counting-word sequences to one hundred and attaching these words to things that characterizes the counting of most young children until the age of six or seven. Another indicator of children's activity is their construction of numerical meanings that can then be read into the counting words. The constraints on children's counting abilities resulting from the gradual nature of this construction process, produce a lag in practical applications of counting in problem-solving situations. These capacities and the constraints on young children's counting will be discussed in further detail in the following sections.

YOUNG CHILDREN'S COUNTING CAPACITIES

To help you appreciate the extent of children's achievements, the following discussion will illustrate the numerous abilities required for successful counting. The difficulties encountered by two- to five-year-old children prior to coordination of the many aspects of basic counting will be illustrated as well.

Rote Counting

Oral recitation of standard strings of counting words characterizes rote counting. These sequences are produced by young children with considerable effort and contain more than a conventional portion that is repeatable from trial to trial. Fuson (1980) reports a general pattern in young children's attempts to learn the counting sequence. Notice the similarities between repeated recitations of two different children.

Child A:	1	2	3	4		9	10	11	8		6	9				
	1	2	3	4		9	10	11	8		10	3	4			
	1	2	3	4		9	10	11	8		4	5	10	11	8	2	4
	1	2	3	4		9	10	11	8		3	8				
Child B:	1	⟶	7		11	12	13		16	11	16	11
	1	⟶	7		11	12	13		14	18	20	
	1	⟶	7		11	12	13		1	6		
	Conventional and Stable			Nonconventional but Stable			Random and Unstable				

Children's counting-word sequences over repeated recitations contain a portion that is a stable sequence of standard names. The sequences also include a portion containing omissions or order changes that are also produced in the same way over repeated trials. Each sequence then ends with a random "spew" of number names representing the children's trial-and-error attempts to extend the sequence. The middle portion may be used consistently by the same child over a period of several weeks although the content of the stable pattern will vary from child to child. The final portion has no pattern, varying for each child on each attempt to extend the sequence.

Beginning at about age two, the acquisition and extension of the conventional counting-word sequence continues for several years. An indication of children's gradual progress is given in a summary of research by Fuson and Hall (1982). They report an average count of 13 for the three-and-a-half to four-year-old group and an increase of the average count to 51 for the five-and-a-half to six-year-old group, with considerable variability within each age group. These researchers also comment on the "decade problem" experienced by these children. In extending their counting-word sequence, young children demonstrate familiarity with the repeating cycle of one to nine within the decades—for example, forty, forty-one . . . forty-nine, but have difficulty in ordering the decades correctly. However, by the end of the first grade most children have mastered the counting-word sequence to one hundred.

Rote counting is the earliest form of counting observed in children. Although they may recite several words in standard sequence, they do not yet use these words for counting objects or events. The children's sing song delivery indicates that this isolated counting-word sequence represents nothing more than a memorized string of words. These number names, often learned in the context of games and rhymes, may signify nothing more than nonsense syllables to the child for some time. It can be argued further that when a child is reciting number names in isolation he is not actually counting; these number names must be placed in correspondence with objects or events before the act of counting is taking place.

Although oral recitation of number names is only one aspect of the counting process, it is emphasized almost exclusively by children's television programs such as "Sesame Street." Furthermore, it is overrated by concerned parents and accepted as a quick indicator of their children's expected brilliance.

Counting Objects and Events

Once rote counters begin to assign counting words to objects or events, they are faced with the challenge of coordination in assigning successive counting words to the items being counted. Only one number name can be paired with each item for an accurate count. The successful assignment of individual number names to each object in a collection requires that the successive oral naming and pointing to each object be perfectly synchronized. For young children, counting things proves to be a formidable challenge requiring years of laborious practice. Some of the coordination demands of this task for young children are reflected in the following partial list of errors a child can make in counting fixed linear arrangements of objects.

1. The child begins the verbal count one step ahead of the pointing action so that four objects get counted as five.

2. The child skips an object so that seven objects are counted as six.

3. The child points between objects as a number name is given, counting eight objects as nine.

4. The child gives two number names to the same object so that five objects are counted as six.

5. The child gives one number name to two different objects so that six objects are counted as five.

6. The child continues reciting the number names beyond the last object as if carried away by the rhythm of the sound sequence. In this way five objects can be counted as six, seven, or eight.

Such problems with children's coordination of different aspects of the counting act account for the gap between the length of a number-name sequence a child can recite and the number of objects he is able to count. For example, a five-year-old child may be able to recite numbers to fifty, yet only be able to count eight objects.

Spatial arrangements of collections of objects to be counted may also affect the difficulty of the counting task. Children who are successful in counting a fixed linear arrangement may make errors when the same number of objects is displayed in a fixed circular

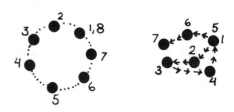

or moveable, random array. Since in both of these arrangements it is difficult for a young child to separate those objects already counted from those yet to be counted, some objects are recounted. Given movable objects, many children will impose their own order on them such as placing them into a linear arrangement or into an array. Furthermore, they may physically partition the collection into successive groupings of counted and uncounted objects.

Young children are most successful at counting small quantities of objects until the collection is exhausted. When asked to count out a prespecified quantity of buttons from a larger collection, they are likely to forget the number of buttons they were asked to count. As a result of the added memory demands of the task the children are likely to overcount, often exhausting the available collection.

Young children's counting is characterized by physical activity, with touching or pointing being an essential component of counting objects. The dependence of very young children on touching the objects being counted has been demonstrated in a study by Gelman and Callistel (1978). Here, three-year-old children showed a marked decrease in accuracy when the collections of objects to be counted were covered with glass. Without touching, these young children had difficulty in pairing up counting words with each object. At the same time that touching and manipulating objects to partition collections are essential in early counting, physical coordination limitations on young children's hands and fingers slows progress to counting larger sets of movable objects. In time, dependence on touching gives way to pointing. Gradually, the physical actions of the counting act become restricted to less apparent nods of the head, or eye movements.

Whereas things to be counted need not look alike, young children initially may get distracted by the physical properties of the objects. Fuson and Hall (1982) report that in the middle of counting a group of objects by successive partition into two subgroups of counted and uncounted objects, two-year-old children may begin to build structures suggested by the physical properties of the objects. The objects to be counted need not look alike, according to Piaget and Inhelder (1969), because by ignoring their physical properties in the process of counting, the child abstracts equivalent units from them. This would explain why it is possible to count a mouse and an elephant in the same set.

An interesting question arises when children repeatedly count a group of objects and obtain the same answer using an unconventional sequence of counting words. The question is whether these children are actually counting.

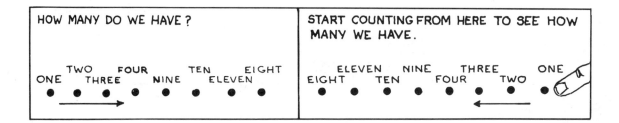

Although most researchers identify the use of a conventional counting-word sequence as a criterion for the counting act, one research team argues against this view. In studying the counting of two-year-old children, Gelman and Callistel (1978) observed them count three objects as "one, two, six" on repeated trials, substitute alphabet letter names for number names, or even invent their own number words. For these researchers, a child's use of a stable nonconventional word sequence would be sufficient to indicate counting.

CONSTRAINTS ON YOUNG CHILDREN'S COUNTING ABILITIES

Thus far, we have looked at two phases of young children's counting that overlap in their development. Young children continue to extend their counting-word sequence while gradually mastering the counting of increasing numbers of objects. In this section we look at a third overlapping phase of counting that is the slowest to develop—the construction of number meanings to assign to the counting words within a relational framework. This slow acquisition of number meanings places constraints on children's counting abilities and on the reliability of counting as a problem-solving tool for them.

Quantification

For adults, number names represent specific meanings in different contexts. These names may represent the number of objects in a collection, a number of units of measure, the relative position of an item in a sequence, or even an identification code, depending on the context.[1] Children gradually construct number meanings for different contexts and integrate them within a relational framework. This construction of meaning to assign to counting words lags far behind the other aspects of counting discussed earlier. The following examples from the work of Churchill (in Ginsburg, 1977) and Ginsburg (1977) illustrate the difficulties young children may experience in the process of constructing these number meanings.

---

1. In the first chapter, the block task presented a measurement context. The number name indicated how many of the smaller units were contained in successively larger units or in the arrangement of mixed units.

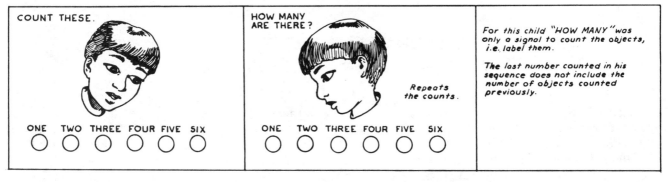

Beyond labeling individual objects in a collection with a name, counting eventually involves a further mental act of relating the individual objects into wholes of increasing size (Labinowicz, 1980).

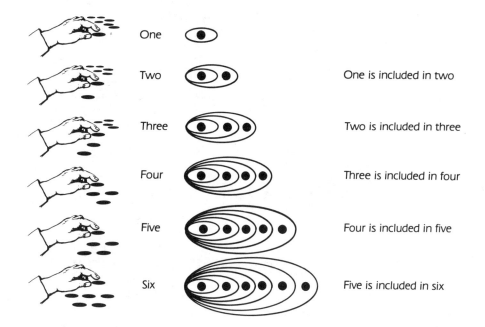

One

Two — One is included in two

Three — Two is included in three

Four — Three is included in four

Five — Four is included in five

Six — Five is included in six

Here, "five" is considered as a whole group. With the addition of one more, "five" becomes part of "six." Numbers come to be understood as whole groups and as the parts that make up the whole. For an older child, "six" indicates not only the counting word attached to the last object counted, but also the "manyness" of the whole group.

Whereas it is clear when a child's response to the "How many?" question indicates his inattention to the "manyness" of a collection of objects, Fuson and Hall (1982) point out the difficulty of determining if a child is attributing this deeper meaning to his counting. In addition to providing a correct response to the "How many?" question, a child may repeat the last counting word in the original counting act, or emphasize the last counting word with a pause or increased volume. A

child's spontaneous adaptation of his counting act to emphasize the last counting word would suggest the construction of a deeper "manyness" meaning. However, in an interview situation it is difficult to determine whether this emphasis behavior was developed spontaneously or whether it is merely the result of training with no underlying meaning. Observing children's behavior on other tasks may be useful in discerning whether they are attributing meanings such as "manyness" to their counting.

If a child is paying attention to the "manyness" of a whole collection of objects, he should have the expectation of producing the same result regardless of his direction in counting the same arrangement. Reiss (in Ginsburg, 1977) reports that children have been observed expressing surprise at this discovery.

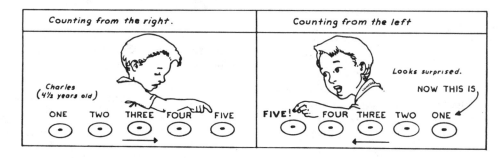

At the same time, other researchers indicate that young children can start their counting at different points of an arrangement in repeated trials as though aware that

order of counting doesn't matter in determining how many. Gelman and Callistel (1978) gave children a linear array of assorted trinkets and asked them to start their count in the middle of the row. Their study showed that four- and five-year-old children behaved as if aware that the order of counting didn't change the count. An episode from an interview with a young child (age: 3 years 8 months) is illustrated below.

It would be interesting to study children's reactions to the order of counting as the quantities and the randomness of the arrangements are increased.

If the child's count focuses on the inclusion of objects already counted to determine the "manyness" of the set, then he should be disturbed if a recount produces a different number. However, Fuson and Mierkiewicz (1980) observed that when three- and four-year-old children produced different answers on a recount of the same collection, they gave little evidence of being disturbed. When the discrepancy of the two counts was pointed out to the children and they were asked which of the two answers was correct, they invariably indicated the last answer as the correct one. Some of the five-year-olds noticed the discrepancy between successive counts and spontaneously recounted the collection a third time. Such behaviors prompted Fuson and Hall (1982) to conclude that much of young children's counting is done for practice in execution of the counting act rather than for determining the number of objects included in a collection.

Also, there is little evidence that young children when extending their string of conventional number names, have constructed number-structure meanings that these names represent for adults. For example, they have little understanding that "twenty-three" is composed of two tens and three ones. For them, "twenty-three" is merely the counting name that comes after "twenty-two." Even though young children are active in extending their conventional string of counting words, their extensions appear to be based more on patterns in language than on the structure of our number system.

This lag in development of number meanings by young children limits their abilities to quantify collections of objects, to develop efficient counting strategies, and to use counting as a reliable problem-solving tool.

Flexible and Efficient Counting

Young children are impressive in their abilities to recite sequences of counting words beginning with "one," but have difficulties in starting their count at other points in the sequence and counting in either a forward or backward direction, in response to questions like the following:

When you are counting, what word comes right $\begin{Bmatrix} \text{after} \\ \text{before} \end{Bmatrix}$ seven?   (sequence)

What number is one $\begin{Bmatrix} \text{bigger} \\ \text{smaller} \end{Bmatrix}$ than seven?
(manyness)

When you are countng, what two numbers come $\begin{Bmatrix} \text{after three and before seven?} \\ \text{before seven and after three?} \end{Bmatrix}$ (sequence)

Tell me two numbers that are $\begin{Bmatrix} \text{bigger than three and smaller than seven.} \\ \text{smaller than seven and bigger than three.} \end{Bmatrix}$
(manyness)

A general lack of flexibility, or inability to see one number in relationship to others—smaller or larger—in young children's counting suggests that their counting-word sequence has been internalized as a total unidirectional entity.

When given addition and subtraction problems young children may be able to solve them by counting, but in ways that are less than efficient. A common solution to an addition task is illustrated on the facing page.

A. HOW MUCH IS FIVE AND THREE MORE?			
ONE, TWO, THREE, FOUR, FIVE.	ONE, TWO, THREE.	ONE, TWO, THREE, FOUR, FIVE, SIX, SEVEN, EIGHT.	Although this method of counting can produce accurate results, it is redundant and time consuming. Both addends are counted twice.
The child counts out five objects for the first addend.	He then counts three objects for the second addend.	Finally, he joins the objects and counts the total.	

However, the following illustrations indicate how some five-year-olds are capable of solving addition tasks in a more efficient manner known as *counting-on*.

Similarly, young children may be able to solve subtraction tasks, but their counting methods vary in efficiency.

This ability of some five-year-olds to count-on and to count-back in such an efficient manner, however, appears to be limited to counting-on or counting-back one, two, or three objects. According to Fuson and Hall (1982), the children appear to be unable to generalize this ability to count-on larger quantities until six-and-a-half to seven years. Hence, although some young children gradually progress in both the flexibility and efficiency of their counting, it is still restricted to small quantities of objects.

Problem Solving

The following example described by Greco (in Kamii, 1982a) illustrates a young child's natural approach to problem solving in the absence of adult direction. In this situation, the child chooses counting to solve a real problem, although he does not do so immediately.

A mother asked her five-year-old to put a napkin on everybody's plate for the main meal every day. There were regularly four people at the table. Jean-Pierre knew how to count to 30 or more. However, he went to the cupboard to take out the first napkin, which he put on a plate, returned to the cupboard to take out a second napkin, which he put on the second plate, and so on, making a total of four trips. At five years three months and 16 days, he spontaneously counted four napkins to take out of the cupboard, and distributed them at the table. He proceeded in this way for six days.

On the 7th day, there was a guest and one more plate than usual. Jean-Pierre took his four napkins as usual, distributed them, and noticed that one plate remained empty. Instead of getting an additional napkin, he collected the four already on the plates, and put them back in the cupboard. He then began all over again and made five trips to accomplish the task.

The next day, the guest was not there, but Jean-Pierre continued to make four trips, and continued the same thing for five more days, until he rediscovered counting. After using this method for ten days, Jean-Pierre was told that there was a guest again. He distributed his four napkins as usual, but this time simply went to get the missing napkin when he saw the empty plate. The next day, when there were only four people again, he counted the number of plates before fetching the same number of napkins. The arrival of a new guest never bothered him after that. (p. 29)

Although Jean-Pierre could recite counting words to thirty and could also count small collections of objects, there were occasions when he either never thought of using counting to solve this problem, or he chose not to rely on his counting abilities.

In another study by Greco (in Kamii, 1981), children were shown nine chips in a random arrangement and asked to put out the same number. All of the children in the study were capable of counting more than nine objects, yet counting was not the predominant strategy of the younger children on this task. It was only in the seven-year-old group that most of the children chose to count. Different levels of responses to this task are illustrated below.

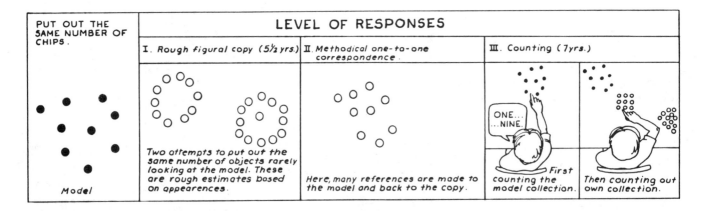

Prior to the age of seven, the children in Greco's study appeared to place more confidence in making one-to-one correspondences or even rough estimations based on appearances than they did in their counting abilities. The findings on this task run contrary to children's general tendency to count as shown on other tasks.

One of Piaget's classical conservation tasks dealing with number provides us with further input on young children's counting as a problem-solving strategy.

To most seven-year-old children, the number of each group is unchanged despite rearrangement, for logical reasons like "You didn't add any more eggs." If these logical reasons were not available, the children could demonstrate the equivalence of the rows by counting. Although many four-, five-, and six-year-old children can count eight in each row, this result does not appear relevant to their conclusion. Some of these young children appear not to anticipate that the counts will turn out the same and therefore do not notice the discrepancy. Often the children become aware of the impending equal count, yet purposely miscount to come out with a larger number for the longer row. Such deliberate miscounts are attempts to preserve their existing ideas.[2] For a long time, making rough estimates based on outward appearances such as length or area of the collection has proved to be a reliable way of comparing the size of groups. In this task the young child's fragile or incomplete concept of "eightness" is pitted against a firmly entrenched perceptual strategy. Young children are convinced that longer or bigger is more, and counting will not shake their conviction for another year or two (Duckworth, 1979).

The following tasks further demonstrate how young children's dependence on perceptual information limits their problem-solving strategies. Both tasks screen off some perceptual information, thereby preventing children from counting in the usual manner. Most young children are unable to invent new counting methods since they are totally dependent on the support of perceptual objects. However, some young children are capable of deriving information that is not perceptually available. A range of responses to the tasks is illustrated below.

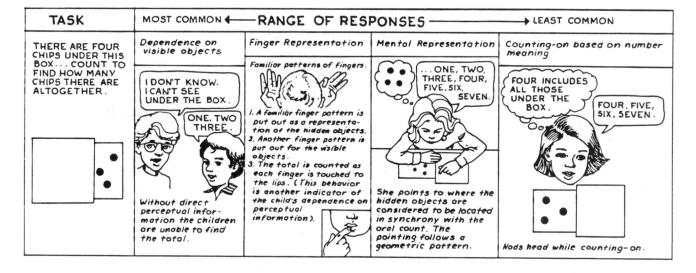

In the absence of visible objects, most young children are unable to count. Some of them progress beyond this direct perceptual dependence by creating representations of the hidden objects that are then countable. In the third response the child's pointing behavior suggests that she deliberately created and counted a figural representation of the hidden objects.

Although the child created a mental representation of the hidden objects, she was still dependent on pointing to their anticipated location. This type of progression in counting has been identified in the work of Steffe, Thompson, and Richards (1982). The fourth response, though least likely to be observed in young children, demonstrated how understanding of the number meaning of the counting word can serve to internalize the counting process and reduce the need for perceptual information.

In an exploration of young children's abilities in obtaining information not perceptually available O'Brien and Casey (1982b) gave variations of the following task to three-, four- and five-year-old children. Some of the response types observed are illustrated overleaf.

---

2. Teachers at the Teachers' Center Project at Southern Illinois University in Edwardsville report that on other tasks, having answered one too many or too few in counting objects, children will deliberately miscount when asked if their answer is correct. This is usually an attempt by the child to preserve himself and his ideas (even his mistakes). It is often not a counting problem.

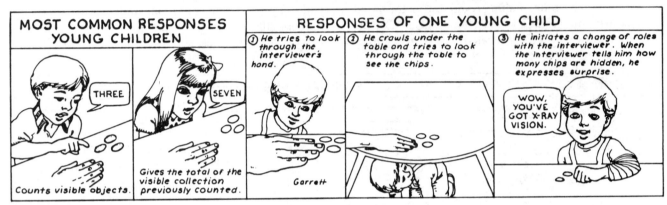

Again, most of the young children demonstrated a dependence on perceptual information, providing answers that were counts of objects currently visible or recently visible (just prior to some being covered). Some of these children were successful when the size of the collection was reduced, when all or none of the objects were hidden, or after changing roles with the interviewer. Yet some children were unable to solve the task regardless of the variations. On the other hand, a small number of children were capable of inferring the number of hidden objects in the initial task. Since these children were not observed using the physical motions characteristic of counting, and they said little in response to "How do you know?" questions, their methods remain a mystery and raise the possibility of strategies other than counting being used. In one exception reported, a child was observed to count in the following manner (O'Brien and Casey, 1982b):

> Child touches the table for each of the showing chips, continues counting up through the total (as though the hidden chips were arranged in a line, not their original position), then counts the touches (again touching) from the number of showing chips to the total. (p. 6)

This child appears to construct his own figural representation of the hidden objects, which is useful in determining how many more counts are needed to reach the total. When the counting is repeated, it is streamlined to counting-on from the visible objects.

Going beyond the perceptual information given is at the core of intellectual activity. For this reason it is important that we explore young children's capacity for such operations in activities like the hidden chips task. O'Brien and Casey (1982b) write on the importance of the activity in mathematics,

> . . . But knowledge—especially mathematical knowledge—is concerned with inference, the deriving of information which is not perceptually available by constructing and coordinating relationships (p. 1–2).

In attempting to understand children's behaviors on such tasks it is also necessary for the researcher to extend himself beyond what is perceptually available and to construct inferences about the child's thought processes. Sometimes by asking clever questions the researcher can gain further information that verifies a preliminary hunch. Unfortunately, because of the young age of these children, further questions were not very productive.

Thus we have seen that young children's counting can be characterized as being based on considerable knowledge that is mainly available externally and on limited knowledge that has been constructed internally. These children acquire a lengthy conventional string of counting words and count small collections of objects when the objects are perceptually available. Despite these accomplishments they have constructed only limited number meanings within a relational framework. Piaget has alerted us to a major reorganization in children's perspectives that marks the initiation of the concrete operational stage at about age seven (an average age). Here, perceptual strategies gradually give way to logical ideas. In Chapter 4 we will explore the counting strategies of seven-year-old children (beginning second graders), and examine the number meanings they have constructed and elaborated to support the development of efficient counting strategies that are reliable problem-solving tools.

COUNTING ACTIVITIES AND TEACHING METHODS

Schools are prone to misjudge incoming children's understanding of counting or of number ideas in two ways. They may assume that a child's rote counting skill (to 50 or even 100) is the equivalent of the child's knowledge of these numbers. Or they may arbitrarily assume that the children have little or no understanding of counting or numbers at all. Regarding the latter case, we have demonstrated in this chapter the considerable knowledge that young children have before entering school. Though the chapter has emphasized general characteristics of young children's counting, it has also provided a glimpse of the wide variations in their abilities. O'Brien and Casey (1982b) write about these variations and their impact in the classroom:

> Of practical importance is the fact that some children within months of one another in age were "light-years" away from one another in inference strategies. And some three-year-olds could do the same inference construction as the most advanced five-year-old. It is clear that a lock-step, single textbook, 'everybody please turn to page 93' sort of schooling is a vast mismatch for young children, not to say anything of older children whose constructions are even more elaborate (p. 9).

In this chapter the glimpses of young children's advanced counting strategies provide us with a preview of some of the counting abilities of the seven-year-olds described in the next chapter. Despite their young age, some of these children appear to be demonstrating thinking characteristic of the concrete operational stage. Conversely, some of the seven-year-olds in the next chapter are likely to exhibit counting strategies typical of younger children.

This chapter will close with a discussion of teaching methods that are congruent with children's counting capacities and constraints. These methods include providing natural problem-solving contexts that encourage counting, observing children at play, and adapting questions for young children.

Providing Natural Problem-Solving Contexts that Encourage Counting

In her book *Number in Preschool and Kindergarten: Educational Implications of Piaget's Theory* (1982), Constance Kamii recommends that teachers of young children take advantage of many natural opportunities for counting that arise both in daily classroom living and in the context of children's play. She also stresses the importance of providing young children opportunities for decision making as well as counting. Rather than tell a child to distribute six cups among his group, allow him to decide whether to include himself and whether to count. In other words, counting is not something merely to be practiced in isolation, but rather something to be increasingly applied in problem-solving situations. Examples of natural contexts for young children's counting are outlined below.

Distribution (Kamii, 1982a)	Distribution of snacks and tools can be undertaken by young children if group sizes are kept manageable. Activities can include the following: - Distribution of cups and juice containers "so there's just enough for everyone at the table" - Distribution of snacks such as nuts or raisins so that everyone gets a "fair share"
Collection (Kamii, 1982a)	Children can also take responsibility for collection of permission slips, work tools, milk money, and so on in small groups of manageable size. The following questions can arise: - How many were turned in? - How many more will we need? - How many slips were brought in yesterday? Today?

(continued)

(continued from preceding page)

Playground Games	Counting the number of swings, bounces, rides, strikes, outs, and so on can be used as a basis for deciding when to take turns. It is also a timing device for seeking in Hide-and-Go-Seek.
Bowling Games (Kamii, 1982a)	In knocking over pins (blocks, empty milk cartons) or knocking marbles out of their indented positions with a rolling ball, the following opportunities for counting arise: - Count the number of pins knocked down or marbles knocked out on each turn. - Count the number of pins or marbles left. - Count the total score on successive turns. (Marbles that are knocked out can be kept until the game is over.) - Count to compare the number of marbles knocked out with that of another child. - Count the number of turns. (Some method of tallying is needed). Allow children the opportunity of deciding how to keep score.

Tower Building
(Fuys & Tischler, 1979;
Lerch, 1981)

Using blocks of uniform size, small groups of children can build the tallest possible towers on a carpeted surface with simultaneous counting. The following questions can guide their activity:
- How many blocks (stories) can you stack up to make a tower?
- How many stories can you stack up if you build two towers side by side?
- Which tower is higher? How much higher?
  • Compare adjacent towers

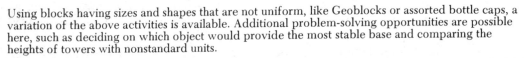

  • Compare nonadjacent towers (or one shielded from another)

  • Compare adjacent and nonadjacent towers built up from different starting levels

Using blocks having sizes and shapes that are not uniform, like Geoblocks or assorted bottle caps, a variation of the above activities is available. Additional problem-solving opportunities are possible here, such as deciding on which object would provide the most stable base and comparing the heights of towers with nonstandard units.

Another variation of tower building involves children pairing up to build the same tower. Each child adds a fixed number of blocks to the tower and counts the total number of blocks in the tower before the next child has his turn. This provides opportunities for counting all the blocks each time or for counting-on from the child's last count. Children with contrasting abilities may be deliberately paired.

A further variation need not involve tower building. Children grab handfuls of loose Unifix cubes and a comparison is made to determine who grabbed the most. The comparison can be made when the objects are in randomly arranged groups or interlocked into towers or roads. Continuous score-keeping is possible from turn to turn by allowing the child with the largest number of cubes to keep the difference

Hiding Games
(Kamii, 1982a;
Baratta-Lorton, 1976)

Hide-and-Go-Seek. This game can be adapted so that the number of players is counted both at the start and as the game progresses. Once one or two of the hiders are caught there is a count of how many are still hiding. Identifying the remaining hiders by name may be helpful for this count.

In another variation of this game, a group of children can be subdivided—one group is to hide a collection of objects that is counted by everyone beforehand and the other group has the task of finding these objects. As each object is found the question of "How many are left to be found?" is raised to the seeking group.

Hidden Chip Games. Children take turns in dropping chips into the slit cut in the bottom of the lower half of a milk carton while the group counts the number of chips. The contents of the carton are stirred while remaining covered. The children take turns in telling how many chips there will be when the carton is lifted. As the chips are exposed they are recounted.

Children pair up and take turns playing the hidden chips task described on page 52. Initially they are restricted to small quantities of chips until some success is experienced.

Observing Children at Play

Careful observation of children during game activities provides the teacher with both a basis for asking questions and for pacing the activities. For example, Kamii (1982a) observes that in bowling activities when a child says, "I knocked over five pins and Freddy knocked over four," he is giving information rather than comparing. Four-year-olds are usually interested in their score alone, while five- and six-year-olds are becoming interested in competition. Here she cautions teachers against insisting that young children compare performances. Kamii has observed that in card games young children usually begin after all the players have stacks of roughly equal size. Young children do not deal the cards in the conventional adult manner. If the children are asked if each of them has the same number of cards, they may show little interest in the question. From their global perspective, this question is of little importance. Kamii asks that teachers not impose this adult concern for equality on the children. Rather, postpone the question for several weeks. Then, if some of the children are interested in the question, allow the children to assert their need for equal stacks of cards.

On the value of encouraging children to exchange ideas, Kamii (1982a) writes,

> When a child is confronted with another child's idea that conflicts with his, he is usually motivated to think about the problem again, and either revises his idea or finds an argument to defend it (p. 37).

The value of group games for socioemotional and intellectual development is discussed further in *Group Games in Early Education: Implications of Piaget's Theory* (1980) by Kamii and DeVries. The book also includes further descriptions of different levels of young children's play in a variety of games.

Observing children at play allows a teacher to notice the different levels of their play and, based on what the children are paying attention to, construct thought-provoking questions. Furthermore, the teacher can choose between allowing the children to benefit from the social interaction springing from the different perspectives within their group, or selecting more challenging activities for those with advanced counting and problem-solving capabilities.

Adapting Questions to Young Children

Sometimes the limitations that we observe in children's counting and other problem-solving behaviors only reflect the limitations in our own abilities to ask the right question that can be understood by them. By rewording questions in terms of children's language and experience researchers recently have uncovered additional children's capacities that previously have gone unnoticed. Two examples of such breakthroughs follow.

Most four-, five-, and six-year-olds have been observed experiencing difficulty in finding exactly how much larger one collection is than another. Yet Hudson (1980) reports that these children demonstrate improved abilities to deal with these situations when the traditional question is rephrased to suggest matching.

We have already described young children's recounting of a collection of objects in response to the "How many?" question. However, Markman (1979) found that this immature behavior can be decreased considerably by rephrasing the question. He switched from the traditional use of class terms (pigs, nursery school children, animals, blocks) to the use of collective terms (pig family, nursery school class, animal party, pile) in asking the question. The use of collective nouns that were already a part of the children's language appeared to help them to pay attention to the "manyness" of the entire group instead of recounting individual items.

Prior to leaving this discussion of uncovering children's capacities by rephrasing questions, here is a word of caution and direction. The caution is against accepting a child's success at a simplified task as an indicator of something more complex. The simplified version of a task may be dealt with successfully by children, but may only require a superficial level of understanding. The alternative direction is to observe the children in a wide range of activities to check your hunches about their level of understanding.

REFLECTIONS AND DIRECTIONS

Locate some young children in your school or neighborhood and engage them in some counting tasks. These tasks can be selected from this chapter or generated on your own. Compare your results with those reported in this chapter. Reflect on the implications of your findings.

4

COUNTING AS A PROBLEM-SOLVING TOOL

PREVIEW

In most classrooms beyond kindergarten, children's counting methods are usually discouraged in favor of the apparent speed of memorized number facts and formal (paper-and-pencil) computational methods. Despite being taught efficient formal methods of computing, children often retain their own counting strategies throughout the grades. The persistence of an individual's own approach into the upper grades was demonstrated in an interview study of seventh graders by Lankford (1974). The most predominant computational strategy employed by the students was counting. It is important, therefore, to look into children's counting abilities further. The added knowledge may allow teachers to use children's natural methods as a basis for further instruction, thereby providing activities that enhance children's counting strategies rather than suppress them.

This chapter will take counting out of the classroom closet and explore the counting of beginning second-grade children (seven-year-olds). Primarily, it will focus on the power and limitations of the children's counting strategies in dealing with problem-solving tasks. The efficiency of their strategies and the understanding of number relationships that the strategies imply will be examined. Counting methods used by the same children and the difficulties they encounter in dealing with more traditional school tasks will be described as well. Finally, the controversy surrounding the use of counting in primary classrooms will be examined along with alternative teaching methods for extending children's counting to more efficient levels.

From the preceding chapter you have gained some understanding of young children's counting capacities and constraints as well as the variability in their abilities within a given age group.

Now for comparison purposes, place yourself in the shoes of slightly older children—seven-year-olds—and anticipate different counting strategies in solving the following tasks. These four tasks will serve as vantage points for observing the counting of second graders.

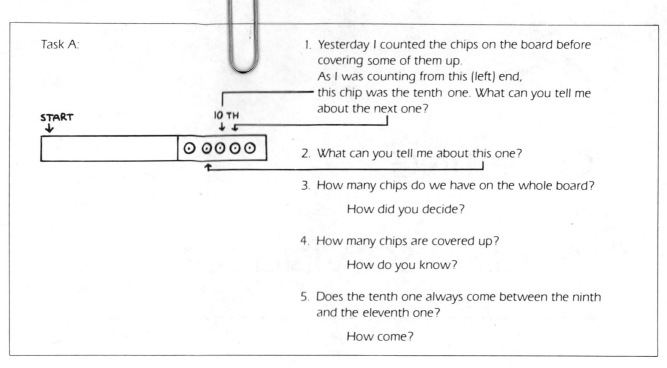

Task A:

1. Yesterday I counted the chips on the board before covering some of them up.
   As I was counting from this (left) end, this chip was the tenth one. What can you tell me about the next one?

2. What can you tell me about this one?

3. How many chips do we have on the whole board?

   How did you decide?

4. How many chips are covered up?

   How do you know?

5. Does the tenth one always come between the ninth and the eleventh one?

   How come?

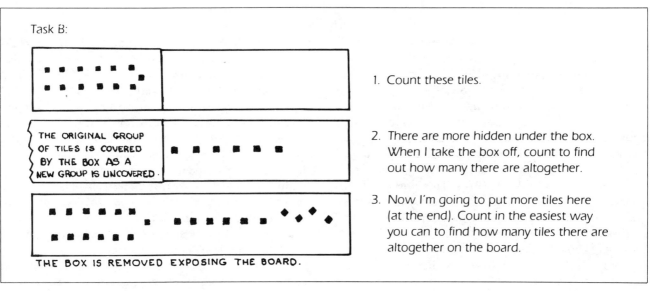

Task B:

1. Count these tiles.

2. There are more hidden under the box. When I take the box off, count to find out how many there are altogether.

3. Now I'm going to put more tiles here (at the end). Count in the easiest way you can to find how many tiles there are altogether on the board.

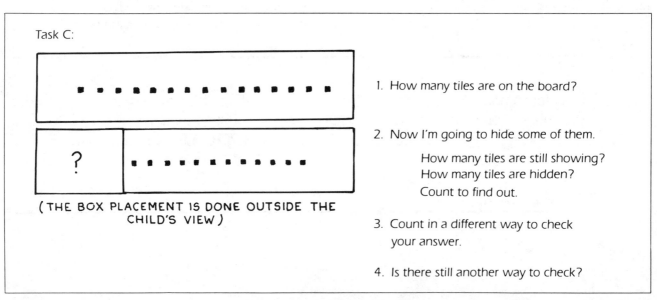

Task C:

1. How many tiles are on the board?

2. Now I'm going to hide some of them.

   How many tiles are still showing?
   How many tiles are hidden?
   Count to find out.

3. Count in a different way to check your answer.

4. Is there still another way to check?

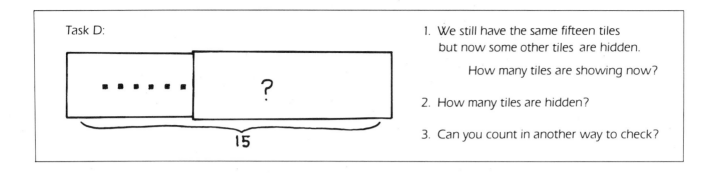

Task D:

15

1. We still have the same fifteen tiles but now some other tiles are hidden.

   How many tiles are showing now?

2. How many tiles are hidden?

3. Can you count in another way to check?

A DEVELOPMENTAL PERSPECTIVE OF CHILDREN'S COUNTING

Prior to looking at the children's responses to these and other counting tasks, an overview of counting within a developmental perspective is in order. This perspective will not only review some of the counting and general thinking abilities characteristic of young children, but will allow us to anticipate further what the interviews with seven-year-old children may reveal. In the last chapter we noticed that major advances in children's counting abilities appeared to coincide with the emergence of logical thinking capacities characteristic of the concrete operational stage. The following table summarizes some of the counting abilities and related logical thinking capacities of children within the context of Piaget's stages of intellectual development.

DEVELOPMENT OF COUNTING AND LOGICAL ABILITIES IN PIAGET'S FRAMEWORK

PREOPERATIONAL STAGE (up to age six or seven)	CONCRETE OPERATIONAL STAGE (from age six or seven)
Perceptual dependence: a) Global estimations of quantities based on outward appearances, such as length, area of collections, or even size of the objects.	A gradual trend away from perceptual dependence as perceptual strategies give way to logical ideas. These logical ideas give support to the construction of more efficient counting strategies and even provide a logical alternative to counting.
b) Dependence on visible objects to count.	Increasing ability to count in the absence of visible objects by generating alternative counting strategies with increased mental involvement, like representing hidden objects internally and/or abstracting relationships from the counting process. The units counted become increasingly more abstract.
Inefficient and inaccurate counting methods	Increased mental activity in the counting process. Reflection on the units counted results in their inclusion in the "manyness" of a collection. Reflection on the counting process allows the child to realize the redundancy in recounting addends and to apply the understanding of "manyness" in the construction of efficient counting-on methods.
Inflexibility in counting, such as difficulty in counting unless starting with "one".	Reversibility and flexibility in counting. Children count from numbers other than "one" and are capable of constructing forward and backward strategies that produce the same result.
Limited meaning of counting words	Number meanings are constructed within a relational framework. "Seven" can now indicate the seventh item counted at the same time that it means the "manyness" of the collection counted.

Younger children are dependent in their counting on what they can see and touch, or perceptually available objects. They gradually become capable of greater internalization of the counting act. Steffe, Thompson, and Richards (1982) found that in this progression children first begin to generate (imagine) mental representations of figural patterns for hidden objects. Yet even here this departure is still dependent on perceptual activity, such as pointing or nodding at expected positions of hidden objects. At a second level, in the absence of visible objects, the researchers report the use of fingers in a particular way that does not depend on direct perceptual input of sight or touch. Here, the children extend their fingers in synchrony with the number names but without looking at their fingers or touching them. It is as if the intentional act of putting up fingers one by one creates the units to be counted. At the third level, children use the counting words as units to be counted; for example, in counting-on 3 from 5: "fi-i-ive, six, seven, *eight*." At this level there are no outward physical indicators of counting activity except for movement of the vocal mechanism.

With each level the units become increasingly more abstract—less dependent on the objects. Further progress in counting results from the children's need to count-on larger quantities. Once again, early methods require perceptual objects for simultaneous tally of the units counted. For example, in counting-on 7 from 8, individual fingers are raised in synchrony with each number named after "eight" until seven fingers have been raised—corresponding with "fifteen." The fingers provide a visual record of the quantity counted on. At a more abstract level, the child simultaneously keeps track of the units counted-on as they are being counted, with increased mental activity. In the preceding task—counting-on 7 from 8—the count would be "e-i-g-h-t, nine (is one), ten (is two), eleven (is three) . . . fifteen (is seven)." Both examples of counting-on large quantities require simultaneous tracking or double counting, but the latter example requires no support from fingers or other objects. This development towards greater internalization of the counting act is also reflected in children's counting-back methods. Based on the variability indicated in young children's counting, it is reasonable to anticipate the entire range of counting levels in the responses of the seven-year-olds interviewed.

Children's increasing reflection on the counting process is accompanied by the construction of an order relation and an inclusion relation within the number sequence. Here they realize that at "six" the sixth item has been counted at the same time that "six" represents the total collection counted. Furthermore, the "sixness" of the collection is retained regardless of physical arrangement. As each number name is assigned in counting forward, the "manyness" of the collection is incre-

mented by one. Here, each counting unit is integrated with those preceding into a new all-inclusive unit.

VISIBLE ACTION	MENTAL ACTION	
One		
Two		Two includes one.
Three		Three includes two.
Four		Four includes three.
Five		Five includes four.
Six		Six includes five.

In mathematics this relationship is called *the cardinal property of number*. Use of this term, however, suggests that this property of number exists "out there" rather than resulting from children's internal construction. By contrast, the terms *progressive integration* (Steffe, Thompson, and Richards, 1982) or *progressive inclusion* remind us of the role of mental activity in the construction of such relationships.

By virtue of the "manyness" of a collection being successively incremented by one in counting forward, a "one more than" relation can be constructed in which six is one more than five, which is one more than four, and so on. Since counting a collection backwards successively decrements the collection by one, a comparable "one less than" relation can be constructed. Eventually the coordination of the "one more than" and the "one less than" relation can result in the child's perceiving an elaborate simultaneous relationship between numbers in the sequence. Here, each number of the series is viewed as one more than its predecessor at the same time that it is one less than its successor: $1 < 2 < 3 < 4 < 5 < 6 < 7 < 8 < 9 < 10$. For example, at the same time that six is one more than five, it is one less than seven. Evidence of understanding of the above relationships will be sought in looking at the children's responses to the counting tasks described earlier.

EXPLORING CHILDREN'S COUNTING IN PROBLEM-SOLVING TASKS

The problem-solving tasks presented earlier (Tasks A, B, C, and D) were given to a group of 31 second-grade children in videotaped interviews conducted in the third month of the school year.[1] Episodes from interviews with different children will be illustrated to describe a range of responses to each task or to each part of a task. Thus, rather than study a child's thinking throughout a series of tasks, we will compare and contrast the thinking of a number of children. Since responses from the same children may be featured for different tasks in this and other chapters, it is possible to compile data for a case study. To facilitate your pursuit of case study information, the episodes of such children are catalogued in the index.

Task A: Constructing Order and Inclusion Relations

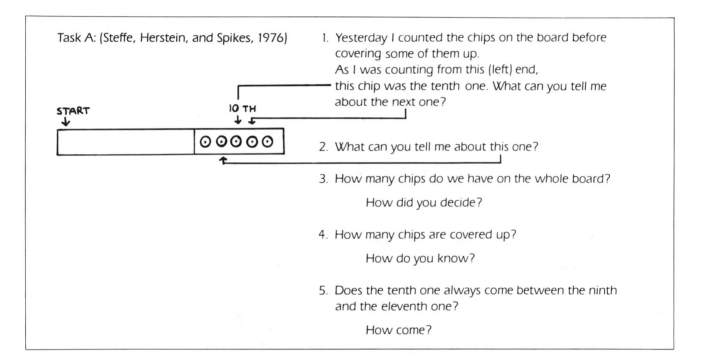

Task A: (Steffe, Herstein, and Spikes, 1976)

START

10 TH

1. Yesterday I counted the chips on the board before covering some of them up.
As I was counting from this (left) end, this chip was the tenth one. What can you tell me about the next one?

2. What can you tell me about this one?

3. How many chips do we have on the whole board?

   How did you decide?

4. How many chips are covered up?

   How do you know?

5. Does the tenth one always come between the ninth and the eleventh one?

   How come?

Task A examines how seven-year-old children take the order information given (tenth) and obtain both the unknown total of the collection and the quantity of hidden objects. It allows us to observe the extent to which these children are able to derive inclusion information from order information and to count-on and count-back. Furthermore, it allows us to find out whether these children have coordinated "one more than" and "one less than" relations.

All of the second graders were able to identify the ninth and eleventh chips as adjacent to the tenth one. It may be worth noting, however, that the responses usually followed long pauses, puzzled looks, and a restatement of the initial question. The following panels illustrate the range of children's responses concerning the number of chips on the board.

1. Interviews were conducted at two schools in an upper-middle class suburb of Los Angeles. Some of the children had experienced traditional instruction in mathematics dominated by a textbook. Since their textbook contained no mention of counting-on methods, it was assumed that none had been taught. The other children had experienced a nontraditional approach to learning mathematics with materials-centered activities taken from a variety of sources. These children had some direct exposure to counting-on visible objects.

Fictitious names have been assigned to the children to protect their privacy.

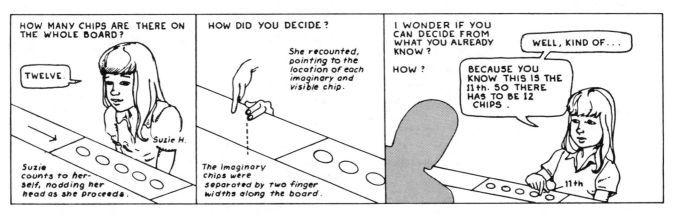

Most of the children were successful in determining the total number of chips on the board, upon reflection. Half of them demonstrated a spatial representation of the hidden chips as shown here by Suzie. They appeared to imagine where the hidden chips would be positioned as they either nodded or pointed to specific locations along the covered section.[2] As they continued to count-on the visible chips, they received feedback on the accuracy of their estimate by the position of the tenth chip. This method of estimation was aided by the regular spacing of the chips in a linear arrangement. Some of the same children also demonstrated the ability to derive the total from the information given. In demonstrating the ability to count-on from ten to twelve, the children showed that counting from one to ten is implicit in the content of "ten" for these children. Both spatial representation strategies and more efficient strategies based on construction of number relations were used by different children to determine the number of hidden chips.

Rita demonstrated that her counting was reversible. By stopping at "seven" and making a sweeping gesture across the hidden chips, she showed that she had integrated the content of that number to include all the hidden chips. It wasn't necessary to count back to "one." Rita also showed no difficulty in using order and inclusion information in context and deriving one from the other. With prompting, some of the children using spatial representation strategies were able to reflect on the task and to derive more sophisticated strategies. When Wendy was prompted to do so, she paused in reflection and then expressed surprise upon realizing the possibility of using the information given

2. Suppose the chips were not regularly spaced and a longer covering was used. How would this affect the effectiveness of this strategy?

in another way. In her justification she used "eight" in an order context. In making a sweeping gesture, Wendy included all the objects counted in a forward direction as being represented by "seven" even though the number was arrived at from the reverse direction in the sequence. In all, twenty out of the thirty-one chil-

dren were able to determine the number of hidden chips.

The last part of this task explored the children's understanding of a simultaneous "one-more-than/-one-less-than" relationship within the number sequence.

The children illustrated above were among the few able to deal with the question. A blank stare was the most common response.[3] Is there a more appropriate question that could have been asked?

On the following page is another task that asks a related question at a lower level of sophistication (Morf; in Kamii, 1982a). It provides information on children's understanding that all numbers are connected by the operation "plus one."

3. Another interesting response was reported by Joni Schaap in an interview with Ola, seven years, eleven months: "Because you count either backwards or forwards, but I don't think you can count sideways. So, if this is nine then there's ten and eleven; that makes ten in the middle."

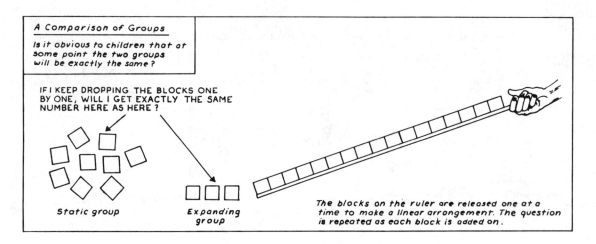

A Comparison of Groups

Is it obvious to children that at some point the two groups will be exactly the same?

IF I KEEP DROPPING THE BLOCKS ONE BY ONE, WILL I GET EXACTLY THE SAME NUMBER HERE AS HERE?

Static group

Expanding group

The blocks on the ruler are released one at a time to make a linear arrangement. The question is repeated as each block is added on.

According to Morf, a common response by children to this task is to keep answering no, until the second group has grown too large. In other words, in their perception the expanding group can go from "not enough" to "too many" without having progressed through "exactly the same." Not until age seven-and-a-half years did children in his study give responses based on a "plus one" relation. Meanwhile, the last question in my interview went beyond the "plus one" relation and explored whether children had constructed and coordinated it with a "minus one" relation. It would be interesting to learn at what age chldren are able to handle the original question.

From the preceding task (Task A: Constructing Order and Inclusion Relations) we have seen that most seven-year-olds in this group were capable of extending order information and deriving related inclusion information in both forward and backward directions. However, half of the children showed a preference for less sophisticated strategies than counting-on or counting-back, such as spatial representation, in solving the problem. Also, we've noted that the children knew which numbers came next in the sequence but had not yet related them as "one more than" and "one less than." Nor had they coordinated these relations as in "ten is in between nine and eleven because it is one more than nine and one less than eleven."

Task B. Counting-On from Hidden and Visible Groups

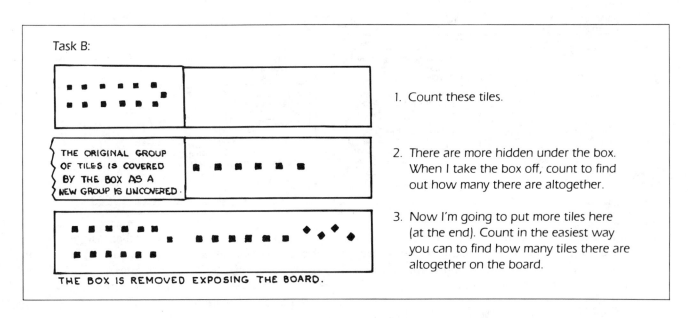

Task B:

1. Count these tiles.

THE ORIGINAL GROUP OF TILES IS COVERED BY THE BOX AS A NEW GROUP IS UNCOVERED.

2. There are more hidden under the box. When I take the box off, count to find out how many there are altogether.

3. Now I'm going to put more tiles here (at the end). Count in the easiest way you can to find how many tiles there are altogether on the board.

THE BOX IS REMOVED EXPOSING THE BOARD.

In Task B notice that some of the objects are hidden, but only part of the time. The box is first used to allow the first addend to be revealed by itself and then the second addend by itself before being removed altogether. Since the first addend is covered once the child has counted it, he is encouraged to count-on.

Actually, by inducing counting-on the covering action may mislead us into overestimating the children's abilities. In the last half of the task, when the box is removed altogether, all of the objects are visible. The child now has the choice of counting from "one", or counting-on from "thirteen" or from "nineteen." In both covered and uncovered situations illustrated below, notice additional prompts to focus the child's attention on the result of the prior count while counting the larger collection. Again, a range of responses is illustrated, beginning with the most efficient counting strategy.

Nearly all of the 31 children counted-on from 13 to 19. This suggests that they progressively integrated the counting acts in the content of "thirteen." At this point they were able to count-on from the all-inclusive larger unit. In the second half of the task it is interesting to note that half of these children reverted to counting from "one." The responses of three other children are illustrated to communicate the range of thinking on this task.

After some confusion, in which he twice acted as if out of sight was out of mind, Nick counted-on from 13 to 19 on the first part of the task. In the next part he reverted to counting-on from a smaller collection despite a prompt aimed at helping him focus on the last number counted. Here, he counted-on from 13 again, resulting in some redundancy in his counting process.

He was not alone, however, as five of the children counted-on from 13 to 23. Despite this redundancy, the method was still relatively efficient. Half of the children reverted to counting all the objects again after having counted-on initially. In one case, a child counted exclusively from "one." Both of these situations are illustrated overleaf.

After counting 13 tiles.

THERE'S MORE UNDER THE BOX....

HOW MANY DO YOU HAVE HERE? **13**

Hazel M.

Just before the box is moved...

COUNT TO SEE HOW MANY ARE ON THE BOARD ALTOGETHER.

1, 2, 3, 4, 5, 6, 7, 8, 9, 10, 11, 12, 13, 14, 15, 16, 17, 18, 19.

From 1 to 13 she did not look at the box. She pointed to each tile beginning with "14."

For the last part of the task she counted from one.

Why did this girl not count from thirteen or nineteen?

In both phases of the task Hazel counted from one whether the objects already counted were covered or not. This makes me wonder whether she could not progressively integrate her count or whether being asked to count aloud triggered an automatic response of counting from one.

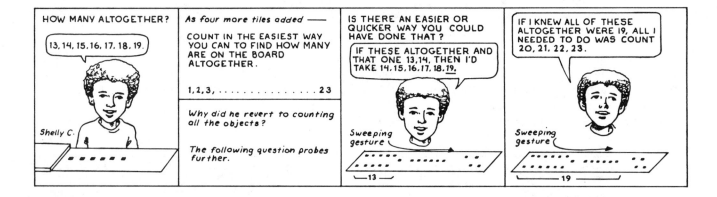

HOW MANY ALTOGETHER?

13, 14, 15, 16, 17, 18, 19.

Shelly C.

As four more tiles added —

COUNT IN THE EASIEST WAY YOU CAN TO FIND HOW MANY ARE ON THE BOARD ALTOGETHER.

1, 2, 3, ... 23

Why did he revert to counting all the objects?

The following question probes further.

IS THERE AN EASIER OR QUICKER WAY YOU COULD HAVE DONE THAT?

IF THESE ALTOGETHER AND THAT ONE 13, 14, THEN I'D TAKE 14, 15, 16, 17, 18, 19.

Sweeping gesture

— 13 —

IF I KNEW ALL OF THESE ALTOGETHER WERE 19, ALL I NEEDED TO DO WAS COUNT 20, 21, 22, 23.

Sweeping gesture

— 19 —

Probing Shelly's earlier response and receiving a dissertation on counting-on in reply reminded me of the limitation of a single observation in making judgments about a child's abilities. In this discussion of counting-on he had no problem in recalling the subtotals involved. Shelly's interview also reminded me of the difficulty of finding a question to ask that all children will interpret in the same way. Children hear what they understand rather than understand what they hear. What could be "easier" than counting from one? After all, this was the first method the children learned. When I experimented with the "quicker" variation of the question, I observed one child revert from counting-on to counting all the objects. In this case, however, he chose to count by twos. Although it is interesting to note the types of responses children gave to the task, it is difficult to make any general conclusions about the children's abilities from them.

To explain children's reversion to counting all from an earlier counting-on strategy, my hunch is that at this age the visibility of all the objects invites children to count from one. When the box (which serves as a prompt not to recount the first addend) is removed, the children's reversion to counting all may be due to an instability in their ability to grasp numbers as inclusive groups. This instability can also be explained by another aspect of the task—the necessity of counting-on from a larger number. The presence of variation in both visibility and quantity limits the value of this task in learning about children's counting-on abilities. Some tasks raise more questions than they answer: this task is a case in point. Other researchers (Fuson, 1982) have observed this reversion phenomenon in more controlled situations, in which the number counted did not change and only the visibility of the first addend varied. While uncovered collections of objects provide a truer test of counting-on ability, covering the first addend appears to suggest the possibility of counting-on to some children. Whether a child is able to benefit from this experience may depend on his ability to integrate numbers at that particular time.

Task C: Forward and Backward Counting Strategies

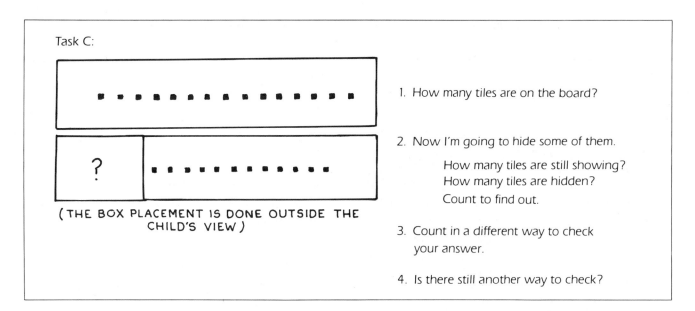

Task C:

(THE BOX PLACEMENT IS DONE OUTSIDE THE CHILD'S VIEW)

1. How many tiles are on the board?

2. Now I'm going to hide some of them.

   How many tiles are still showing?
   How many tiles are hidden?
   Count to find out.

3. Count in a different way to check your answer.

4. Is there still another way to check?

The essence of Task C is similar to that of the hidden chips task in the last chapter, although it does vary in its details. A black box rather than a hand is used to cover the objects. The total collection of objects used is now larger, although the number of hidden objects remains small. Also, the collection is in a fixed linear arrangement rather than in a random, movable one. Although most of the children were able to solve this problem, they approached it with considerable variation in counting strategies. Forward counting strategies will be illustrated and discussed first.

When asked to solve the problem by counting, the children must infer the number of hidden tiles by counting-on. For some children, like Carol, this appears to be a purely mental process since no motor activity is evident. They may envision the hidden objects as they are counted-on verbally, tune into the beats of the word pattern as counting-on begins, or envision numerals 13, 14, and 15 as things to count. In other words, some children can count-on 3 more units from 12 to reach 15 and keep track of this counting-on without use of a physical assistance. Other children appear to create units to count by each touch of the box, nod of the head, or extension of a finger. The latter activity has the advantage of leaving a record of the counting. Another forward counting strategy that was an effective, though somewhat inefficient, method of solving the problem involved recounting the visible objects from one. Once these children had exhausted the visi-

ble objects, the box served to separate their counting act so they were able to keep track of their counting as they continued to 15. Nate touched the tiles as he recounted them, but once he reached the box his fingers served to simultaneously keep track of his count. This strategy, which has aspects of both counting all and counting-on, will be referred to as the counting-all-on strategy.

Since the focus of the exploration was to determine the limits of the children's counting abilities, they were asked to check their answers with alternative strategies. One of the strategies invented by the children was designed specifically for checking answers. They counted-on from their original answer for the number of hidden tiles (3) and counted-on the visible tiles to reach 15. Once again, there were variations within this strategy.

The complexity of the task for children can be appreciated best by following Vince's efforts to coordinate the number inclusion of the starting point, the

recording of the counting-on process, and the termination of the count at an appropriate point.

After making two adjustments of his strategy based on feedback from the task and prompts from the interviewer, Vince successfully coordinated his strategies on the third attempt.

Half of the 31 children solved the problem by

counting-on. Nine children used the less efficient counting-all-on strategy. Five children used a forward counting strategy to check their original response. Thus far, the data indicate the children's preference for forward counting strategies in solving the mystery box problem. An examination of their backward counting strategies may explain this preference.

In solving the mystery box problem, only four of the seven-year-olds spontaneously used counting-back methods. Five additional children were able to invent a counting-back strategy when encouraged to check their answer by another method. In contrast to the forward-counting strategies used, the counting-back methods were less highly developed.

Owing to a general lack of clarity in the children's counting-back methods, alternative counting-back strategies will be illustrated and discussed prior to returning to illustrations of the children's methods.

Since the number of hidden objects is small, the task of counting-back from the total (15) to the known amount (12) and keeping track, of the count (3) can be accomplished mentally. The same mental representation of three units by visualizing objects, numerals, and so on can be used as for counting-on. Since most seven-year-old children still have some dependence on physical action and perceptual support, they are likely to keep track of the count by some visible physical means such as extending fingers or making tally marks. Two alternative approaches to use of fingers to keep track of the counting-back process are illustrated below.

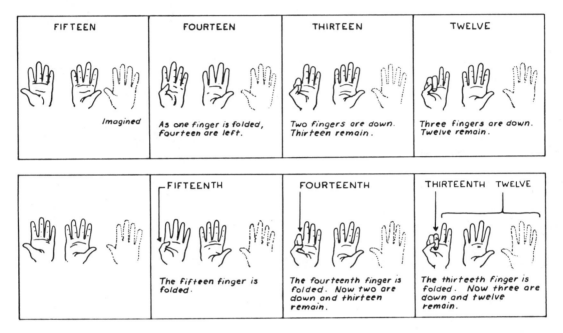

As these methods are generalized to larger quantities, the fingers are used only for keeping track of the counting-back process. Thus this mystery box problem can be solved by using the fingers of only one hand. However, now that the total quantity cannot be represented directly on the fingers, the number of fingers remaining has no significance. As the method becomes further abstracted from a concrete representation of a "take away" subtraction, the fingers may be extended in the tracking process rather than pulling the fingers down.

While the first count-back strategy emphasizes inclusion information, the second strategy emphasizes order with an eventual conversion to inclusion information. Children's lack of clarity in focus is indicated by their use of counting words, implying number inclusion in the context of an order strategy where *th* endings should be used. They appear to vacillate in their focus and may combine aspects of both strategies unsuccessfully, producing answers that are off by one unit.

Until these strategies are coordinated in their thinking, they will experience concerns about what to call the first finger and how to interpret the last counting word.

Now let us look at the children's counting. The first two children illustrated, Carol and Marsha, spontaneously chose to count-back in solving the problem. Examine their strategies and decide whether they have an inclusion or an order focus.

The difficulty of counting-back strategies is reflected both in the small number of children who spontaneously chose them and in the attempts of other children to check their original answers by reverse methods. Some of these attempts are illustrated below.

Initially, Vic had a unique interpretation of counting backward. He just counted forward in a direction opposite to his last counting-on-to-check method. Since the clarification of the direction resulted from the rocket reference, I wonder if counting backward means the same as counting down to all children. Although Vic was able to count down, he wasn't able to apply this to the problem at hand. Stopping his countdown at "three" coincidentally with gesturing across the box would have been an appropriate counting-back-to-check strategy. Since he exhausted the counting-word string and did not attach any special significance to the last three counting words, he was not applying a reverse equivalent of a counting-all-on strategy.

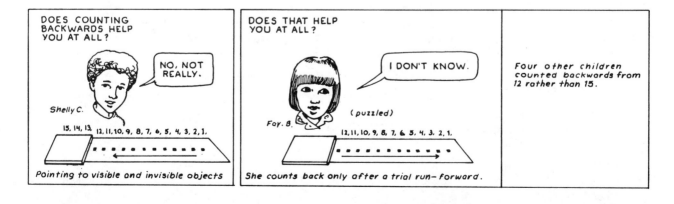

From these examples of second graders' counting it is obvious that their counting-back strategies lag far behind those of counting-on. These children's backward counting abilities fall into the following categories:

1. Unaware of counting backward; counts forward from the opposite direction

2. Recites the number words in a backward sequence only after a practice run in the forward direction

3. Recites the number words in a backward sequence but cannot relate this to problem-solving tasks

4. Unsuccessfully relates backward counting to problem-solving situations by inventing counting-back strategies that combine aspects of two distinct perspectives.

5. Relates backward counting to problem-solving situations by invention of effective counting-back strategies

The extent of this lag between children's forward and backward counting strategies appears to vary with individual children. An explanation for the existence of such a lag would be a relative lack of practice in the reverse counting-word sequence and of modeling its application outside of the rocket countdown. Furthermore, Piaget's theory of intellectual development predicts a lag since most children are unable to reverse actions mentally before the age of seven. We take a further look at children's counting strategies in the next task.

Task D: Tracking Systems and Other Innovative Strategies

Task D, an adaptation of Task C, hides a larger number of tiles, presenting a new challenge for the children. Whereas the previous task could be solved without need for a visible means of keeping track of the counting-on or the counting-back, now all the children must find a way to do so. Many of the children are able to respond to this need by inventing innovative methods.

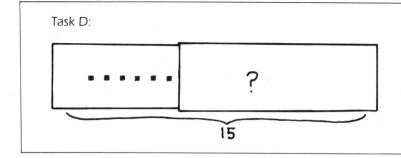

Task D:

15

1. We still have the same fifteen tiles but now some other tiles are hidden.

   How many tiles are showing now?

2. How many tiles are hidden?

3. Can you count in another way to check?

You will see children use their fingers extensively in the absence of objects or paper and pencil for tallying the count. Since these methods have the stamp of unique personal construction, be prepared to slow down your reading. You will find no uniformity in the children's methods, despite there being an American social convention of representing "one" with a particular finger. Arrows will be used to help you locate the varied starting points in the children's counting.

COUNTING-ON

SO WE STILL HAVE 15 ON
THE BOARD AND WE'VE
COVERED SOME UP
LEAVING 6 SHOWING.
HOW MANY ARE HIDDEN?
COUNT TO FIND OUT.

THERE'S
NINE
HERE.

15
14
13
12
S-I-X

7
8
9
10
11

Marsha J.

She makes a sweeping gesture over the six tiles.
Then each finger is touched by the thumb of the
same hand in counting-on.

COUNTING-ALL-ON

11 (thumb)
7 (index finger)
8
9
10

15
14
13
12

Sal S.

He points to 6 tiles individually with thumb
then continues the count on his fingers.

In a clear example of progressive integration, Marsha makes a sweeping gesture over the known quantity of tiles as she says, "S-i-x." She then begins counting-on units and keeping track of them with folded fingers until she reaches "fifteen." At this point she has 9 fingers folded. By contrast to Marsha's efficient counting-on method, Sal uses a method that is less efficient—counting-all-on. Although Sal counts from "one to fifteen," he separates his counting act into two distinct phases. First he recounts the visible tiles using his thumb as a pointer. Upon reaching the box he counts-on 9 more units while simultaneously extending individual fingers. He doesn't appear to be paying much attention to his fingers during this time. Once he reaches "fifteen", however, he looks at the pattern of his extended fingers and immediately says, "Nine." In Sal's approach it is possible that he is progressively

integrating his initial count of the visible tiles, but the result is still too unstable to use as the basis for counting-on in the next part of the task.

One way to solve the problem without constructing a tracking system for counting-on is to estimate the number of hidden objects and then count-on from the estimated answer to the known total. If counting-on the visible tiles produces a result that coincides with the total (15) then the child has feedback that the estimate is correct. If "fifteen" is reached before all the tiles have been counted-on, or if all the tiles have been counted-on at "fourteen", then the child has a basis for readjusting the original estimate. Notice that this method can solve the problem without developing a tracking system. This method can also be used to check answers obtained through tracking strategies. Beth demonstrates counting-on from an estimated answer.

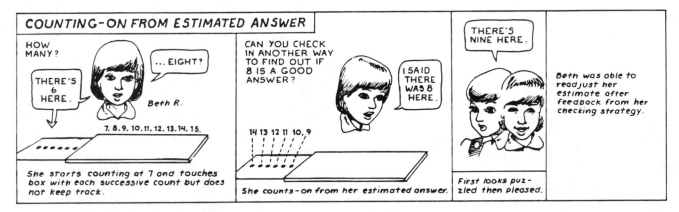

COUNTING-ON FROM ESTIMATED ANSWER

HOW
MANY?

THERE'S
6
HERE.

... EIGHT?

Beth R.

7, 8, 9, 10, 11, 12, 13, 14, 15,

She starts counting at 7 and touches
box with each successive count but does
not keep track.

CAN YOU CHECK
IN ANOTHER WAY
TO FIND OUT IF
8 IS A GOOD
ANSWER?

I SAID
THERE
WAS 8
HERE.

14 13 12 11 10, 9

She counts-on from her estimated answer.

THERE'S
NINE HERE.

First looks puz-
zled then pleased.

Beth was able to
readjust her
estimate after
feedback from her
checking strategy.

Some of the children have little awareness of the need to keep track of their count as they count-on from six to fifteen while touching the box along its length at each successive count. In counting along the length of the box they appear to be visualizing the contents and their spatial location. While engrossed in this task, the children may lose track of the target number and count beyond fifteen. Research (Steffe, Thompson, and Richards, 1982) indicates that such

visualization methods are effective only when five or fewer objects are hidden. The lack of success experienced in using familiar counting methods served to motivate many of the children to search for other possibilities. Beth generated a counting-on-from-the-estimated-answer strategy, while other children integrated a tracking component into their counting with on-the-spot constructions. Two such unique constructions are illustrated at the top of page 73.

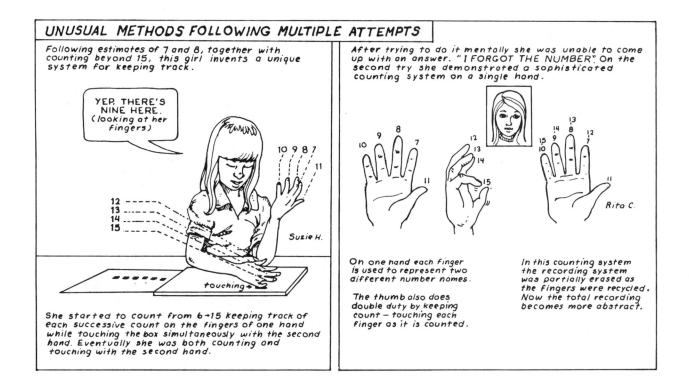

UNUSUAL METHODS FOLLOWING MULTIPLE ATTEMPTS

Following estimates of 7 and 8, together with counting beyond 15, this girl invents a unique system for keeping track.

YEP, THERE'S NINE HERE. (looking at her fingers)

10 9 8 7
11
12
13
14
15

touching

Suzie H.

She started to count from 6→15 keeping track of each successive count on the fingers of one hand while touching the box simultaneously with the second hand. Eventually she was both counting and touching with the second hand.

After trying to do it mentally she was unable to come up with an answer. "I FORGOT THE NUMBER." On the second try she demonstrated a sophisticated counting system on a single hand.

10 9 8 7   12 13 14   13 14 8 12   15 9 7   10   11

Rita C.

On one hand each finger is used to represent two different number names.

The thumb also does double duty by keeping count — touching each finger as it is counted.

In this counting system the recording system was partially erased as the fingers were recycled. Now the total recording becomes more abstract.

Despite the added complexity of the mystery box's containing a larger number of hidden objects, 21 out of the 31 children were able to solve the problem. Of this group, 16 of the children used counting-on strategies. Ten children were unable to invent a new counting strategy in response to the added demands of the task.

CAN YOU KEEP TRACK THE NEXT TIME YOU COUNT?

NO, BECAUSE I CAN'T SEE THROUGH THE BOX.

Ken M.

Four of the children generated counting-back strategies for their initial solution of the mystery box problem. These strategies involved counting-back a different number of units while simultaneously tracking them:

- Counting-back the known quantity (6) from the total (15) to reach the unknown (9)
- Counting-back from the total (15) until the known number (6) is reached while tracking 9 units.

The following two examples of children's counting illustrate these strategies. Both girls use an ordering approach and experience difficulty in coordinating the order and inclusion aspects of the method. They count from the fifteenth through the tenth or from the fifteenth through the seventh, leaving the consideration of how many remain.

COUNTING-BACK

Nancy J.

9 8
10 7 14 13 12
6 15 11
↑
Start

9 8
10 7 14 13 12
15 11

For her original solution, she counts back to the 6 tiles showing.

At this point she hesitates, looking at her fingers, she readjusts, tucking in the thumb and says, "NINE."

COUNTING-BACK

After attempting a similar strategy and experiencing the same problem, this girl tries another strategy.

She counts back 6 from fifteen (counting her thumb first, then using it as a counter-touching other fingers).

10
14 15
13
12
11

Marsha J.

TEN...I MEAN NINE. THERE'D BE NINE IN THERE.

In the physical context of the mystery box task, the visible tiles can serve as a tracking system for counting back from fifteen. Here, "nine" would be reached coincidentally with arrival at the box. With progressive integration of the contents of nine, no further counting is necessary. Sal demonstrates this method in the following example. However, rather than

stop at "nine" with the understanding that it incorporates the results of any further counting, he continues to "one" while tracking nine units. Despite Sal's need to count-all in forward and backward counting strategies (shown earlier in this section), he generated three different solutions to the mystery box problem.

COUNTING-BACK TO CHECK

Checking his fingers.

3 4
2
1 5
9
6 7 8

15, 14, 13, 12, 11, 10

Touching the tiles.

IT'S STILL NINE.

Sal S.

Following a long hesitation at "10", he continued the count, representing the hidden tiles with his fingers and keeping a running tally of his count.

After the "10th" he could have said "9 are left." His failure to do so is surprising, particularly since he had already determined that there were 9 hidden chips.

Perhaps he had a problem similar to the girls and resolved it in a different way.

In spite of the complexity of solving this mystery box problem by counting, 21 out of the 31 seven-year-olds solved this problem by either counting-on or counting-back. Since these children counted on request, no information was gathered on their ability to generate a solution by strategies other than counting. Alternative, noncounting strategies will be examined in the next chapter.

Exploring the Strength of Children's Convictions

The children's long pauses prior to using alternative strategies to check their original answers to Task D suggested two things to me. First, the children need time to generate a response. Second, they are unaccustomed to checking their answers with alternative strategies. Despite the apparent novelty of the request, six of the children generated a second strategy to check the accuracy of their first answer. A seventh child, Sal,

generated three different strategies. Since the seven children derived information that was not perceptually available by at least two different methods, I wondered about the strength of their convictions. I explored it by giving the children the option of peeking under the box to check or moving on to something else.

Sometimes when I do something, I'm so sure that I don't need to check it. At other times, even though I may be right, I'm still not sure and need to check it.

I'm wondering how sure you are. If you're sure there are nine things in here we can put this away and do something else. If you're not so sure, you can peek to check. So, what would you like to do?

Of these seven children, four of them wanted direct perceptual feedback.

The children's preference for looking under the box to count the hidden tiles invites speculation about their attitudes. Perhaps some of these children didn't have good self-concepts as problem-solvers and backed down when presented with this novel challenge. Maybe the children were attracted by the fun of peeking. Perhaps the children wanted to expose the tiles in an "I told you so" gesture. Even the rejection of the offer to expose the hidden tiles is open to speculation. Could

any of these responses be interpreted as an expression of disinterest in further discussion of the mystery box task? This is another aspect of the exploration of children's counting that raises more questions than it answers.

Despite not providing any clear answers regarding the children's strength of conviction in their solutions, the test provokes some real concerns in need of further exploration and reflection.

All of us are sure of aspects of our knowledge due to logical necessity. What size must the hidden quantities be before all children are sure?

Which situation is more likely to result in increased numerical understanding— children doing fifteen problems lacking conviction and awaiting teacher feedback or doing four problems with conviction?

Rather than defining conviction as sureness resulting from reflecting on the reasonableness of answers and cross-checking answers through alternative methods, schools universally recognize conviction in terms of

quick answers obtained through single, standard methods. This contrast in ways of defining conviction raises questions not only about methods of teaching but also about the goals of education.

EXPLORING CHILDREN'S COUNTING IN TRADITIONAL SCHOOL TASKS

At this point you may not recognize the preceding problem-solving tasks as being school related and, therefore, may not appreciate the relevance of counting in the second-grade classroom. After all,

these tasks focused on thinking processes rather than right answers, they were presented outside the context of workbooks, and they actually encouraged finger counting in some cases. In shifting to more familiar classroom tasks in this section your prior exposure to children's counting methods will allow you to either anticipate or to recognize their approaches and to gain insights into children's common difficulties with such tasks.

Using Objects in Addition Situations

The same children who did Tasks A, B, C, and D were asked to determine the answer to an addition situation presented as an open sentence, 8 + 3 = ___. They were asked to count aloud and reminded that they could count on their fingers or use blocks. The children selecting the blocks for counting used a variety of approaches ranging from time-consuming methods characteristic of young children to very efficient methods.

Larry G.

1. Counts 8 blocks silently.
2. Picks up 3 blocks.
3. Counts-on, "8", 9, 10, 11." simultaneously joining individual blocks to larger group starting with "9".

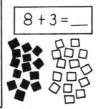

$$8 + 3 = \_\_$$

Rita C.

1. Says "EIGHT" immediately.
2. Takes 3 blocks.
3. Counts-on, "9, 10, 11," touching each block, simultaneously.

Fay B.

1. Counts 8 blocks.
2. Counts 3 blocks (different colors)
3. Joins 2 sets.
4. Counts-all from one.

Suzie H.

1. Counts 8 blocks.
2. Counts 3 blocks.
3. Joins 2 sets.
4. Counts-all silently.
5. Asks, "DO YOU WANT ME TO TELL YOU THE ANSWER?"
6. Recounts all.

Nick S.

1. Counts 8 blocks.
2. Shows 3 more blocks.
3. Ignores the materials and counts with fingers.

Rita demonstrates the most efficient strategy based on the number inclusion represented by the numeral 8. She was able to anticipate the results of counting to "eight" without actually doing so, and counted-on from that number. By contrast, Larry needed to count from "one." Once he reached "eight" he was able to continue his count to include the additional blocks. Because Larry did not recount both addends, his method is more efficient than Fay's, Suzie's, or Nick's. Although Nick counted out two sets of objects, he preferred to count-all using his fingers. We have seen Suzie exhibit earlier a preference for global, inefficient strategies. Although her counting strategies are not highly developed, she also has shown a capacity for constructing more efficient strategies as demanded by the tasks. Here, the availability of materials seemed to invite her to count-all.

Using Fingers in Addition Situations

Once sums exceed 10, addition becomes more difficult for children to represent by counting on their fingers. This is amply illustrated by a worksheet completed by a second grader.[4] Using blocks she was able to complete the worksheet accurately. However, when

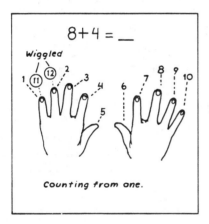

$5 + 3 = \underline{8}$

$7 + 4 = \underline{10}$

$5 + 6 = \underline{10}$

$9 + 3 = \underline{10}$

$8 + 2 = \underline{10}$

A 2nd grader's worksheet.

these materials were unavailable and she needed to resort to her fingers, she was unable to extend her counting beyond her own ten fingers. Some children adapt to this situation by reusing or imagining additional fingers to represent numbers beyond ten. It is not too difficult to imagine or to wiggle one or two fingers beyond ten as in solving $7 + 4 = \underline{\quad}$ or $9 + 3 = \underline{\quad}$.

$8 + 4 = \underline{\quad}$

Wiggled

Counting from one.

However, this method becomes too cumbersome with larger totals. At this point, many children invent either more efficient counting strategies or even efficient

noncounting strategies. Alternative strategies to counting will be discussed in the next chapter.

In determining the answer to $9 + 6 = \underline{\quad}$, several of the seven-year-olds attempted to represent both addends on their fingers before realizing that another method was needed. At this point, following some hesitation, many were able to count-on 6 more units from

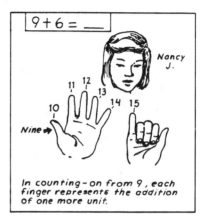

$9 + 6 = \underline{\quad}$

Nancy J.

Nine ➡

In counting-on from 9, each finger represents the addition of one more unit.

9. In keeping track of the counting-on process the fingers represent the second addend. The total quantity is no longer represented directly on the fingers. Once again we see the importance of a child's anticipating the counting acts implied in the contents of a number so that a finger representation and recounting of the first addend is no longer necessary.

Using Fingers in Subtraction Situations

Subtraction situations involving quantities greater than 10 are also difficult for children to represent on their fingers. However, many children invent representations to supplement their ten fingers, find ways of recycling their fingers, or use their fingers only for tracking the counting-back process. The number of visible remaining fingers not only loses its significance now but can be misleading as well.

---

4. This worksheet was completed by a second grader at another school during the first month of the school year.

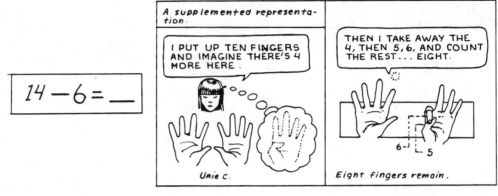

14 − 6 = ___

A supplemented representation.

I PUT UP TEN FINGERS AND IMAGINE THERE'S 4 MORE HERE.

Unie C.

THEN I TAKE AWAY THE 4, THEN 5,6, AND COUNT THE REST... EIGHT.

Eight fingers remain.

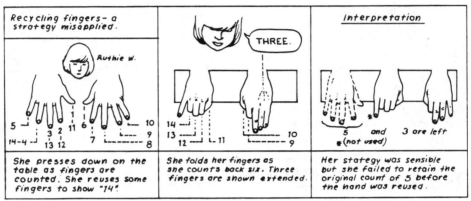

Recycling fingers— a strategy misapplied.

Ruthie W.

She presses down on the table as fingers are counted. She reuses some fingers to show "14".

THREE.

She folds her fingers as she counts back six. Three fingers are shown extended.

Interpretation

5 and 3 are left
*(not used)

Her stategy was sensible but she failed to retain the original count of 5 before the hand was reused.

No strategy.

I CAN'T DO IT. I DON'T HAVE 14 FINGERS.

Rita C.

Despite highly developed counting-on strategies.

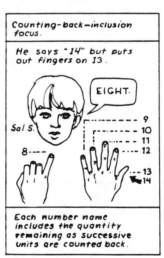

Counting-back—inclusion focus.

He says "14" but puts out fingers on 13.

Sal S.

EIGHT.

Each number name includes the quantity remaining as successive units are counted back.

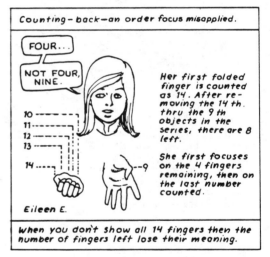

Counting-back—an order focus misapplied.

FOUR...

NOT FOUR, NINE.

Her first folded finger is counted as 14. After removing the 14th thru the 9th objects in the series, there are 8 left.

She first focuses on the 4 fingers remaining, then on the last number counted.

Eileen E.

When you don't show all 14 fingers then the number of fingers left lose their meaning.

Strategies for Missing Addend Situations

We have already seen children's counting strategies in determining missing addends in the physical context of the mystery box tasks. Here, we'll look at the range of responses to missing addend situations presented in open sentence contexts.

I wrote down three numbers on this card and two of these numbers add up to eleven/nine.

Then I covered up one of these numbers so you couldn't see it.

7 + ▨ = 11

▨ + 3 = 9

Count to find the hidden number.

An inefficient counting strategy was exhibited by Hazel despite using abstract units. She first counted from one to eleven. Then she counted from one to seven, paused and counted, "Eight, nine, ten eleven . . . four." No

visible tracking system was evident. This is another example of the counting-all-on strategy. The predominant strategy among the second graders in their third month of school was counting-on.

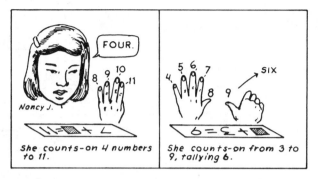

She counts-on 4 numbers to 11.

She counts-on from 3 to 9, tallying 6.

He focuses on remaining fingers before focusing on fingers counted.

For ___ + 3 = 9, some of the children counted-on from an estimated answer or guess. They started with "six" and counted-on three more. Although counting-back three from nine would be the most direct and efficient strategy, none of the second graders used it in this context. Researchers (Steffe, Thompson, and Richards, 1982) posit that as children are able to count-on and count-back with keeping-track methods they gradually integrate the two processes and come to see that addition and subtraction are related inversely, so that $4 + 5 = 9$ and $9 - 5 = 4$. Accompanying this integration, children acquire the flexibility to make a selection

from among the counting-on and counting-back strategies according to the demands of the problem. Although some of these seven-year-olds are using counting-back strategies to check the result of counting-on, their failure to select the most efficient strategy here suggests that they have yet to fully integrate the forward and backward processes.

A unique response constructed by a seven-year-old to a missing addend situation is illustrated next. It serves as a springboard to a discussion of thinking, learning, and teaching.

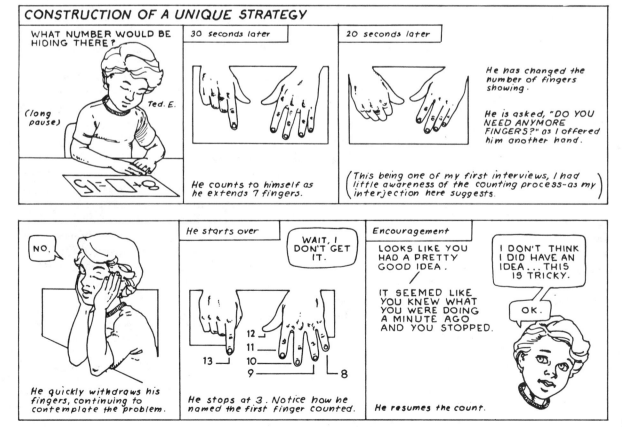

CONSTRUCTION OF A UNIQUE STRATEGY

WHAT NUMBER WOULD BE HIDING THERE?

(long pause)   Ted. E.

30 seconds later

He counts to himself as he extends 7 fingers.

20 seconds later

He has changed the number of fingers showing.

He is asked, "DO YOU NEED ANYMORE FINGERS?" as I offered him another hand.

(This being one of my first interviews, I had little awareness of the counting process—as my interjection here suggests.)

NO.

He quickly withdraws his fingers, continuing to contemplate the problem.

He starts over

WAIT, I DON'T GET IT.

He stops at 3. Notice how he named the first finger counted.

Encouragement

LOOKS LIKE YOU HAD A PRETTY GOOD IDEA.

IT SEEMED LIKE YOU KNEW WHAT YOU WERE DOING A MINUTE AGO AND YOU STOPPED.

I DON'T THINK I DID HAVE AN IDEA . . . THIS IS TRICKY.

OK.

He resumes the count.

Following six minutes of intense involvement with the missing addend problem, Ted constructed a unique, sophisticated solution. He counted-on and simultaneously counted his counting as it progressed by generating a second series of numbers. Here, each verbalized counting word, "nine, ten . . . fifteen" was taken as a countable item. Yet the things he counted from "one" to "seven" did not exist before he counted them. Then he must have "counted the acts of creation" (Bridgman, cited in Steffe, Thompson, and Richards, 1982) of these abstract units. Similarly, it may be argued that counting from nine to fifteen involves the progressive integration of numbers—putting them into inclusion relationships, where these relationships do not exist externally in his fingers (or blocks if available). Thus "counting the acts of creation" applies to both series in the double-count strategy. However, the second (counting the counting) is dependent on the first and is the more abstract of the two.

Despite the high-level construction, Ted also simultaneously extended his fingers as if to keep track of the counting-on in still another way. At one point in his final counting sequence he lost track of which finger was "twelve" and asked where he had left off on his fingers. Actually, since he was already double counting, he could have managed without his fingers. The role played by his fingers throughout the episode is unclear to both Ted and this observer. Although he used his fingers, he seemed to distrust the feedback they gave him.[5] Perhaps he became disoriented when he started to extend his fingers at "nine" rather than at the customary "one." His invention of the sophisticated double-count strategy with abstract units may have been the result of the need to resolve this discrepancy. In his new construction, "Nine is one."

Ted's invention came near the end of a thirty-five minute interview. Following the interview he volunteered that his head was spinning. His invention came on the fifth visible attempt to coordinate a solution. Prior to this time he was unable to recognize the apparent solution on his seven extended fingers. My attempts to get him to look more closely at his fingers weren't productive as he was intent on looking at the problem differently. Throughout the task his actions were quite deliberate, suggesting considerable reflection during the pauses. A good indication of his level of concentration was his ability to brush off my misguided offer of another set of fingers and immediately return to his own two hands.[6]

---

5. Ted's distrust of his fingers may be the result of classroom experiences in which counting blocks or fingers was discouraged.

6. The interview with Ted was conducted in a pilot study that preceded the explorations reported in this chapter. Other opportunities to follow Ted's thinking are presented in Chapter 7.

Apart from my embarrassing intrusion in the child's problem solving, I was satisfied with my role as interviewer. The blunder was the direct result of limited knowledge of counting strategies during some preliminary interviews. My general manner was one of interest, patience, and support. The encouragement I provided seemed appropriate. The suggestions for concluding something about his seven extended fingers also seemed appropriate at the time. Prior to his invention, I had no way of knowing that he was viewing the problem from a more sophisticated level. All of this points out the difficulty of conducting such interviews and following the direction of children's thinking. This episode also indicates the unpredictability of the time required by children in making higher-level constructions.

The prevailing view of teaching is one of helping—intervening quickly in times of confusion to rescue the child by giving the correct answer or straightening out his thinking. From the preceding episode, however, it is apparent that help is not always helpful since learning involves personal construction. Ted was intent on doing his own thinking and anything other than general acceptance and encouragement seemed to only add static to his system. Teachers accustomed to quick, ready answers would be likely to make wrong inferences about Ted's thinking abilities. After all, he didn't give an answer for several minutes. It would also be possible to conclude wrongly that Ted had arrived at an answer after first extending seven fingers. This episode can serve as a reminder of the importance of restraining urges to intervene quickly and to straighten out children's thinking, despite the obvious time pressures in the classroom.

COUNTING IN DEVELOPMENTAL PERSPECTIVE—A CONTINUING LOOK

Now that you have viewed the counting strategies of seven-year-olds on several tasks, you are invited to return to the developmental perspective of counting earlier in this chapter. You may be surprised at how much more meaningful that section is to you now.

In children's transition to more efficient and more internalized strategies, they exhibit alternative strategies to counting, based on logic.

Logical capacities that unfold in the transition to and during the concrete operational stage both support and provide alternatives to counting in certain situations.

Matching objects in adjacent rows is not an easy task for preschoolers. Although they line up a row of objects within the boundaries set by the other row, they tend to lose sight of the density of the objects within the row. Once children recognize that two sets are matched, there is no need to count the sets to find out if they are equal. If one of the rows is counted, the older child may be able to infer the number in the second row with no further counting needed.

Again, the young child will focus on the boundaries of the rows to determine "sameness." Although the rearrangement is carried out in full view, he will focus on the final outward appearance rather than on the process of rearrangement. The child's ability to count each row may not overcome his dependence on perceptual strategies. Most seven-year-olds have constructed capacities for logical reasoning that replace perceptual dependence and provide an alternative to counting in deciding on the equivalence of the two rows.

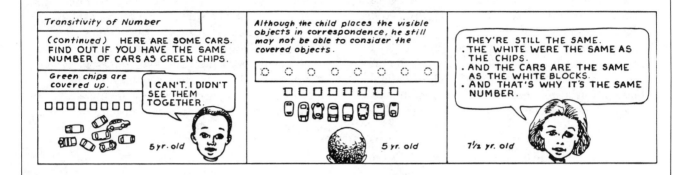

Seven- to eight-year-olds can place multiple rows of objects into one-to-one correspondence with the power of their logic, even though all the rows are not seen together.

Row of white blocks = Row of green chips

Row of yellow cars = Row of white blocks

Row of yellow cars = Row of green chips

Again, by counting the number of objects in one row the child can infer the number in each of the other rows.

ALTERNATIVES FOR TEACHING COUNTING IN THE CLASSROOM

Having taken counting out of the classroom closet and examined the complexities of children's counting methods in problem-solving situations, you may be ready to reconsider the role of counting in the child's development of number understanding and to examine both alternative approaches to the teaching of counting and existing classroom research. A discussion of these areas follows.

Is Counting Appropriate in the Classroom?

Conflicting statements on the virtues of counting have been made by researchers and mathematics educators since the 1920s, and still persist in the 1980s. This continuing controversy is reflected in the following statements.

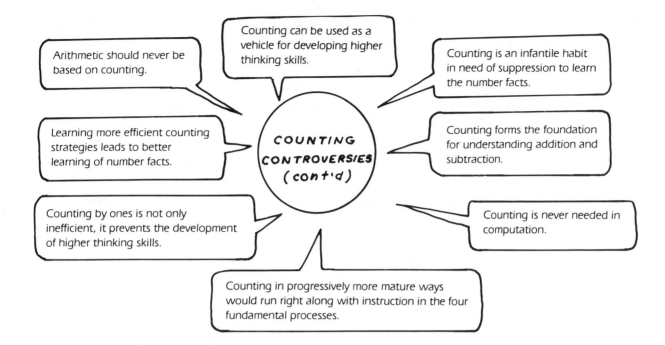

The controversy raises questions about the basis for these statements. Are these experts all looking at children? Could they be looking at children differently?

An in-depth look at children's counting provides considerable support for placing it at the core of the early primary mathematics curriculum. Yet mathematics textbooks give no emphasis to children's development of advanced counting methods. Since the textbook defines the curriculum in most classrooms, counting is deemphasized or, in some cases, even suppressed. To ignore what children already know is simply wrongheaded! Let us examine the perspectives behind the contrasting positions.

One of the primary objectives of the school math program is to have the children know the number facts for later use in more complex computations. The criterion here is an automatic response on cue; for example, when a child sees (or hears) 4 + 3, he immediately says "seven." In teaching for memorization, the teacher does not allow a child time to figure out an answer by counting or by another thinking strategy. Rather, the teacher provides immediate feedback on the correct answer as well as additional drill on that number fact. Although the method is virtually mindless, it does result in a veneer of isolated number information. If one views mathematics as consisting of these number associations rather than of relational knowledge, then observing young children's time-consuming, redundant counting methods would be both frustrating and misleading. Suppressing children's inefficient counting methods makes it impossible to observe their construction of more efficient ones that rely less and less on action on blocks and fingers and increasingly incorpo-

rate mental involvement. According to Piaget, such "interiorized actions" constitute thinking. O'Brien (1982a) argues that by suppressing children's counting actions, teachers are depriving children of thinking. There is research evidence that children who count-on perform better on tests of addition facts than do children who do not use that strategy (Leutzinger, 1979).

While one opposing perspective may result from limited knowledge of children beyond their capacity to function at levels similar to laboratory animals, another may be the outcome of closer observation of another aspect of their intellectual functioning. A math educator could notice that some children invent alternative thinking strategies that don't involve counting, and could then set the development of these alternative, noncounting strategies as the objective for all children to the exclusion of counting. The next chapter will examine these alternative thinking strategies and consider whether they should be taught in isolation or together with counting.

Indirect Teaching Methods

Throughout this chapter you have seen evidence that if you place children in problem-solving situations in which old counting methods don't work, they will invent new ones. From this knowledge of children's capacities it is possible to generate teaching methods that promote/support their natural abilities. My proposal for teaching methods is a bold one—flying in the face of what currently takes place in classrooms. My proposal is to take advantage of children's ability to adapt to new situations by providing counting and problem-solving activities in the context of searching for more efficient alternative methods. Examples of such gamelike situations for pairs of children are illustrated next, together with examples of the teacher's role in facilitating this constructive process.

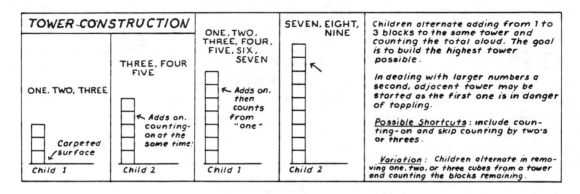

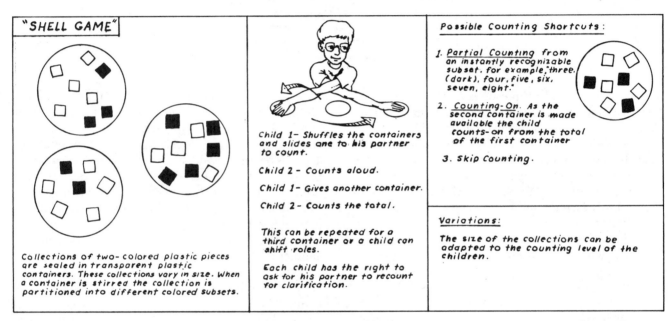

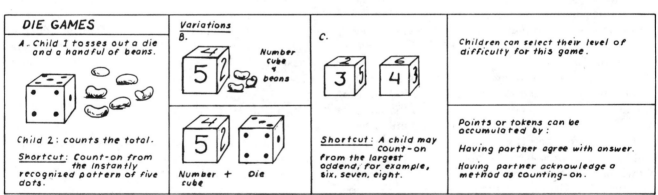

MYSTERY BOX

Total collection is displayed.

Child 1

Child 2

Child 2 counts the collection.

I'M GOING TO HIDE SOME. TURN AROUND.

OK, NOW COUNT TO FIND HOW MANY CHIPS ARE HIDING.

The child 2 may

- Count-on from quantity showing.
- Count-on from an estimated answer.
- Count-back from the total to the number showing.
- Count-back the number showing.

These may or may not involve tracking methods, depending on the size of the hidden quantity.

Possible Variations

- Method of hiding: box, cloth, hand.
- Arrangement and fixedness of objects: linear / fixed random / moveable.
- Number of objects available.
- Location of hidden quantity: e.g. left/ right side of linear array.
- Request for level of difficulty: hard/ easy problem.
- Test of conviction: use / non-use (to be discussed).

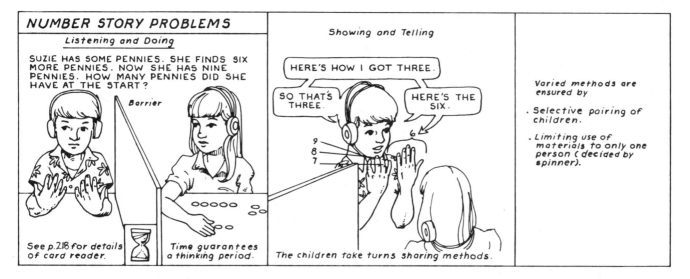

NUMBER STORY PROBLEMS

Listening and Doing

SUZIE HAS SOME PENNIES. SHE FINDS SIX MORE PENNIES. NOW SHE HAS NINE PENNIES. HOW MANY PENNIES DID SHE HAVE AT THE START?

Barrier

See p. 218 for details of card reader.

Time guarantees a thinking period.

Showing and Telling

HERE'S HOW I GOT THREE.

SO THAT'S THREE.

HERE'S THE SIX.

The children take turns sharing methods.

Varied methods are ensured by

- Selective pairing of children.
- Limiting use of materials to only one person (decided by spinner).

Rather than to suppress children's counting, and thereby their thinking, the teacher's role is to facilitate children's natural constructive processes. The children's involvement in the game activities does not minimize the teacher's role. Instead, it provides numerous opportunities for the teacher to learn from children through firsthand observation and to base decisions and interactions on this information. Different aspects of the teacher's role are outlined below.

Pairing Children Selectively. By observing children and noting their counting methods, the teacher can pair up children in a variety of ways. Some of the ways have potential for maximizing interaction that may provoke invention of new strategies. Other ways may reduce the interaction between children and allow them to practice a newly-acquired strategy without being exposed to anything new. When children of unequal counting abilities are teamed, the less-advanced child may benefit from the other child's modeling of a

more efficient method. At the same time, by displaying and verbalizing his method, the more-advanced child may become more conscious of his own thinking. When children using advanced strategies are paired, they may benefit from the exposure to an alternative advanced strategy. Children who are just beginning to count-on may profit from being paired up so they can practice this method in different situations without the pressure of adapting their method again. The organization of ideas supporting their newly-acquired method may still be too fragile for the introduction of additional novelty. When handled with sensitivity, selective pairing has considerable potential for either facilitating change or for supporting continuity.

Requesting Clarification of Counting Methods. Asking a child to repeat his method for clarification not only can inform you of his process, but also can do the same for the child's partner. Furthermore, asking a child, "Would you show and tell me what you did so I can understand your counting?" can give the second child permission to ask for his own clarification at another time.

Encouraging a Search for Shortcuts in Counting. Periodically, at the end of a game period, the teacher can get the children together and ask them to show and tell about any shortcuts they invented in their counting. The teacher can request volunteers or can ask specific children to display and explain their methods: "Jenny, I noticed you were doing some different counting today. Would you be willing to show the others how you did it?" Having the game materials handy may provide the contextual support for the child's demonstration. However, for some children, ideas in the process of construction are still too fragile to be shared publicly, and this need for privacy must be respected to provide psychological safety for children's efforts. When a child's method has been shared the teacher may ask the child to repeat his method for clarification—for example, slow down the finger work. Alternatively, the teacher can model a child's unique method while thinking aloud: "Let me see if I understand how you counted . . . " Eventually, other chil-

dren may be able to undertake this clarifying role that has been modeled by the teacher. The length of this summary session may depend on the children's capacity to consider alternative methods. Initially, children may repeat a method already displayed, as applying this method in a different context may be considered by them as another method.

Testing Inferences About Children's Understanding of Counting-On. Although children may be able to count-on from the number of objects in a first set, it is possible that they are only imitating other children and have no understanding of number inclusion. Such an inference can be made from the children's lack of sweeping hand gestures over the first set or lack of emphasis on the number of that set ("f-o-u-r"). For an advanced counter, the lack of such behaviors might indicate that the child had automatized the procedure. A method for testing the inference is reported by Russian researcher V. Davydov (in Fuson, 1982). He attempted to provoke both a reflection on the counting-on process and rethinking of the meaning of the first number word by the following intervention.

For some children this timely question provokes both a disequilibrium and a reconstruction of the meaning of "four" to one of progressive inclusion, according to Davydov. However, he does not consider the possibility that any of them might be only clarifying their counting for the adult. At the same time, other children responded in ways that indicate lack of understanding or confusion with the question. Other responses included referring to another single object in the first set as "four," counting-all, or no response. The sensitive application of this method with a child who may be merely imitating others may provoke a rethinking and construction of understanding to support the method. It also may confirm your hunch of lack of understanding and influence your selection of his partners and activities. This method may also provoke advanced

counters to reflect on their process so that they are able to clarify it for other children.

Modeling an Alternative Strategy as a Possibility. On occasion, the teacher can take a turn at a game and model an alternative method while thinking aloud. The purpose of this modeling is to suggest a possibility rather than to show repeatedly so that a child can make a carbon copy. With this germ of an idea, children who are close enough in their thinking can construct a similar method in their own minds and begin to apply it. Others may attempt to do so and construct something entirely different. Application in different tasks provides the children with feedback on the effectiveness of these methods. Some children whose thinking is not close enough to consider the teacher's method will simply ignore it. It is important to note here that despite the suggestion, the child must still reorganize his thinking to some extent to make the method his own.

Encouraging Children to Generate and Compare Alternative Counting Methods. A mystery box problem can be presented to a group of children (possibly using an overhead projector), and generation and comparison of alternative methods can be encouraged by the following questions:

1. In what different ways can you count to solve this problem?
2. Which way do you like best? How come?
3. Is this way to count always best or is it best for this problem? Explain.

The mystery box can be uncovered to expose the hidden objects to provide additional feedback on the children's methods.

Encouraging Children to Check Answers by Alternative Methods. In a group presentation of a mystery box problem the teacher can follow up each demonstrated method with a test of conviction:

1. If you think eight tiles are hidden by the box, raise yor hand.
2. If you're sure enough of the answer so you don't need to peek, raise your hand.
3. What could you do so that you would be sure (so that you don't need to peek)?

The initial problems presented in this manner should involve less than five hidden objects. As children grow in both competence and confidence, the size of the hidden quantity can be increased. With these more challenging problems the mystery box situation can remain on display until the following day, allowing time for further generation of methods. Providing sufficient experiences in cross-checking answers by alternative methods can eliminate the children's need for peeking entirely. One such experience can be adapted to a game situation initially role-played by the teacher with a child as partner. Once the children grasp the details of the game they can pair up and play with each other.

If a child declares that he is "SURE" and can demonstrate two alternative solutions he collects 2 tokens from the bank.	If a child declares that he is "SURE" but can only demonstrate a single solution he collects 1 token from the bank.	If a child declares that he cannot demonstrate a correct solution, he gives up 2 tokens to the bank.	If a child declares that he is "UNSURE" but can demonstrate a correct solution, he collects 1 token from the bank.	If a child declares that he is "UNSURE" and is unable to demonstrate a correct solution, he gives up 1 token to the bank.

Each child starts with the same number of tokens. The winner is the child with the most tokens after a given period.

The children take turns in setting up a mystery box situation for their partners. Prior to his turn, a child can request an easy or difficult problem.

Mystery box problems simultaneously arouse curiosity and caution. The children are not only intrigued by the hidden objects but they are now less likely to accept their first inference of the number of objects hidden. They are enticed to check their inference by alternative methods, thereby gaining an ap-

preciation of the value of cross-checking. The children not only become more conscious of their counting strategies, but also of their level of confidence in their inference. They gain the ability to intervene when feeling unsure and to do something to change that feeling. Once children's cross-checking ends in the rejection of the need to peek under the box, they experience the power of their thinking abilities. Eventually, the children are able to do autonomous thinking without dependence on either the teacher or the hidden objects for feedback. Although the children don't see all of the objects, logical necessity based on relational knowledge constructed internally provides the basis for their conviction. Since children's construction of number meanings proceeds gradually (in what Piaget calls *progressive arithmetization*) their convictions are limited to that range of understanding. The mystery box activity has long-range appeal to children for the very same reason.

Considerations for Implementation. The above suggestions are based on knowledge gained from interviewing seven-year-olds and observing their ability to adapt to problem situations by constructing unique and efficient counting methods on the spot. Similar constructive abilities of children have been reported in Moser's interview study of their solutions to orally presented number story problems (1980).[7] Since the available evidence comes as a result of interviews with individual children rather than from classroom studies, implementing these methods requires considerable trust in children's capacities to adapt.

In considering implementation of these methods, a natural question to ask is whether all children are capable of their own construction of efficient counting methods. The question is basically a developmental one. It is reasonable to expect that all normal children are capable of such construction and will not get stuck at a level of counting-all. However, it is not reasonable to expect all of them to be able to do so on cue before the end of the first grade. Some of them may not develop the intellectual capacity for such construction for another year. Providing for learning within the constraints of intellectual development requires more time than today's schools are prepared to give. Are you willing to provide that time?

Because of the degree of mental engagement required of the children and their teacher, the method described above might be more appropriately named reflective teaching. The primary goal of this method is to use children's counting as a vehicle for developing thinking abilities and related positive attitudes. Knowledge of the number facts is a secondary outcome.

Direct Teaching Methods

Research evidence also exists that counting-on methods can be taught directly. This approach to teaching differs from the methods already described in several ways. Some of these differences are outlined in the following table.

COMPARISON OF DIRECT AND INDIRECT METHODS OF TEACHING COUNTING-ON

	INDIRECT (REFLECTIVE)	DIRECT
Goals	Thinking is basic (with counting as a vehicle). Learning number facts is secondary.	Learning of number facts is basic (with counting a tool for this purpose).
Planning	Game activities are designed to provide opportunities for adapting counting methods and interacting with others. Selective pairing of children is based on careful observation of children's counting. Follow-up discussions are based on spontaneous occurrences during the game activities. Scheduling of more challenging games is based on observation of children. The transition from concrete to abstract is gradual over a period of several weeks. Worksheets are not emphasized, even avoided.	The act of counting-on is broken down into small skill steps for teaching. A tight sequence of activities is designed beginning with the simplest skill, such as counting from points other than one. Nothing is left to chance. Decisions made are often independent of careful observation of children. Game activities may be scheduled for small groups, but only following explicit teaching in large-group situations. The transition from concrete to abstract is rapid over a period ranging from one day to several days. Worksheets are emphasized.
Teacher	Interacts with children on the spot following careful observation of counting methods. Helps children to verbalize and clarify their methods by serving as model for these processes.	Explicitly models counting-on and counting-back methods.
	Encourages shortcuts in counting, generation, and evaluation of alternative methods. Any modeling done here is not explicit; it is merely for the purpose of suggestion.	May involve children in the demonstration to introduce the notion of counting-on but one specific method is modeled. (This may include pauses and gestures of emphasis for the number of the first addend.)
	Encourages cross-checking of problems solutions by alternative methods.	Corrects children's worksheets and repeats modeling as needed.

7. Children's interpretations of a variety of number story problems will be explored in Chapter 8.

Child	Inventor, clarifier, and evaluator of his own multiple methods.	Imitator of teacher's single method. (Teacher is the inventor.)
	His counting is a powerful problem-solving tool.	His counting is a limited tool for generating answers to exercises.
	His locus of evaluation is internal. The child is not dependent on the teacher. Although there is feedback from the materials and from other children available, he becomes increasingly reliant on internal consistency of cross-checking.	His locus of evaluation is external. The child is dependent on the teacher for determination of correctness.

It is important to note that approaches to direct teaching vary in their details. They may even combine some features of indirect teaching. Also, despite being taught a single method of counting-on, there are always some children who will invent their own alternative methods and who will not be dependent on the teacher. Thus, the main value of the table is in making some broad distinctions in methods.

Deciding whether to include counting activities not emphasized by textbooks and then deciding on the type of teaching to use may depend on your view of how children think and learn, external pressures to conform to a certain schedule, class size, and your view of yourself as a professional teacher. Chapter 16 of this book discusses these issues. If you are interested in teaching counting-on and counting back by direct methods, you can adapt the variety of tasks and activities previously described and design your own instructional sequence. Information from available research also may be useful to you for this purpose, so we review some of it here.

A clear example of direct teaching is found in the research of Leutzinger (1979). His study, conducted with 29 first-grade classrooms, indicates that counting-on can be taught directly. Following an instructional unit of ten days' duration during the fall of the school year emphasizing both underlying skills and counting-on, these results were reported: The total counting-on instructional group performed significantly better than the total set-strategy group on 40-item tests containing addition facts (sums 6 to 10) studied. Moreover, it performed better on a 40-item test containing more difficult addition facts *not covered* in class instruction. A pretest together with process observation had identified children who already demonstrated counting-on abilities prior to the counting-on instructional unit, and these children also benefited from this instruction. They demonstrated greater gains than children of similar abilities exposed to set-strategy instruction. Also, children who had no counting-on abilities prior to the counting-on instruction outperformed children in a control group (no instruction) who had already developed counting-on abilities on their own. From the perspective of group comparison, Leutzinger's direct teaching approach appears to have been quite effective.[8]

Although there was a significant difference in performance by eleven classrooms in Leutzinger's study on the addition facts tests in favor of the total counting-on instructional group, this general conclusion does not apply to all children exposed to this instruction. There were a few children in each classroom who failed to show counting-on following ten days of instruction. Another interesting observation was that some children in the counting-on instructional group viewed the strategy as a game and didn't attempt to count-on during the test situation. When questioned later they replied that they didn't know that they could use it on the test. Also, in the weeks following the instructional unit many children who had demonstrated the counting-on strategy following instruction and posttest needed reminders for continued use of the strategy. By shifting the perspective from a group comparison on tests administered soon after instruction to observation of individual children during instruction, testing, and in the weeks following, some important concerns are raised. Fortunately Leutzinger included both group comparison and observation of individual children in his study. These observations provoke the reader to ask why the instruction did not reach some children initially and why others reverted to inefficient counting-all methods.

Again, individual rates of intellectual development can account for all children not being able to count-on on cue. Furthermore, since the number meanings underlying the counting-on process develop gradually, it is reasonable to expect instability in children's

8. Another bit of evidence in support of direct teaching comes from the children interviewed in this chapter. Since the incidents of counting-all-on occurred almost exclusively in a subgroup of the children, I tried to learn more about their prior instruction. This subgroup (half the original group) had experienced traditional textbook instruction. The other half had been given explicit modeling of counting-on and practice in game situations involving visible objects. For many of these children, this experience provided a basis for inventing efficient tracking methods and counting-back methods. Although the textbook learners could invent strategies to solve the problems, these were indicative of incomplete progressive integration. Since these differences are not the result of a controlled comparison study, no firm conclusions can be drawn.

use of more efficient methods. Vacillation back to counting-all may be viewed as a natural part of the equilibration process. On the other hand, the reversion may be due to children's imitation of a method when modeled, with no underlying understanding (in what Piaget calls false accommodation to adult authority). Leutzinger recognized the time limitations of the study and recommended both a delay in the onset of instruction and a longer instructional sequence. This recommendation is supported by the five weeks reported by Steffe, Herstein, and Spikes (1976) for the less-advanced children to make the transition to counting-on. This study took place during the winter of the first grade. Fuson's research (1982) suggests that scheduling such a unit would be most appropriate when the children are six-and-a-half years of age. Despite care in scheduling of large-group instruction, there will always be some individuals that cannot adapt to the schedule. Is that their problem or ours?

Another recommendation of Leutzinger's study is to extend the period of emphasis on experiences with concrete materials before moving into worksheets with abstract representations like those illustrated here. This recommendation is based on feedback from the teachers participating in the counting-on phase of the study.

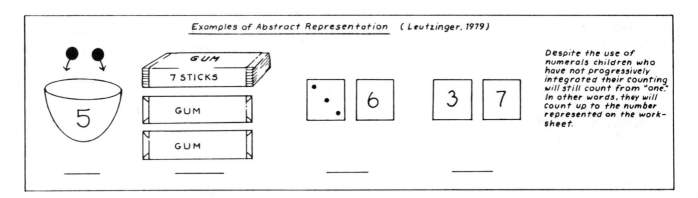

The importance of concrete experiences for introduction of counting-on and counting-back methods is supported by Fuson's study (1982). Her study explored first graders' abilities to count-on and count-back in the absence of objects. The children had little difficulty responding to "Start with five and count-on three more." However, when the task became, "Start with eight and count-on five more," one-third of the children demonstrated consistent errors in keeping track. These children would extend their first finger at "eight" rather than at "nine," resulting in answers that were consistently one less than the accurate answer. Even greater problems were experienced when counting-back and keeping track in the absence of objects. From this study, Fuson recommends that advanced counting methods be learned first in an object context where the object arrangement supports the coordination of the counts for each addend.

In contrast to the preceding recommendation, Steffe, Herstein, and Spikes (1976) suggest that objects not be made available in order not to delay the development of counting-on. In their instructional study of first graders, these researchers observed many first graders counting-on when the first addend (already counted) was covered, yet counting-all when both addends were exposed. Yet some of these same children counted-on using their fingers in the absence of objects. Covering the first set of objects (already counted) serves to suggest to children that the first addend need not be recounted. The regular arrangement of fingers may offer a similar suggestion. Here the ability to obtain number information from instantly recognizable patterns of fingers eliminates the need to count from one. Thus, fingers may be considered a special set of regular objects.

A follow-up study by Secada, Fuson, and Hall (1983) analyzed first-grade children's transition from counting-all to counting-on using tasks with and without objects showing the first addend. In all cases the quantities were labeled with numeral cards.

HOW MANY DOTS ARE THERE ON BOTH THESE CARDS ALL TOGETHER ?	
First addend not visible	*Both addends visible (with a hint)*
	THIS CARD TELLS HOW MANY DOTS THERE ARE HERE 13 SO YOU DON'T NEED TO COUNT OVER AGAIN. 6
⬚ ⸺ · · · · · ·	· · · · · · · · · · · · · ⸺ · · · · · ·

Three subskills essential for making this transition to counting-on were established. Several children who previously counted-all on the above tasks were induced to count-on through the process of individual subskill assessment itself. Other count-all children were able to count-on in the posttest when individually taught the lacking subskills. Details of both the assessment and the direct teaching sequences are given in Appendix C. Whereas extended time was required in the group teaching methods reported by Leutzinger (1979) and Steffe et al. (1976) while working with small numbers, this study reports 88 percent success in a single teaching session with individual children, while working with large numbers.[9] An important consideration in this success rate is that Secada, Fuson, and Hall worked with more mature first graders (mean age: 6.9 years). In accounting for the ease of teaching the lacking subskills, the researchers conclude that the layout of materials and the subskill questions helped the children to focus on relevant word meanings already in their possession and to coordinate them in counting-on applications. They regard their teaching procedure as "a first step toward inducing a flexible and full use of counting-on."

This chapter has explored children's counting as a problem-solving tool and has justified its inclusion in the classroom as a vehicle for development of children's thinking capacities. Alternative methods for teaching advanced counting strategies have been examined as well. At the same time that the chapter reflects considerable advances in our understanding of children's counting processes, it also reveals some gaps in this knowledge. For example, we still need to account better for differences in understanding between children who can count-on only in a single, modified task—for example, when the first addend is covered after counting—and children whose underlying understanding generalizes across a range of counting-on tasks—for instance, when both addends are exposed in an object context, when no objects are available, when tracking is required, and so on. Accounting for differences between children's understanding on single, modified tasks and their understanding on a variety of tasks is a major goal identified for research by Riley, Greeno, and Heller (1982). This critical goal applies not only to counting-on but to all areas of mathematics learning.

REFLECTIONS AND DIRECTIONS

1. Locate some six-to-eight-year-old children in your school or neighborhood and engage them in advanced counting activities and problem solving with mystery boxes. The tasks can be selected from this chapter or generated on your own. Compare your results with those reported here. Reflect on the implications of your findings.

---

9. For all the tasks in this study the first addend ranged from 12 to 19 and the second ranged from 6 to 9.

2. If you are unable to locate any children for interviewing, select three of the following children for a closer study of their counting behaviors on different tasks.

Marsha: pp. 70, 72, 74, 75   Sal: pp. 63, 72, 74, 75, 78, 79
Rita: pp. 62, 71, 73, 75, 76, 78   Nancy: pp. 68, 74, 75, 77, 79
Suzie: pp. 62, 63, 73, 76   Eileen: pp. 71, 75, 78

Reflect on any consistencies or inconsistencies uncovered. Try to account for them.

3. Reflect on the extent to which thinking is basic in your teaching of mathematics.

5

DERIVED THINKING STRATEGIES: BASIS FOR A NETWORK OF NUMBER RELATIONS

PREVIEW

Traditionally, a primary goal of elementary school mathematics has been to teach children instant recall of number facts such as addition and subtraction combinations. For example, by the end of the second grade, children are expected to have mastery over 121 addition combinations (with sums up to 20). Although traditional methods of drilling with endless flash cards and worksheets have had limited success, many schools have opted for even more drilling rather than searching out better methods of teaching. One of the failures of the drill method is that when a number combination is forgotten there may be no way to reconstruct it, since it was memorized in total isolation from other combinations. This method is so mindless that active learners may suddenly require external rewards to participate in the repeated drill that is so common. While many children succumb to this rote system, others remain active by inventing their own strategies for relating combinations in solving unfamiliar ones.

Alternative thinking strategies that are not primarily based on counting for determining sums and differences will be the focus of this chapter. These strategies are characterized by the use of a known sum or difference to derive an unknown sum or difference. In addition to being powerful shortcut problem-solving tools that are alternatives to counting, these derived strategies also provide a basis for constructing an intricate network of number relations. This relational network not only serves as a long-term memory system, but also gives meaning to addition and subtraction combinations. Evidence of such strategies is found in the thinking of primary-grade children, despite teaching methods that don't encourage their development.

A brief glimpse of the capabilities of beginning second graders to generate such alternatives to the advanced counting strategies of counting-on and counting-back was provided by the children's responses to the following task.

$9 + 6 =$ _____ $14 - 6 =$ _____   (Presented in the absence of physical materials).

Would you figure out the answer to this number sentence?

(If the answer was obtained by counting, a follow-up question was asked:)

Suppose you couldn't count and didn't remember the answer.

Is there another way you could figure it out?

Think aloud so I know what you're doing.

In the absence of materials and specific requests to count, 5 of the 31 children approached at least one of the tasks with a derived strategy. Two additional children generated a noncounting alternative to their initial counting method. The responses of four children are illustrated below. Notice how the children transform the tasks in terms of known addition or subtraction combinations to derive solutions for the unknown combinations.

$9 + 6 = $ _____

TEN PLUS SIX IS SIXTEEN, MINUS ONE IS FIFTEEN.

Marsha J.

Marsha increases one of the addends by one and compensates by decreasing the sum by one.

SIX PLUS TEN IS SIXTEEN. ONE LESS IS FIFTEEN.

Shelly C.

Shelly also compensates an increase in one addend by an equivalent decrease in the sum.

TEN PLUS FIVE IS FIFTEEN.

Vic J.

Vic compensates the increase of one of the addends with a decrease of the other addend by the same amount.

$14 - 6 = $ _____

SEVEN AND SEVEN IS FOURTEEN. TOOK SEVEN AWAY— LEAVES SEVEN. IF YOU TOOK SIX AWAY IT WOULD LEAVE ONE PART OF THE OTHER SEVEN... EIGHT.

Larry G.

EIGHT, BECAUSE SEVEN PLUS SEVEN IS FOURTEEN. IT MUST BE ONE MORE—IT'S LESS THAN SEVEN YOU TOOK AWAY.

Shelly C.

Larry and Shelly view subtraction in terms of known addition combinations in which the increase of one addend is compensated by an equivalent decrease of the other addend.

FOUR OFF—TEN, AND TWO MORE, THAT MAKES EIGHT.

Vic J.

Vic breaks up six into four and two, taking six away in two stages (first four, then two) while using ten as an intermediate unit.

Since the purpose of the preceding chapter was to explore the limits of the children's counting abilities, they were asked to respond to a variety of tasks using counting methods exclusively. Yet the preceding tasks illustrate that some of the same children may have preferred to respond by using noncounting methods. Thus, by not restricting responses to a counting method we may have observed both counting and noncounting strategies being used alternately on different tasks. This was demonstrated by both Marsha and Larry. Also, we may have observed greater success by some children, such as Vic, as effective derived strategies replaced ineffective counting methods. Let's summarize the methods we have observed four of the children using:

Marsha: A fast, efficient counter on a variety of tasks in the last chapter, she chose a derived strategy for the first task. On the second task she counted back but did not generate an alternative strategy.

Larry: He counted-on for the first task, generating no further alternative. For the second task he invented a derived strategy only after experiencing difficulty with counting-back.

Shelly: He showed a preference for derived strategies on both tasks despite an ability to follow them up with counting-on methods. On other tasks in the preceding chapter he showed efficient counting methods, though inconsistently.

Vic: He showed a strong preference for derived strategies. Although an inefficient counter on a variety of tasks, he was able to generate two distinct derived strategies.

Reflection on this limited data generates several questions, about the origins of these abilities to operate on numbers with derived strategies, their relation to counting, the existence of distinct learning styles, and the teachability of derived strategies.

This chapter will present the results of other research on child-invented derived strategies that don't depend primarily on counting. It will also examine the origins of children's abilities to operate on numbers using noncounting methods, and their relations to counting. Furthermore, it will illustrate classroom activities for developing derived strategies and discuss related classroom research.

RESEARCH ON CHILD-INVENTED DERIVED STRATEGIES

Two research studies will be described which provide evidence that children not only are capable of using derived strategies to obtain answers to number combinations but also of inventing these strategies. The latter conclusion is based on the lack of any direct instruction in such non-counting methods. Particularly notable is the fact that the children in one of the studies did so in spite of instruction to the contrary.

A Longitudinal Study

In the absence of exposure to teaching methods that focus specifically on strategies for relating combinations and applying them to finding unknown sums and differences, a child's demonstration of such strategies is assumed to be the result of his own construction. The preceding glimpse of these capacities of beginning second graders (5 out of 31) may suggest that they are manifested by only a select number of superior children. Yet Carpenter's longitudinal study of 100 children (1981) provides evidence to the contrary. Beginning in the first grade, the same group of children was interviewed every few months and presented with sets of 12 story problems to solve. By the end of the first grade over half the group had used a derived strategy at least once during the interviews. By the middle of the second grade over three-fourths of the children had done so. Since the children had received no classroom instruction in these strategies, their existence was attributed to the children's own inventiveness.[1]

Carpenter (1980) reports that most of the observed strategies were based primarily on known combinations having a sum of 10 or on known combinations of "doubles". Examples of such strategies are illustrated below, together with schematic representations of the possible steps required in children's thinking (Carpenter, 1980).

ADDITION

Decomposition

4 + 7	"Seven and three is ten, so I put one more on there and got eleven."

$$7 + 4 \longrightarrow 7 + \square = 10 \longrightarrow \underline{4 = 3 + 1} \longrightarrow 7 + 3 = 10 \longrightarrow 10 + 1 = 11$$
(Unknown)   Decomposition   (Known)

6 + 8	"Six and six is twelve and two more is fourteen."[2]

$$6 + 8 \longrightarrow \underline{8 = 6 + 2} \longrightarrow 6 + 6 = 12 \longrightarrow 12 + 2 = 14$$
Decomposition

Compensation

4 + 9	"Ten add on four is fourteen and one less than fourteen is thirteen."

$$4 + 9 \longrightarrow 9 \,\boxed{+\,1} = 10 \longrightarrow 10 + 4 = 14 \longrightarrow 14 \,\boxed{-\,1} = 13$$

6 + 8	"I took one from the eight and gave it to the six. Seven and seven is fourteen."

$$6 + 8 \begin{array}{c} \nearrow\; 8\,\boxed{-\,1} = 7 \searrow \\ \\ \searrow\; 6\,\boxed{+\,1} = 7 \nearrow \end{array} 7 + 7 = 14$$

---

1. It is still possible that some of these strategies were learned from other children in the classroom or from adults at home.

2. The child's response to 6+8: "Six and six is twelve and two more is fourteen," can also be interpreted as a compensation by simultaneously reducing one of the addends by two and increasing the sum by two.

Decomposition

9 – 6	"Nine take away four is five; take away two is three." $9 - 6 \longrightarrow \underline{6 = 4 + 2} \longrightarrow 9 - 4 = 5 \longrightarrow 5 - 2 = 3$

Compensation

11 – 3	"Three from ten is seven. It will have to be more because it's eleven." $11 - 3 \longrightarrow 11 \boxed{- 1} = 10 \longrightarrow 10 - 3 = 7 \longrightarrow 7 \boxed{+ 1} = 8$
14 – 8	"Seven and seven is fourteen. Eight is one more than seven, so the answer is six." $14 - 8 \longrightarrow 8 + \square = 14 \longrightarrow 7 + 7 = 14 \begin{cases} 7 \boxed{+1} = 8 \\ 7 \boxed{-1} = 6 \end{cases}$

Another strategy used by children (Carpenter, 1981) is to relate successive tasks, using the result of the first to derive the result of the second. At one point in the presentation of the tasks at the beginning of this chapter, the child's answer to the first task was written down with the open sentence for the next task immediately following.

$$9 + 6 = \underline{15}$$
$$14 - 6 = \underline{}$$

This method of presentation provided the children an opportunity for relating the two tasks in the following way.

> "Nine plus six is fifteen. So, fifteen minus six is nine.
>
> Since fourteen is one less than fifteen, fourteen minus six must be eight."

Yet none of them did so.[3] Although 31 beginning second graders treated these as independent tasks, some of Carpenter's children did construct relations between other successive tasks. They did so despite the use of numbers on successive tasks that were not readily related.

Although Carpenter (1981) reports a large proportion of children demonstrating derived strategies on at least one occasion, these strategies were applied inconsistently. Both counting and noncounting methods were used side by side in successive problems. Although 75 of the 100 second graders had used derived strategies on at least one of the twelve problems by midyear, only 12 of these children had used three or more of these strategies. The ability to apply a derived strategy may depend on specific number combinations.

Other related research indicates that children derive strategies as early as kindergarten. Blume (in Carpenter, 1981) observed similar shortcuts to counting employed by 20 percent of the kindergarten children in his study. Furthermore, in comparing these strategies to ones used by first-grade children, Blume noticed a distinct shift in preference. Whereas kindergarteners using derived strategies favored compensation, their counterparts in the first grade tended towards decomposition methods. Although this evidence suggests the possibility of a developmental trend, further research is needed to establish its existence.

An Indirect Outcome of Direct Drill

The traditional goal of instant recall of addition and subtraction combinations has been most frequently

---

3. One notable exception in this group of second graders was observed in another context. Hazel M. was able to relate successive number story problems. Rather than clear the table of blocks following each problem, she used the ones available from the last problem—adding to, separating from, or transfering between subgroups. In one sequence she related problems of the type $6 + \underline{} = 10$, $\underline{} + 3 = 9$ and $9 - \underline{} = 5$. Also notable, is that Hazel was an inefficient counter on the tasks described in the preceding chapter.

pursued in the classroom through the accumulation of independent and unrelated combinations by drill and practice. The drill method described by Brownell in 1935 still dominates the educational scene, decades later.

> . . . the combinations were exposed, a few in each lesson, in random order. A combination, such as 5 + 3, was exhibited, and the answer 8 was supplied by the teacher or some child who knew it. Individual and group practice followed, in the form of oral, silent, and written exercises—flashcards, games, and the common rapid-exposure devices being employed with a view to establishing mastery over the combinations. If, at a later time, a child did not know a sum or remainder, he was not allowed to find it for himself. Instead, the answer was immediately provided. "Learning" consisted in repetition of the appropriate verbal formulas. Simple one-step problems then provided use of the combinations so taught. (p. 20).

Research employing written tests of speed and accuracy provide support for this method. Yet studying only children's written responses under timed conditions can provide misleading information. The 1935 research of Brownell and Chazal extended beyond paper-and-pencil tests to individual interviews, producing startling results that both questioned the effectiveness of drill and uncovered the existence of alternative thinking strategies invented by children. This research studied both the speed and accuracy of 63 beginning third graders on a written test of 100 addition combinations, and the process of obtaining their answers in individual interviews. Prior to the third grade, these children had been exposed to 200 addition and subtraction facts by repeated drill over a period of two years. If this method of direct drill was effective, children would provide accurate and automated responses even following a summer vacation. If not, a brief period of further drill would reactivate their capacity for instant recall. Following the written test, a subgroup of 32 children was selected for a follow-up interview. This subgroup was composed of children having the poorest, average, and highest test scores. The interview encouraged the children to think out loud when given each of 16 items of average and greatest difficulty on the test. Despite two years of drill, less than 40 percent of the interview responses were automated. Actually, the children's counting methods (23 percent) and derived strategies (14 percent) accounted for almost as many combinations (37 percent) as did instant recall. The remaining responses were categorized as wrong guesses. The researchers pointed out

that drill does not guarantee instant recall by children, and therefore does not achieve what it sets out to do. Unbeknownst to the teacher, who assumed the children were repeating the combinations, the children had taught themselves other ways of thinking in an attempt to make sense out of the combinations.

In a second phase of the study, the beginning third graders received a further month of drill on addition combinations through daily five-minute sessions. The drill, in the form of oral, silent, and written practice, was so distributed that each combination was presented at least twice daily. Following the month of drill, the same written test was given to all 63 children and interviews were conducted with the same 32 children on the same 16 combinations. At this point only 48 percent of these combinations were known as memorized facts, while the percentages of counting and derived methods remained about the same, for a total of 85 percent accuracy. The average time to complete the written test was reduced from 17 to 11 minutes. From the interviews, the researchers found that children's established methods were employed with greater proficiency—for instance, children who tended to count did so more rapidly. In other words, rather than markedly alter the thinking used by a child, the additional drill tended merely to speed it up. Furthermore, as proposed by Steffe (1979b), there is a possibility that the percentage of responses attributed to drill was exaggerated when some of the children switched from counting-on or derived strategies by "meaningfully habituating the combination." Following two years of drill on addition and subtraction facts, more of the same methods failed to produce the desired type of mastery. Yet if teachers do not find the time to listen to children, this discrepancy goes unnoticed.

In discussing this research, Brownell (1935) concludes:

> Under such conditions if arithmetic becomes meaningful, it is absurd to assign the credit to drill. It is nearer the truth to say that if under these conditions arithmetic becomes meaningful, it becomes so in spite of drill. (p. 12).

He argues that children must be capable of reasoning such as that involved in derived strategies in order to give meaning to addition and subtraction combinations. For Brownell, combinations such as 8 + 7 = 15 and 9 + 6 = 15 are not adequately learned until the child has become aware of how they are related. He argues further that time must be provided for such meanings to develop prior to practicing for proficiency.

TOWARD A NETWORK OF NUMBER RELATIONS: A THEORETICAL PERSPECTIVE

Numbers are melded by arithmetic operations into new numbers. The earliest method of putting numbers into relation involves incrementing or decrementing numbers in steps of 1 to establish a ±1 relation between adjacent numbers in the series.

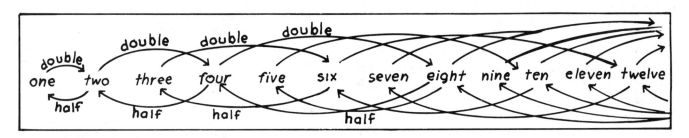

Although building on the number-name chain of oral counting, this ±1 relation is not necessarily established by the process of counting, according to Steiner (1980). (An alternative basis for this relation will be described later in this chapter.) He further points out a certain reversibility of thought required to reason that if seven is one more than six, then six is one less than seven. As well as relate adjacent neighbors in the series, it is possible to relate near neighbors by steps of ±2. These increments and decrements introduce addition and subtraction into number learning.

Whereas incrementing and decrementing in small equal-sized steps provide for the construction of relations between neighbors and near-neighbors, doubling increases the distance between numbers being related. Steiner (1980) points out that this process of adding an equal amount also includes a certain reversibility: six is double three, so three is half of six.

These two processes of relating numbers provide a basis for initiating the construction of a network of number relations. For example, if double three (3 + 3) and double five (5 + 5) are known, then 4 + 3 and 7 + 5 can be inferred as being one or two more than the respective doubles combinations.

Just as doubles become building blocks in the construction of small numbers, tens serve a similar function in the construction of larger numbers. Initially, knowing the constituents of ten—8 + 2, 9 + 1 and so on—enables children to determine the sums of 8 + 5 and 9 + 7. This is accomplished by the decomposition of one of the addends to "build ten and some more."

$$8 + 5 \text{ is the same as } \underbrace{(8 + 2)}_{10} + 3$$

$$9 + 7 \text{ is the same as } (9 + 1) + 6$$

The gradual construction of larger numbers using multiples of ten as units or building blocks is the topic of Chapter 10.

As indicated by Brownell (1948), "the child does not come, suddenly and once and for all, to a meaningful grasp" of seven, or even one of its combinations, 3 + 4 = 7. Rather, this involves both a reorganization of thinking and a continuous elaboration of relations over extended periods of time. The elaboration process will be illustrated first. Neither seven nor any one of its combinations can be understood in isolation. Rather, they acquire meaning in relation to other numbers. The elaboration of relations in the construction of a steadily-growing network is suggested in the following illustration. Since new and different relations can be constructed in all directions, the number of possible relations is virtually unlimited. According to Steiner (1980), once relations have been developed in several directions, that part of the network is ready to be condensed into a single concept of seven and used in development of further relations. This continuous constructive process eventually produces a coherent system of relations. Furthermore, the high level of organization resulting from linkage of new relations to previously constructed ones also ensures that the knowledge will be retained in long-term memory. This tightly knit network can also be unfolded to gain access to constituent relations of seven condensed in the concept.

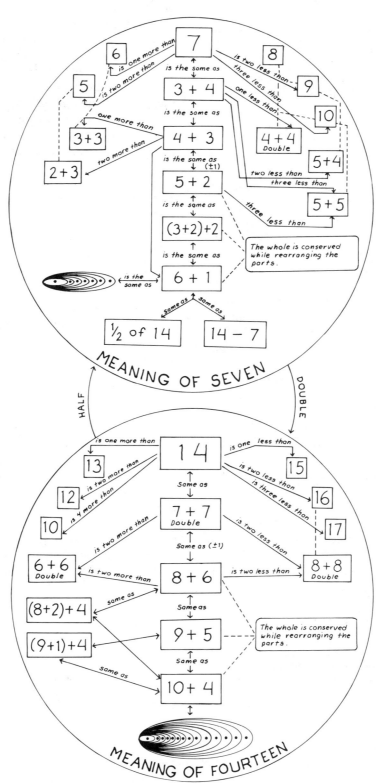

Brownell (1935) argues that children must be capable of reasoning such as that involved in derived strategies in order to give meaning to addition and subtraction combinations. These potential meanings are clarified by Carpenter (1980).

> Understanding addition and subtraction requires that children have some basis for relating the magnitude of different combinations, that they know how different sums and differences can be composed and decomposed, that they understand the relation between addition and subtraction, and that they understand basic properties of addition like commutativity and associativity. (p. 326).

Carpenter continues to point out that these basic concepts are all involved in the derived strategies for building numerical networks.

One of the striking aspects of numerical networks is the flexibility of thinking required to construct them. Yet Piaget's work (1965) shows that the thinking of young children lacks this necessary mobility. Most five-year-olds are unable to mentally reverse a physical action just completed. Thus in combining or decomposing groups of objects, they are unable to simultaneously consider the whole and its parts. Having combined two groups (3 + 4) to make a whole (7), they lose sight of the parts (3 + 4) and are unable to mentally reverse the joining action to regain them. For them, the parts no longer exist.[4] Without the ability to simultaneously consider the whole and its parts they are unable to relate quantities as in 3 + 4 = 7. Similarly, a lack of reversibility in thinking prevents young children from constructing a system of ±1 relations between numbers. However, a major reorganization of thinking resulting in increasing flexibility of thought, such as reversibility, characterizes the child's entry into the concrete operational period at about age seven. A further source of flexibility is the realization that the order of combining numbers in addition is irrelevant (commutativity and associativity).[6] The latter flexibility contributes to the capacity for mental detours and construction of multiple relations.

In using derived strategies, children deal with numbers without the need to count up to them, even in the absence of objects. Yet in examining children's development of counting, we noticed the initial importance of objects, the intellectual breakthrough involved in counting-on, and then the gradual decrease in dependence on objects. In such a comparison, it is tempting to assign greater maturity to noncounting methods. Indeed, Brownell considers noncounting methods to be at a higher developmental level. However, a consensus is lacking among researchers. Having observed the coexistence of counting and noncounting strategies in the same children, Carpenter (1980) avoids the temptation of relating them developmentally. On the other hand, Davidson (1982) focuses on children's preference for one type of strategy or the other, thus positing two distinct learning styles. A final interpretation is that these strategies represent two distinct avenues to number understanding, developing simultaneously and interacting at times to extend this understanding. The latter interpretation will be expanded later in this chapter. Another topic for extended discussion will be an alternative process of dealing with numbers without counting up to them. First, we will detour to discuss existing research on teaching derived strategies.

RESEARCH ON TEACHING DERIVED STRATEGIES

Although Carpenter (1980) reports a high percentage of children using derived strategies in the absence of any instruction, he also indicates that such strategies are used inconsistently. This inconsistency was illustrated by Marsha and Larry in the introduction to this chapter.[5] These observations support Steffe's (1979b) contention that although there is evidence of children's inventiveness in constructing derived strategies, many children either do not develop such strategies or have developed only limited strategies. This inconsistency in appearance and sophistication of derived strategies raises questions about methods of furthering children's development of such strategies. Schemes for teaching addition combinations by means

---

4. Yet they are quite capable of focusing on the result of the combination. This behavior will be illustrated in the next chapter as children are asked to represent relationships as equality sentences.

5. This issue of consistency in the use of derived strategies deserves closer attention, in day-to-day observation over an extended period rather than in a single interview.

6. Whereas commutativity is the order-irrelevant property of combining pairs of numbers, associativity is the order-irrelevant property of the paired combination of three or more numbers. For example, 8 + 2 + 4 = (8 + 2) + 4 or 8 + (2+4), where the combination of any pair is added to the remaining number.

of derived thinking strategies have been identified by several mathematics educators, including Thornton (1978, 1982), Rathmell (1978), Taylor (1980), and Wirtz (1982b). Each scheme is designed to assist children in organizing their thinking about the combinations in order to reduce the number to be memorized directly. One such scheme is illustrated below and the results of its effectiveness are reported.[7]

Thornton (1978) organized the addition combinations into clusters for easier learning. Each cluster of combinations used similar strategies for derivation from already known combinations. Following an initial mastery of the doubles (5 + 5, 6 + 6, and so on) the children studied related facts such as "doubles plus one" (5 + 6 is one more than 5 + 5). Next they studied the cluster of combinations that could be derived as "doubles plus two". Here the compensation strategy taught involved converting these combinations into another double by sharing the two (7 + 5 becomes 6 + 6, a known combination). Another cluster isolated for study involved adding nines to other numbers by considering them as one less than tens combined with those numbers (9 + 6 is one less than 10 + 6). The final cluster included the following combinations together with their commutatives: 2 + (5, 6, 7, 8), 3 + (6, 7, 8), 4 + (7, 8), and 5 + 8. Strategies for dealing with these combinations included counting-on and decomposing one of the addends to make a "ten and some more." For subtraction combinations, children were encouraged to think in terms of related addition combinations. The following table illustrates similar clustering of combinations (Thornton et al., 1982).

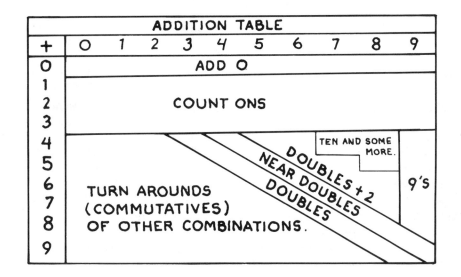

Although the teaching involved explicit modeling of the selected strategies, followed by immediate practice on combinations within specific clusters, children's alternative constructions were accepted and encouraged. For example, despite being taught a specific strategy of converting doubles plus two into another double by sharing, some children generated a more direct approach (7 + 5 is two more than 5 + 5). In this method there is a heavy emphasis on children relating the combinations either through methods modeled by the teacher or through those of their own creation. Yet this experience is followed by drill, in the form of motivating games and worksheets, to provide further practice for speeding up the processing and ensuring the retention of known combinations. As suggested by Brownell (1935), this method provides some time for meanings to develop prior to practicing for proficiency.

Thornton (1978) studied two groups of 25 beginning second graders pooled from multigraded classrooms and assigned to either experimental or traditional teaching. Beginning in September the children were exposed to eight weeks of direct instruction (twenty minutes per day for three days per week). These groups were taught by regular classroom teachers after undergoing training. In the experimental group the children were taught according to the method of relating combinations described above. The traditional group was taught addition combinations, primarily with a mathematics textbook supplemented by drill worksheets and games. A paper-and-pencil test was used to determine how many combinations the children could complete in three minutes. The same instrument was administered as a pretest, a posttest at the end of eight weeks of teaching, and a retention test two weeks later. Following the retention test all of the children were interviewed using a blank copy of the

---

7. A second scheme devised by Taylor is described in Appendix D.

test, to determine their thinking processes. They were asked to think back to a time when first learning the combinations (2 + 6, 4 + 7, 6 + 7, 9 + 4, 8 + 6) and to identify any tricks, shortcuts, or other easy ways they used for remembering them. They then were invited to do the same for any other combination on the test.

The children's performance on addition and subtraction combinations, as well as on forty harder addition facts (where both addends were greater than 3 and at least one was greater than 7) is graphically represented in the following graphs as average group scores.

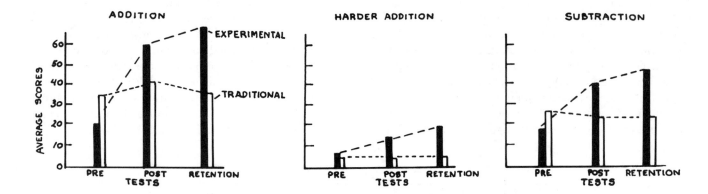

A comparison of group test performance indicates significant gains and retention by the experimental group. Follow-up interviews of the upper and lower thirds of each group revealed that about three-fourths of the experimental group used the derived strategies that had been taught explicitly. In a comparison of the thinking strategies used by high- and low-achieving children, it was found that methods used by the high achievers accounted for 62 percent of the use of derived strategies tallied during the interviews, while the low achievers accounted for the use of the remaining strategies. However, retention of facts as shown on the posttest was consistent for both groups. The study showed that both high and low achievers in the exper-

imental group learned, although to different degrees, the strategies taught explicitly, and also retained the ability to do so in the absence of further teaching. On the strength of these results, and the results of other studies by Thiele (1938), Swenson (1949),[8] and Rathmell (1978),[9] Thornton recommends that such derived thinking strategies be developed systematically prior to any drill on number combinations. Children capable of derived strategies are better able to interrelate the combinations and remember or retrieve them. Children who forget combinations learned as isolated facts in drill have never constructed the relations and therefore have no basis for reconstruction.

AN ANALYSIS OF METHODS FOR TEACHING CONSTRUCTION OF A COHERENT NUMERICAL NETWORK

Thornton's program (1978) for direct teaching of derived strategies achieved a degree of success, particularly with high-achieving second graders, in terms of the traditional goal of rapid recall of addition combinations. Yet examined in light of broader goals of mathematics education—children's development of independent thinking abilities and construction of a coherent network of number relations—it fell short of its potential. Without question, Thornton's program does make a contribution towards the latter goals. Yet an analysis of further possibilities will clarify how this contribution might be extended.

Explicit modeling of derived thinking strategies by the teacher minimizes children's capacities for

developing their own thinking abilities. Some of the children may be left with the impression that these efficient shortcuts originate only in the teacher's mind and that their job is only to imitate the teacher. Similarly, placing the primary emphasis on the recall of number combinations restricts the construction of a network of number relations somewhat. Steiner (1980) distinguishes between placing the emphasis on constructing relations and computing correct answers.

---

8. Both Thiele and Swenson demonstrated that children were able to transfer these strategies to combinations not studied previously.

9. Rathmell applied a more stringent test of this method of teaching derived strategies by retesting the children following summer vacation. The retention level remained high.

Constructing a numerical network does **not** primarily mean to find the correct answers to numerical tasks but to find relations to other numerical tasks or propositions (statements of relation). That is why the teaching and learning process, the program in the classroom looks different from other programs: Students are not looking for the outcome of 5 + 5 but for different propositions relating 5 + 5 to 6 + 5 or 5 + 6, to 5 + 4 or 4 + 5 or even to 4 + 6, and examples where a compensation of +1 and –1 relates to the original 5 + 5! Or 10 – 5 will be related to 10 – 4 or 10 – 6, to 9 – 5 or 11 –5 or to 9 – 4 or 11 – 6! Each time the primary task is to put the first task or proposition where the solution is known into relation to another one and tell exactly what the relations between the two are and what can be inferred from this relational knowledge. (pp. 6–7)

This shift in emphasis will be clarified further through a discussion of classroom activities.

Rather than explicitly model derived strategies, the teacher can encourage children to search for shortcut alternatives to counting to find correct answers to combinations. Their shortcut methods can be shared in group discussion with the teacher's role being to encourage the exchange of ideas by helping children verbalize and clarify their ideas, to help children refine their methods by asking questions that provoke comparison and evaluation, and to encourage the cross-checking of outcomes through alternative derived strategies. Once the children have begun to adapt to the restriction of not counting by construction of their own derived strategies and by learning shortcuts from other children, the focus can be shifted further to the construction of a network of relations. The assorted small-group activities illustrated below can serve as springboards for further construction of relations.[10]

---

1. 2   4   In what different ways are these combinations related?
    +2   +2

    (If necessary, this question can be made more specific.)

    3   5   Which combinations are related by:
    +3   +3   one more?

                      one less?

    4   6   two more?
    +4   +4   two less?

    Write more number combinations across or down the page to make the series longer. Have your partner figure out how they are related to their neighbors.

2. Which of the following combinations is the easiest to add up? How come?

7	4	3	7	4	3
4	7	7	3	3	4
+3	+3	+4	+4	+7	+7

3. What double combinations are related to 3 + 4? How are they related?
    What double combinations are related to 6 + 8? How are they related?
    How is it that we only have two doubles for 3 + 4, but we have three doubles for 6 + 8?

    How many related doubles do you think you would find for these combinations:

        4 + 4?   5 + 7?   6 + 4?
    Check your prediction.
    Keep trying different combinations until you are sure of the number of related doubles without checking.

---

10. Initial activities were adapted from Taylor (1980).

4. Find all the combinations of two numbers that are the same as 14.
   Find a way to put these combinations in order, discarding any that you don't need.

   Decide how each combination is related to its next-door neighbor.
   Decide how each of these combinations are related to their neighbors:

   a. 7 + 7   b. 7 + 7   How can you explain the different
       9 + 5   10 + 4   relations between 7 + 7 and its different
      11 + 3   13 + 1   neighbors?
      13 + 1

5. Find some three-number combinations that are the same as 14.
   Find all the three-number combinations for 14 having the first two numbers combining as 10.

6. What are all the possible relations for the number 15?

The preceding activities encourage children to use their derived strategies as problem-solving tools, thereby extending these abilities and contributing towards the construction of a network of relations and the development of attitudes of independence and confidence in problem solving. Again, Steiner (1980) helps to clarify the potential outcomes of such an emphasis in the classroom.

> The student has not primarily learned to find outcomes —this is what the calculator can do and is made for. He has learned to find relations and by doing so he got accustomed to knowing that he can always find a solution, that every problem is solvable. He has even learned to find whole sets of solutions—a plurality of solutions—whereby two of them suffice to arrange a self-control over whether he has done a good job or not. This in turn enhances the student's independence from the teacher as well as from the program. More-over, finding a whole plurality of solutions is a kind of divergent thinking, a kind of creativity. Being able to give reasonable, explainable, prediction by "putting into relation" certain numerical propositions is a capability that cannot be replaced by any calculator. (pp. 7–8)

In contrast to the mindless method of memorizing dissociated number facts, where active learners may suddenly need rewards to participate in endless drill, generating a multiplicity of interesting solutions can serve as its own reward. Rewards such as praise are not only unnecessary, they may actually interfere with the constructive process. Once children begin to focus on relationships, they can actively quest to find further relationships between other numbers by asking their own questions. A more specific analysis of the contrasting methods follows.

COMPARISON OF CONTRASTING METHODS FOR TEACHING DERIVED THINKING STRATEGIES

	DIRECT METHODS	INDIRECT METHODS
Goals	Learning number facts is basic. Derived strategies are specific tools for this purpose.	Independent thinking is basic. Construction of a coherent network of number relations is basic.
		Learning number facts is secondary.
Teacher's Role	Models selected derived strategies explicitly. Clusters of combinations are studied at a particular time since they can be determined by a specific method. This is followed by immediate practice of that method.	Encourages children to find noncounting computing methods for combinations. Facilitates process of sharing, clarifying, and comparing methods. Encourages generation of multiple solutions. Provides activities that shift focus from computing answers to putting numbers into relation.
	Accepts incidental invention of alternative strategies by children.	Encourages invention of alternative strategies by individuals and in collaboration.
	Provides for large group lessons with small group or individual practice activities.	Provides for small group interaction.

Teacher's Role (cont'd)	Encourages speed of computing answers to combinations. Once meanings have been developed, provides a variety of practice—worksheets, games, and so on.	Provides for practice games that shift back and forth from focus on computing accurate answers to putting numbers into relation. Time for reflection is included.
	Corrects worksheets, and repeats modeling of strategies as needed.	Encourages children to correct computational worksheets by cross-checking with alternative methods. Feedback on putting numbers into relation is provided through interaction with peers and teacher in group discussions.
Child Outcomes	Imitator of teacher's selected methods. (Possible inventor of alternatives.)	Inventor of multiple methods, imitator of other children's methods.
	Derived strategies—a limited tool restricted to looking for the answer to specific combinations.	Derived strategies—a powerful problem-solving tool producing satisfying multiple solutions. Also a critical crosschecking device for computing accurate answers.
	External locus of control—although beginning to think for himself, the child is still reliant on the teacher for feedback on accuracy of answers, and as source of strategies to apply.	Internal locus of control—the child is not dependent on the teacher to check accuracy of computation or as the source of strategies.

Thornton's program of direct teaching of derived strategies is a major improvement over current classroom practice. Yet further development of children's mathematical knowledge and attitudes of confidence and independence is possible by shifting the emphasis from computing correct answers to generating multiple solutions.

PREVIEW II

Piaget's theory has led us to expect that the origins of children's abilities to deal with numbers without counting up to them, even in the absence of objects, are to be found in earlier concrete experiences. The remainder of this chapter will discuss both an origin of these abilities and how it can be extended to make derived strategies accessible to more children, including the low achievers, and to contribute towards the construction of a coherent network of number relations. Ways in which the origin of this alternative, non-counting pathway to number relations supports the child's development of the counting process also will be discussed.

SUBITIZING AND VISUALIZING AS A BASIS FOR DERIVED STRATEGIES

At the same time that humans have a universal ability to count, they also have the capacity to instantaneously identify the number of objects in a group without counting. This process of instantaneous recognition of number patterns without counting is known as *subitizing*. To experience this curious phenomenon and explore how it is affected by the size and configuration of the groupings, try the next activity. Following you will find rows of groupings to be viewed individually as you would if each grouping were flashed on the screen for one second with a pause between groupings. Mask rows with 3-by-5 index cards, reveal each new grouping in a row by unmasking, then look away after one second of viewing, to approximate such an experience.

1. Read off the number of dots in each group without counting.[11]

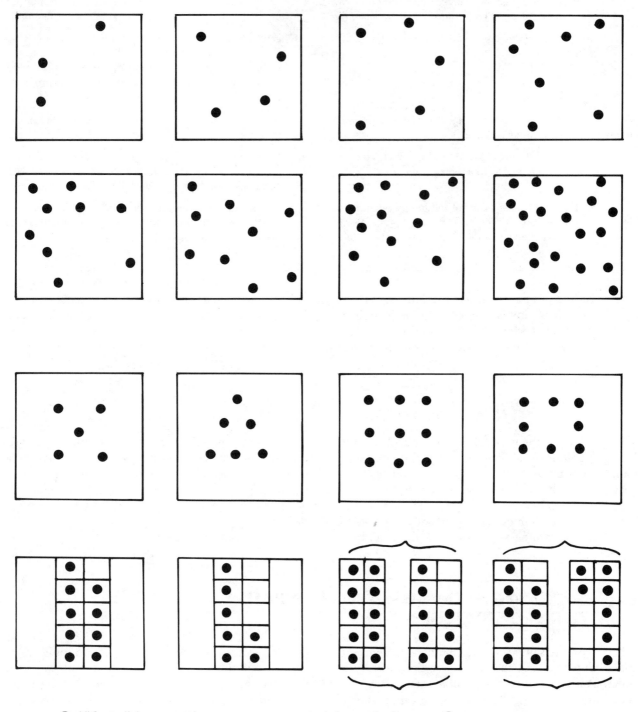

2. What did you notice as you proceeded through the page?
   How come?

11. These activities are adapted from Wirtz (1980).

Origins of Recognizing and Visualizing Number Patterns

Man's natural tendency to physically and mentally organize objects into symmetrical groupings contributes to the curious ability to instantly recognize and give a number name to groups of objects without counting them one by one. Lacking such organization, it is possible to recognize about five randomly grouped objects. It is also possible to estimate larger random groupings by mentally clustering the objects into recognizable subgroups and subconsciously adding the quantities. Organizing the objects into patterns facilitates our ability to recognize and give number names to groupings larger than five. The process of instantaneous recognition of number patterns without counting is known as *subitizing*. The choice of arrangement, familiarity with certain arrangements, and prior understandings brought to the subitizing process determine the limit of what is seen, as well as the speed and accuracy of the process. Although treated merely as a curiosity for many years, subitizing plays a fundamental role in the genesis of the concept of number, according to von Glasersfeld's theoretical analysis (1980).

Piaget's theory (1966) indicates that young children experience a breakthrough in abilities to visualize or otherwise re-present ideas and events internally beginning at age two. By the age of four, they are able to instantaneously recognize groups of one, two, three, or four objects.[12] Young children are surrounded by patterns at home and at play and soon develop stable images of these patterns—they are mentally able to re-present them for themselves. Having acquired some number names for oral counting, they become attuned to hearing number names used in context. Soon they begin to hear these names together with configurations of objects, fingers, and dots. Once the same name is heard repeatedly with the same pattern, the children make a pattern-name association. On seeing a pattern

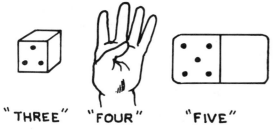

"THREE" "FOUR" "FIVE"

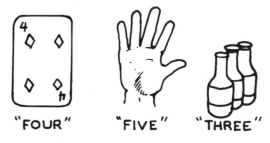

"FOUR" "FIVE" "THREE"

in their environment, they compare it to mental images of known patterns, and even mentally rearrange a nonstandard pattern in the attempt to name it. Now they are able to call out the number names when they see associated patterns, much to the delight of their parents. Also, upon hearing a number name, they are able to mentally visualize its pattern or refer to a pattern of actual objects. However, just as children take considerable time to develop number meanings through the counting process, these meanings also develop gradually through subitizing.

Although a young child may be able to assign appropriate number names to patterns of objects or dots, the quantitative aspect of the pattern is not likely to be evident. This limitation of young children's abilities is demonstrated by Churchill's five-year-old Steven (in Ginsburg, 1977). He does not realize that five refers to any arrangement having the same quantity. A

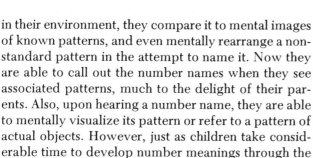

I SEE YOU'VE GOT FIVE DUCKS IN YOUR PICTURE.

IT ISN'T FIVE. THERE ISN'T ONE IN THE MIDDLE.

For Steven the particular arrangement seems more important than the number.

---

12. Piaget avoided using small quantities (up to five) in his conservation of number task because of children's abilities to recognize their *perceptual number* despite rearrangement.

similar restricted view of patterns is revealed by some of Wirtz's kindergarten children (1980) while attempt-

ing to match their own patterns to standard domino patterns. With more experience, they will notice that

more than one configuration is called five, yet they may still not have the concept of "fiveness." For example, following the ordering of patterns from one to

five the children's understanding of five can be explored by removing the other patterns and asking the children to show four. Some of the young children will

not yet see four as a part of five or see five as one more than four. Although children acquire a variety of representations in association with the same number names,

these initially serve merely as interchangeable labels conveying limited quantitative information.[13]

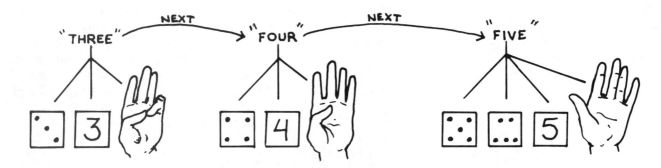

To explain this delay in development of meaning, von Glasersfeld (1980) reminds us that numbers are not concrete. Piaget considers number as a mental construction of relations that is not perceptually available

in either the objects or perceptual patterns themselves. It is this mental construction that requires considerable time for development.

According to von Glasersfeld (1980), the child's

---

13. Just as small numbers can be represented by visual configurations in space that are readily recognized, they can also be represented by auditory configurations in time—rhythmic patterns (von Glasersfeld, 1980).

simultaneous interaction with the previously separate process of counting objects and subitizing patterns of objects results in the first major breakthrough in numerical understanding of patterns. Whereas the child's initial focus in counting is on individual countables, his primary focus in subitizing is on the group as a whole—a global configuration rather than a collection of count-ables. Once a child begins to count objects, he is willing to count everything, including known dot patterns. When the last number name of his counting of the elements of the pattern coincides with the resulting name of the subitizing process, the child may react with surprise. This may provoke him to reflect on the dual processes leading to the same result, and thereby,

to reconstruct his five at a higher level of number meaning—a unit of units integrated by "one more" relations. Once this compositional aspect of five is understood, patterns of five need no longer be counted to confirm this. However, further coincidence of counting and subitizing processes for other number patterns is necessary for the construction of a system of relations. Just as progressive inclusion in the counting-on process results in a consideration of the first addend as a whole through the integration of its parts, the interaction of counting and subitizing results in an understanding of the parts encompassed in the whole. Both pathways eventually arrive at an understanding of number as a unit of numerable units despite coming originally from different directions.

Just as the coincidence of counting and subitizing processes provides the experiential foundation for constructive abstraction of a compositional view of number, joining and separating perceptual patterns provides an experiential foundation for future constructive abstraction of a coherent network of relations. Some relations are being established even prior to a child's development of a concept of number. Duplicating a "two" pattern produces a "four" pattern without reference to number. Similarly, by examining a "five" pattern a child may notice that it can be separated into a "three" pattern and a "two" pattern. Although these manipulations resemble addition and subtraction, von Glasersfeld argues that they are not yet truly numerical operations without further constructive abstraction. He goes on to clarify the process of constructive abstraction and distinguishes it from simple abstraction of patterns of objects and visual images of them.

Abstraction is here taken literally: It lifts the construction pattern out of a sensory-motor configuration, leaving behind the actual sensory material—in other words, holding on to the connecting, relating, integrating operations and disregarding the stuff that was connected by them. (p. 23)

Then and only then do we come to have "number" as an abstract unit of units that are themselves wholly abstract from sensory-motor experience. We can then operate with the relations between these abstract entities and can represent them symbolically by means of numerals and letters. But whenever we want to comprehend them or visualize their specific numerosities, we switch back to the figurative level and *re*-present to ourselves spatial configurations that we are able to see. . . . (p. 25)

This theoretical perspective establishes number as a mental construction extending beyond its experiential and perceptual foundations in patterns.

Classroom Activities to Encourage Pattern Recognition

This section describes classroom activities that can provide a rich variety of experiences with patterns and thereby establish foundations for the derived strategies described at the beginning of the chapter. These activities have been adapted from two classroom programs by Wirtz: *New Beginnings: A Guide to Think • Talk • Read Math Centers for Beginners* (1980) and *Thursday Math Sampler* (1982b). The activities use a 2 x 5 pattern frame since its base of 10 is valuable in helping children build a decimal system. Initially the pattern frames can be made from shallow plastic lids taped together, or egg cartons shortened to hold up to ten objects. Objects can also be placed in frames outlined on paper. Variations of the activities use sketches of object patterns and visualization of patterns and their transformation.

Some introductory activities for small group discussion are suggested by the following questions.

1. Without counting, read the number of spaces filled.

   Read the number of empty spaces.

   In how many different ways do you see six?

   How are these patterns related?

2. Fill seven spaces in your frame in any way you want. Tell how you see seven.

   Now move the beans around so that it's easier for you to read without counting.

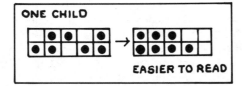

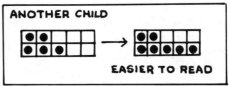

3. Add one more bean to your frame. Read the number of spaces filled . . . empty.

   How do you see eight? What's the easiest pattern to recognize as eight?

   Does anyone see any doubles in their pattern? Which ones?

4. Take one bean from your frame. In how many ways is seven related to six and eight?

5. Close your eyes and picture a pattern of eight. How do you see your eight? Add another bean to your pattern. What do you see now?

As pointed out by Wirtz (1982b), regularity and pattern in the grouping of objects invites discussion by children. Both the use of patterns and the sharing of different perspectives in discussion serve as springboards to the construction of a multiplicity of relations, as illustrated on page 111.

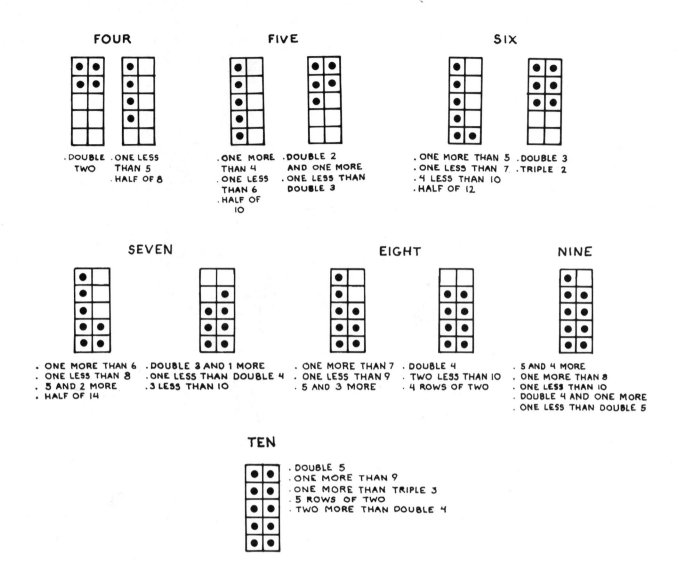

FOUR

. DOUBLE TWO
. ONE LESS THAN 5
. HALF OF 8

FIVE

. ONE MORE THAN 4
. ONE LESS THAN 6
. HALF OF 10

. DOUBLE 2 AND ONE MORE
. ONE LESS THAN DOUBLE 3

SIX

. ONE MORE THAN 5
. ONE LESS THAN 7
. 4 LESS THAN 10
. HALF OF 12

. DOUBLE 3
. TRIPLE 2

SEVEN

. ONE MORE THAN 6
. ONE LESS THAN 8
. 5 AND 2 MORE
. HALF OF 14

. DOUBLE 3 AND 1 MORE
. ONE LESS THAN DOUBLE 4
. 3 LESS THAN 10

EIGHT

. ONE MORE THAN 7
. ONE LESS THAN 9
. 5 AND 3 MORE

. DOUBLE 4
. TWO LESS THAN 10
. 4 ROWS OF TWO

NINE

. 5 AND 4 MORE
. ONE MORE THAN 8
. ONE LESS THAN 10
. DOUBLE 4 AND ONE MORE
. ONE LESS THAN DOUBLE 5

TEN

. DOUBLE 5
. ONE MORE THAN 9
. ONE MORE THAN TRIPLE 3
. 5 ROWS OF TWO
. TWO MORE THAN DOUBLE 4

The construction of relationships such as these forms a solid foundation for the study of addition, subtraction, multiplication, and division. Since such relationships are not constructed in a linear fashion, activities must be experienced over an extended period of time (for example, K–2), returning with questions that evoke a slight shift in focus.

The following page from Wirtz's *Thursday Math Sampler* (1982) involves reading combinations of patterns to ten. Notice how each sketch becomes a conversation piece.

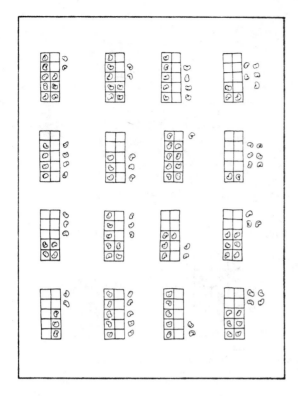

A second page from Wirtz's *Thursday Math Sampler* (1982) provides activities in reading combinations of patterns to 19.

Lets's read the page. How many beans are inside each ten-frame? Outside?

How many empty spaces are there in each ten-frame?

If the outside beans were moved inside, how many would there be inside? How many empty spaces would be left?

How many beans inside, outside, and altogether? (8, 2 and 10, and so on)

How many beans outside, inside, and altogether? (2, 8 and 10, and so on)

How many are there altogether, outside and inside? (10, 2 and 8, and so on)

Make up three addition sentences (oral) for each sketch.

Make up a subtraction sentence (oral) for each sketch.

Read the number of beans altogether as a double (including a double and some more and a double and some missing).

Rather than simply focusing on determining answers to combinations, the preceding questions also ask the children to consider whole-part relations from different perspectives.

Let's read the page. How many do you see inside each ten-frame? Outside?

Describe each sketch as ten and some more. (10 and 2 more, and so on)

How many beans are in each sketch? (12, 16, and so on)

How many inside, outside, and altogether? (6, 6 and 12, and so on) Outside, inside, and altogether?

How many are there altogether, inside and outside?

Make up three addition sentences (oral) for each sketch . . . a subtraction sentence.

How many would be left if you took away ten beans from each sketch?

Suppose you added one more bean inside and one more outside; how many would there be inside, outside, and altogether?

How many beans would you need in each sketch so it would have just twenty beans?

If the beans in each sketch were arranged in as many groups of five as possible, how many groups of five would there be in each and how many left over? (2 groups of 5 and 2 left over, and so on)

Combining activities are best introduced with actual materials that can be transferred physically to fill a ten-frame. Use of different colored objects inside and outside the frame prior to the combination facilitates a child's recognition of both the parts and the whole. Otherwise, the child may focus on the result, 10 and some more, and lose sight of what has taken place. In these activities, the results are important within the context of the relationship of the parts and the whole. Later, transfer of objects to fill ten-frames can be made mentally while looking at the patterns. Eventually the original patterns are visualized and are transformed mentally as needed. Although most activities in Wirtz's programs are designed for group discussion, they may be adapted to game situations as follow-up activities for pairs of children. For example, the combinations can be randomly generated by rolling a pair of dice made from one-inch blocks covered with such different number patterns on each face. Children can take turns mentally combining the patterns and identifying the result and the relationship of parts and whole.

Throughout Wirtz's program *New Beginnings: A Guide to Think • Talk • Read Math Centers for Beginners* (1980), object and pictorial patterns serve as catalysts for considerable small-group discussion of relations and different perspectives of patterns. Having observed the mindless manipulation of marks on paper by primary-grade children in many classrooms, Wirtz chooses to shift the emphasis to mental activity and postpones the introduction of conventional mathematical notation such as the equality sentence. (Our next chapter examines the difficulties experienced by children in this area.) Yet in the latter stages of the program (near the end of the first grade), he agrees that equality sentences can be meaningfully introduced in the context of discussion of pattern combinations as a shorthand record of relational statements made by children. Initially, only the teacher does the recording in these contexts.

Another adaptation of the preceding activities is to focus on the generation in small groups of multiple strategies for approaching a single combination. Such strategies can be generated by questions like: "In how many different ways can we see/think of the combination of eight and six?" See illustrations of such strategies on the following page. Since the strategies applied may depend not only on the size of the quantities represented but also on the specific patterns chosen to represent them, using more than one pattern for each number also can facilitate the generation of a set of multiple solutions in collaboration. Although children usually have preferred ways of focusing on patterns, hearing other perspectives expands their awareness of different possibilities.

Such activities provide a rich experiential foundation for future reflection and reconstruction of derived strategies at a level of abstract number concepts where visualization of patterns and their transformation is no longer necessary because these concepts are in a purely abstract network of relations. Although all children will not achieve this level of abstraction, derived strategies may still be accessible to them at a lower level of perceptual patterns.

In the economy of the equality sentence, different approaches to a combination are all represented in the same way, as $8 + 6 = 14$, for example. Any attempt to record the underlying thinking using mathematical conventions requires the introduction of another symbol (for example, $8 + 6 = 8 + 2 + 4 = (8 + 2) + 4 = 14$), resulting in enough complexity to grind creative activity to a halt. Yet having a record of the group's set of multiple strategies is desirable because it encourages reflection by children. A practical recording of the strategies can be done pictorially on sheets having prepared outlines of paired ten-frames. Having a complete record allows group consideration of such open questions as: Which way of looking at the combinations is the easiest? Hardest? The best? How come? Although no group consensus is likely to be reached, the ensuing discussion encourages critical thinking and can reveal more about individual children's interpretations of these patterns.

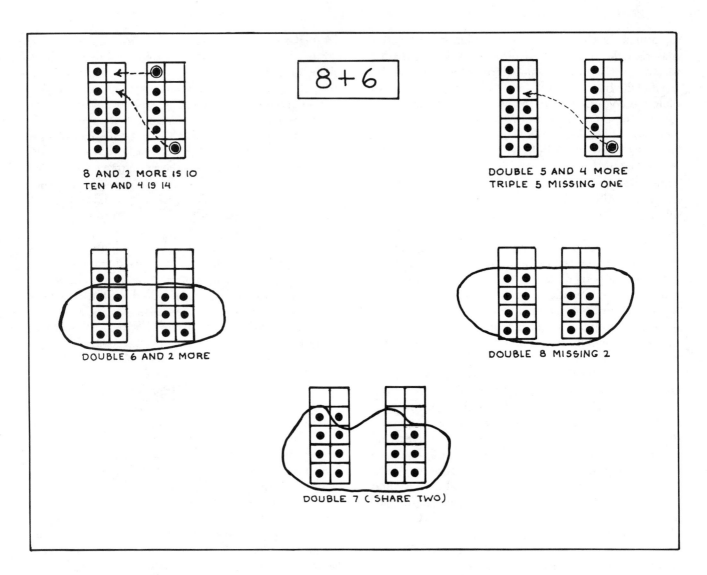

8 AND 2 MORE IS 10
TEN AND 4 IS 14

8 + 6

DOUBLE 5 AND 4 MORE
TRIPLE 5 MISSING ONE

DOUBLE 6 AND 2 MORE

DOUBLE 8 MISSING 2

DOUBLE 7 (SHARE TWO)

A third adaptation of number pattern activities is to vary patterns and materials. Variations in current use are the following:

1. *Domino patterns* are a part of children's natural environment and can be used in ways similar to the ten-frames. According to Hendrickson (1981), one drawback of domino patterns is that they are not always simply additive—sometimes when a child adds one more to an existing pattern it must be transformed in the process of forming a standard pattern for the next number name. Wirtz (1980) includes many domino pattern activities in his program. The value of domino patterns is limited to representing small numbers.

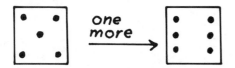

2. *Interlocking cubes* are used by Jordan (1979) to simulate ten-frame patterns.

The physical union of five stresses its use as an intermediate unit, with two fives forming the next intermediate unit. In the absence of the outline of the ten-frame, children must construct mental outlines to remind them of the complements of ten. The physical union of five discourages the consideration of other numbers less than ten as doubles.

3. *Tiles* forming patterns similar to the interlocking cubes, except using an unsegmented unit for five, are the most popular classroom model in Japan, according to Hatano (1982). Time must be given

for the construction of meaning of the continuous five-unit from the individual one-units through measurement. Again, these materials are useful for thinking in terms of fives and tens, but discourage thinking about other doubles for number patterns less than nine. Since they can be cut from cardboard, the tiles may be more practical than interlocking cubes.

4. *Bead strings* that are color-coded to highlight the fives and tens are also useful in visualization of number patterns. Weinzweig (1982) uses 25 beads strung on a stiff wire. Each group of ten is painted a different color, with each group of five within the ten painted a different shade of that color.

A paper clip can be used as a marker to indicate the subgroups of the combination. Like the tiles and interlocking cubes, the fixed nature of the materials encourages the development of strategies emphasizing fives and tens as intermediate units. For example, the combination 7 + 6 can be transformed into 10 + 3 by one's action and the break in color. A doubles focus on small numbers is discouraged by the single arrangement of materials. However, all three types of materials allow a doubles focus for combinations totaling 10 or more.

5. *Structureless materials* such as collections of objects (beans, blocks, or counting discs) without any frame provide both flexibility in pattern construction and the opportunity to observe children's natural approaches to grouping objects in patterns. Here, they are invited to arrange groupings in ways that would make them recognizable the next day without the need for counting. Wirtz (1982c) reports that children prefer to play domino games using nonstandard pattern variations of their own construction.

A Cautionary Note

It is easy for adults to get carried away with clarity, efficiency, and potential creativity of the pattern visualization method. Our prior understanding of number relations contributes to this apparent clarity and efficiency. Yet children are not likely to consider efficiency during their first exposure to combinations of patterns. Furthermore, they may experience difficulty in visualizing the transformations involved. For example, I interviewed four end-of-the-year first graders who were acquainted with number patterns but had no experience with pattern combination, and I found that three of the children experienced difficulties on the following task.

After being asked to visualize the transformation, two of them ignored the direction and physically moved blocks. Rather than complete a ten-frame in a single move, they took several moves before doing so and arriving at an answer of 15. They did not focus on the ten-frame as a potential intermediate unit initially. Efficiency comes with reflection on experience, sometimes prompted by questions such as, "Could you have done that in fewer moves?" Working with other children is another source for shortcut ideas. The third child arrived at an answer of 16 through a mental process. In trying to visualize the transformation she apparently failed to coordinate the filling of one ten-frame with the removal of the block from the second frame. The performance of these three children suggests that visualization of transformations is not an easy process and that they don't appreciate the power of grouping by tens as yet. The fourth child's facility with visualization stood in sharp contrast to the difficulty of the first three and serves to remind us of the range of individual abilities found in the same classroom.

The work of Piaget and Inhelder (1966) in the area of mental imagery indicates that children have the ability to visualize static images at an early age. However, this research also indicates a strong relationship between the capacity to visualize transformations of objects and the onset of logical thinking at about age seven. Thus, asking children to mentally anticipate the result of a transformation without actually carrying it out may be asking too much of them, and they may be unable to respond. Since the tasks employed by Piaget's research may not be comparable to those described here, his conclusions may be regarded as a general caution to pace the activities according to the children's abilities. Although Wirtz (1982b) includes many "What if . . ." and "Pretend that . . ." questions for his activities, he suggests that any reluctance to respond to these questions may be a signal for their postponement.

COUNTING AND VISUALIZING NUMBER PATTERNS: SEPARATE OR OVERLAPPING PATHWAYS?

The chapter began with a consideration of derived strategies as noncounting alternatives. Yet upon examination of an origin of derived strategies, we noted the role of counting in a child's construction of meaning for the recognition and visualization of number patterns. Similarly, upon closer examination of the counting process we find that subitizing and visualizing contribute to children's ability to count-on. For example, instead of counting a single collection from one, a child may identify an easily recognizable subgroup and count-on from there. This combination of subitizing and counting is known as *partial counting*. According to Steffe and Thompson (1979), the visualization of patterns also appears to play a role in children's counting-on up to 5 more without keeping track on their fingers. In totaling 8 + 5, children are thought to visualize a configuration of five and use it as an internal tracking system. Children have also been observed to touch the table according to this configuration in synchrony with their oral count.[14]

---

14. Tapping a rhythm (tap, tap, tap . . . tap, tap) in synchrony with counting-on has also been observed by Steffe and Thompson (1979) as a tracking system for counting-on small numbers.

Object patterns also play a role in counting-on addends greater than 5—for example, 8 + 7. The child starts counting with "eight" (the first addend), extends the first finger at "nine", and continues counting and simultaneously extending fingers until seven fingers are extended. A way of knowing when to stop is by recognition of the pattern made by the extended fingers. Based on this information, it may be argued that the total integration of subitizing, visualizing, and counting methods can result in a higher level of number understanding. School programs, therefore, should provide children with experience with both methods of processing.

But despite a strong argument that subitizing, visualizing, and counting are mutually supporting processes resulting in higher levels of number meaning, there is tendency for recent innovative classroom programs to emphasize one process or the other. For example, Wirtz's classroom program (1980) uses subitizing and visualization of domino and ten-frame patterns to the exclusion of counting. Children are asked to count only if unsure of the name of the pattern. No time is devoted to counting-on strategies. Although this program appears to have considerable potential for providing all children with access to derived strategies, research is still needed to establish its effectiveness. Coincidentally, a study by Hatano (1982) reports the success of a program in Japan that stresses subitizing and visualization of tile and domino patterns while using counting only as a means of checking answers. Testing was done in the absence of these patterns, so the study bridges the gap between subitizing and visualizing patterns and derived strategies with numerals. Hatano's study indicates that following exposure to direct teaching methods using these patterns, Japanese children rely mainly on decomposition strategies using ten as an intermediate unit. Although not providing much experiential foundation for derived strategies, Thornton et al. (1982) and Rathmell (1978) do combine direct teaching of these strategies with counting-on in their programs. Here, counting-on is restricted to small quantities, up to three more, where, according to Rathmell, memory is not overloaded and the process is most efficiently used. In their programs, the decision to limit the focus of counting-on appears to be motivated by a greater interest in teaching sums of number combinations than in developing thinking abilities and a more integrated concept of number.

Decisions on the extent to which classroom activities should attempt to help children integrate the two processes are influenced by many considerations, such as available knowledge of how children learn, your goals for the mathematics program, external definers of the curriculum—district objectives and textbooks—and time available in the school day for teaching mathematics. Although considerable knowledge is available currently on children's counting processes, their derived strategies have been relatively unstudied. Important additions to available knowledge would include research comparisons of the effectiveness of classroom programs that provide an experiential foundation with patterns and those that do not. Another area in need of further exploration is the relative effectiveness of direct and indirect teaching methods. A detailed description of classroom interactions over an extended period of indirect teaching of derived strategies would be extremely valuable in itself.[15]

Finally, further research on the processing of individual children is needed to clarify whether there is enough evidence for a consideration of different learning styles. The global, spatial nature of visualizing object patterns as wholes and the more analytical focus of counting-on parts suggest different intellectual demands on children despite overlaps in pathways. The possibility of distinct learning styles occurred to me when interviewing Vic, a beginning second grader with highly developed derived strategies but inefficient counting methods (see his activity early in this chapter). Davidson's research with learning disabled children (1981) is being analyzed from this perspective. Although it may be argued that the integration of the two processes results in a higher level of number understanding, teachers find that some children experience difficulty with the single process of counting. The latter children may benefit from the subitizing and visualizing of patterns. Not only may the novelty of the new process spur their involvement, but its spatial nature may also be more compatible with the way they see the world.

---

15. Chapter 7 provides such a detailed description of interaction during twelve days of indirect teaching in another area of mathematics.

REFLECTIONS AND DIRECTIONS

1. Interview some children on derived strategies with tasks providing both patterns and numerals. Compare your results to those reported in this chapter.

2. Summarize your most significant learnings from your interaction with this chapter and indicate their implications for your work with children.

3. Available knowledge of how children learn indicates the importance of integrating both counting and the subitizing/visualizing process leading to derived strategies. Yet textbooks do not emphasize either process. Reflect on this dilemma and decide how you would resolve it as a K–2 teacher.

6

ARITHMETIC NOTATION AND ITS INTERPRETATION

PREVIEW

The preceding chapters have described children's construction of number concepts and relations, and problem solving through counting and derived strategies. Contrary to the common classroom emphasis on written symbols, the activities described usually took place in physical or pictorial contexts having no requirements for reading and writing numerals or equality sentences. The primary emphasis was on construction of number ideas and relations prior to their representation in written notation. Yet written notation is important in keeping a record of relations constructed and in communicating these relations to others. This chapter will examine children's notations, and will probe the underlying ideas being represented.

Conventional number notation, the generally accepted language of arithmetic, involves an economy of representation. For example, a single numeral, 5, is used to represent a collection of five objects. This arbitrary symbol neither maps directly the quantity of the collection in a one-to-one correspondence, nor does it denote the kind of object in the collection. To different children, 5 may represent a shorthand for the counting word *five*, a loose collection of five objects, "fiveness" of an integrated group, multiple relations of five, and so on. In other words, number ideas are not built into symbols. Rather, they are constructed mentally and read into the symbols. In the historical development of written notation of number, primitive systems used a multiplicity of symbols, such as tallies, to denote number in a direct symbol-to-object correspondence.

With the advent of the existing conventions of notation, earlier forms have been eliminated from the school curriculum and children become immediately exposed to numerals, with other conventions of arithmetic notation soon to follow. What meanings do children read into these symbols?

In addition to representing number, numerals are used in combination with other symbols according to specified rules to represent situations and relations in equality sentences ($3 + 2 = 5$) and in statements of inequality ($5 > 3$). For adults, $3 + 2 = 5$ represents a considerable amount of information such as number, order, operation, and relationship. Its shorthand form uses a minimum number of symbols and bears no direct resemblance to physical situations represented. It is important to note here that the equality sentence is only one means by which number relations can be communicated in writing. Other possible forms include the following:

$$(3, 2) \xrightarrow{+} 5 \qquad \textcircled{3} \xrightarrow{+\,2} \textcircled{5} \qquad \begin{array}{r} 3 \\ +2 \\ \hline 5 \end{array}$$

The adoption of a standard convention facilitates communication among mathematicians. Are other forms of notation more suitable for children? Again, in Piaget's theoretical framework, the relationships are not automatically built into the notation of the equality sentence. Rather, they are mentally constructed by people and read into the marks on paper. From this perspective, it is the people who are representing the ideas, and not the symbols themselves (C. Kamii, 1980a). Interviews with children will explore whether they read the same meanings into equality sentences as adults.

Most teachers, guided by textbooks, introduce the formal mathematical convention of equality sentences to first graders by the end of the first month, almost simultaneously with an introduction to addition and subtraction. The symbols are introduced arbitrarily as children are provided with little time or experience for developing related meanings. This is followed by months of practice in addition and subtraction combinations in which children complete blanks in open sentences, like 5 + 2 = ___. Assumptions made in this method are:

1. Children are ready for an early introduction to this formal mathematical convention.

2. Children have no ideas of their own for informal written notations that communicate.

3. Children understand equality sentences and their implied relations once they successfully fill in the blanks of open sentences in worksheet exercises.

A review of interview research involving children from preschool through the first grade, both prior to and following instruction in conventional notation will allow you to examine these assumptions. The chapter will conclude wth a consideration of alternatives for introducing children to equality sentences.

The following notations of five objects (houses) were made by kindergarten children prior to instruction. You are invited to:

 —interpret them for underlying number meaning

55555 —suggest other notations made by young children

YOUNG CHILDREN'S ARITHMETIC NOTATION

Children are exposed in many areas of their everyday graphic environment to written number notation, as indicating quantities of objects, prices, game scores, length markers, and coding of addresses, phones, cars, buses and game players. Do these conventional notations influence the notations of young children prior to instruction? Does instruction in conventional notation of arithmetic influence the notations of kindergarten and first-grade children? Do these conventional notations have the same meaning for chil-dren as for adults? These questions have been addressed by the research studies to be described.

Notation of Number

A study by Sinclair, Siegrist, and Sinclair (1982) examined how children represented numerical quantities prior to being taught the alphabet and numerals in school. Forty-five Genevan kindergarten children (ages four to six, fifteen of each age group) were asked to respond to the following tasks.

Could you write about what is on the table?	Could you write down two (or three, four, and so on) houses?
(The collection of identical objects—pencils, balls, toy cars, and houses—varied in quantity from one to eight).	(Numerical quantities were given orally in the absence of objects).

The children were asked to explain their notations through the following questions.

1. What did you write down?

2. Can you read it? What does it say?

3. Can you show me where three is written? And where is "balls" written?

4. What is written here (pointing)? And here?

All but one child, a four-year-old, produced interpretable notations from which the researchers identified six distinct notation types (described in the following table). The types of notation used most by the youngest children did not appear to be influenced by the existence of numerals in their environment. Once their notation included conventional numerals, these were used in a redundant way that overlooked the potential economy of representation. Older children used individual numerals as notation for arrays of objects. Although primitive attempts at written notation clearly precede correct usage of conventional symbols, the details of a developmental sequence were blurred by usage of a variety of notations by most children.

YOUNG CHILDREN'S IDEAS ABOUT THE WRITTEN NUMBER SYSTEM (SINCLAIR, SIEGRIST, AND SINCLAIR)

TYPES OF NOTATION	EXAMPLES OF NOTATION		
	THREE BALLS	TWO BALLS	FIVE HOUSES
1. *Global Representation of Quantity.* The line of hooks, bars, or squiggles doesn't represent the quantity correctly and doesn't represent the type of objects.	/ \ \ \	/ \ \ \ \	/ \ \ \
2. *Representations of the Object-Kind.* While attempting to represent the kind of object, its nature, shape, or name, this type of notation ignores quantity.		B	
3. *One-to-One Correspondence with Nonnumerals.* Each object is represented by a graphic symbol. The type of symbols used and their combinations vary with quantity and objects represented for some children. Other children use only one type of symbol.	P P P T (⌐ A E I 0 0 0	φ p ⌐ T 0 I 0 0	7 7 7 7 7 I J T T P 9 A Ǝ 0 I ◻◻◻◻◻
4. *One-to-One Correspondence with Numerals.* Each object is represented by a numeral in one of two ways. Each object may be assigned a different numeral in a "counting" sequence. An alternative is to repeat the numeral indicating the "manyness" of the array.	1 2 Ɛ 3 3 3	1 2 2 2	1 2 Ɛ 4 ᘔ 5 5 5 5 5
5. *A Single Numeral Indicating Manyness.* Manyness is represented with one written symbol. When asked, many children show an alternative representation in the form of the appropriate word.	3... TRO (trois)	2...D3 (deux)	5...sin (cinq)
6. *Representation of Manyness and Object-Kind.* Most children writing numerals alone, sometimes specify the object-kind.		deu bal	3 mèzone

A curious observation made by the researchers was that young children's usage of one-to-one correspondence notation with nonnumerals was unrelated to their counting abilities. Although unable to produce a number-name sequence or count a small collection of objects accurately, some four-year-olds were able to produce notations in correspondence with objects. At the same time, other children showing counting abilities also used this method of notation. This interesting observation suggests the primitive nature of this type of representation. The history of mathematics indicates an early use of tally sticks and marks, even before verbal counting systems had been developed.

In notation-type 4, even though children used conventional numerals in their representation, they adapted these numerals to their existing number ideas. One-to-one correspondence of counting words to objects is a major focus in young children's counting, with considerations of "manyness" developing more gradually. Observations such as these support Piaget's posi-

tion that symbols themselves do not automatically convey meaning.

The researchers (Sinclair, Siegrist, and Sinclair, 1982) also report that the older children used numerals in both nonconventional and conventional notations (types 4, 5, and 6) on different trials of the tasks. This may be interpreted as a vacillation between different perspectives of number and its representation, as in a state of disequilibrium prior to the integration of different perspectives. One-to-one correspondence is a powerful idea in children's early view of number, which eventually becomes integrated in the higher-level view of the "manyness" of collections.

Other studies conducted by these researchers (in M. Kamii, 1981) have explored children's interpretation of numerals in a variety of environmental contexts. It is important to note that some of these contexts are not truly quantitative, yet use numerals in labeling or coding. Some interpretations by four-year-olds follow.

Given their simultaneous exposure to letters and numerals and quantitative and nonquantitative contexts for numerals, it is not surprising that children's interpretations of the different systems of notation and their applications interact as they attempt to make sense of their complex world.

A similar study of Allardice (1977) placed young children's number notation more clearly in the context of communication by asking them to send a message to a stuffed animal, Snoopy. Her interviews were conducted with groups of twenty 3½-, 4½-, 5½-, and 6½-year-old children.

All of the children accepted her request for a written message as a reasonable one, although their ability to carry out the request in a communicative way varied.

The following children showed little awareness of the need for the sender (encoder) and receiver (decoder) to use a shared system of notation.

Yet most of the children's representations did communicate quantity. Even half of the three-year-olds was successful in nonnumerical one-to-one notation. Most five- and six-year-olds used numerals, although three-quarters of the younger group did so in nonconventional ways.

Allardice suggests that children's early exposure to numerals in kindergarten may interfere with their experimentation with tally marks or other nonnumeral notation. Since numerals are taught as the preferred method, the kindergarteners may use them even when inaccurately applied and may resist their earlier informal methods such as tallies. Such an influence is suggested in the following illustrations.

An alternate explanation for the latter children's responses is that they have acquired an understanding of the "manyness" of small groups and now find their earlier notations to be inefficient.

Young Children's Notation of Addition and Subtraction

Since young children usually interpret addition and subtraction in terms of the actions of joining and separating, order of actions to be represented becomes important. Allardice first examined the children's abilities to represent order on paper.

(Three objects are presented individually and removed following each exposure.)

Put something on paper to show Snoopy which one I showed you first, and which one I showed you next, and which one I showed you last.

To denote temporal order, preschoolers drew objects on a horizontal plane from right to left, reproducing the correct order. Once completed, however, the written representation gave the reader no clue as to the sender's convention. The older children used a left-right convention for their pictures more often. Some first graders persisted with right-to-left despite school cues from reading and writing. Many of the older children also failed to give the reader an indication of the direction, although some children introduced arrows and numerals as indicators of temporal order. These varied ways of denoting order were also found in the children's representation of addition and subtraction situations.

Space, circles, and lines are devices used by chil-

dren for representing separate groups or subsets. The younger children often represented the individual objects of one set, paused, then continued to represent the objects of the second set with no space between the markings. Although they represented that pause in their action, it was not preserved on paper. Similarly, the older children using numerals without operator signs failed to separate them, so that 1 + 3 would be written as 13. Other children left obvious spaces between numerals even though no signs were placed between them.

The children also were asked to represent addition and subtraction in situations involving joining and separating actions.

The preschooler's representations showed only the end result of the joining or separating action. Older children tended to represent more information and to represent the physical action of joining and separating. Representing these dynamic situations proved to be very challenging and a variety of methods were devised. The joining or combining action was shown by using lines, dots, hands, words, or physical actions. Some of the children were aware of the need to repre-

sent this dynamic aspect of the demonstration but were not able to preserve this action on paper. Although some children used arrows or words such as in or out to clarify the direction of lines or hands, others were less specific. Another successful method of notation was writing or drawing the original amount and crossing out or erasing (but leaving traces of) all or part of it.

Addition

Subtraction

Although only 3 of the 15 first-grade children spontaneously wrote complete equality sentences in writing messages to Snoopy, others produced communicative informal written representations of addition and subtraction situations. It is interesting to note that Allardice's first graders responded in this manner

following an entire year's exposure to the conventional notation of the equality sentence. The next section will examine the responses of another group of end-of-the-year first graders in more detail and will offer some explanations as well.

FIRST GRADERS' REPRESENTATION OF ADDITION

I interviewed thirty-three first-grade children in the last month of their school year to explore their written representations. During the year these children had experienced the typical introduction to the symbolism of the equality sentence as well as considerable practice in filling in blanks for addition and subtraction combinations in this format. The interview tasks went beyond the typical worksheet to gain a better perspective of first graders' abilities to deal with this formal convention and to read meaning into its symbols. Not only were children asked to encode addition in written notation, but also to decode existing addition sentences in demonstrations with blocks. Similar tasks were given for subtraction. The interview shifted from the typical classroom focus of finding and writing answers to representing the entire situation/relationship.[1]

Below is the range of first graders' written notations of the illustrated demonstration of combining two groups of objects. You are invited to examine these responses and decide if they are typical of first graders.

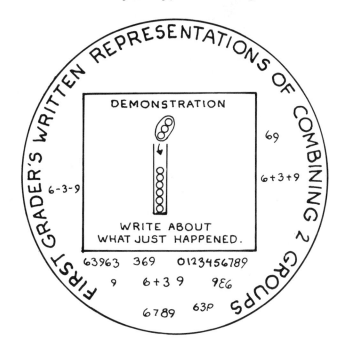

How would you explain these responses?

---

1. The interview research to be described is based on the earlier work of C. Kamii (1980) with inner city children in Chicago and children in Geneva, Switzerland.

Task A: Notation of Addition

I demonstrated an addition situation by combining two groups of objects. Prior to making a written notation (encoding) of the situation, each child was asked to describe what had taken place. The child was also asked to read his own notation.

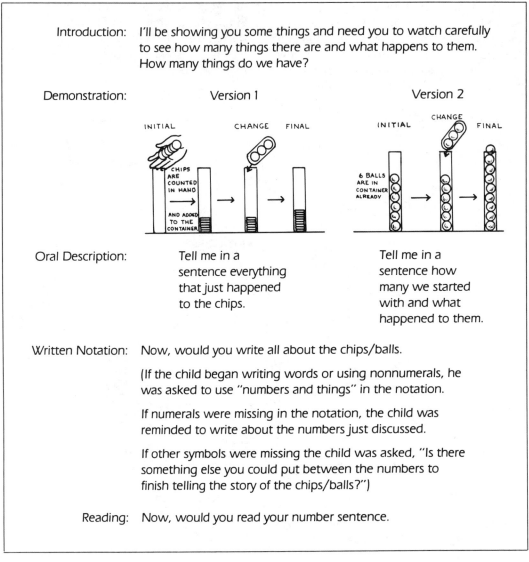

Introduction: I'll be showing you some things and need you to watch carefully to see how many things there are and what happens to them. How many things do we have?

Demonstration: Version 1   Version 2

Oral Description: Tell me in a sentence everything that just happened to the chips.

Tell me in a sentence how many we started with and what happened to them.

Written Notation: Now, would you write all about the chips/balls.

(If the child began writing words or using nonnumerals, he was asked to use "numbers and things" in the notation.

If numerals were missing in the notation, the child was reminded to write about the numbers just discussed.

If other symbols were missing the child was asked, "Is there something else you could put between the numbers to finish telling the story of the chips/balls?")

Reading: Now, would you read your number sentence.

Two variations of the task were used, each with about half the children. The second version was an attempt to improve the visibility of the objects and to sharpen the focus on the initial group. Notice how the change in both the objects and the wording accomplished this end.

Oral Description. Of the 33 children interviewed,[2] only four were able to give a detailed description of the demonstration without needing any prompts. The other children needed up to three prompts to elicit the details. Even with the aid of focusing questions, a small number of children still missed some of the details. Problems with written notation of addition may be anticipated from the children's difficulties in giving detailed descriptions orally. A range of their descriptions is illustrated.

DETAILED DESCRIPTIONS

TELL ME IN A SENTENCE EVERYTHING THAT HAPPENED TO THESE CHIPS / BALLS.

HAD 6 PUT 3 MORE THAT'S 9.

Stan R.

SIX, THEN YOU PUT 3. THAT'S 9.

Felicia

GLOBAL DESCRIPTIONS

YOU PUT ALL OF THEM IN THERE.

Steven M.

THEY'RE ALL PUT TOGETHER NOW.

Kathy L.

Prompts followed to coax out further information.

INCOMPLETE DESCRIPTION

STARTED WITH 6 AND THEN IT WAS 9.

Miguel C.

WHAT HAPPENED IN BETWEEN?

I DON'T KNOW . . . . . . THREE.

This child focuses on initial and final groupings.

INCOMPLETE DESCRIPTION

NINE

Celia C.

This child focuses on final result.

HOW MANY WERE PUT IN?

THREE

HOW MANY DID WE START WITH?

SIX

In examining the incomplete descriptions and the number of prompts required to draw out further information, two questions come to mind. Are these isolated responses to the interviewer's focused questions related in the children's minds? Would better questioning draw out more complete relational statements?

Writing and Reading Own Notation. The first graders' written notation of the physical combination of two groups of objects was rarely what one would expect after a year of related instruction. Only 4 of the 33 children wrote a complete equality sentence. The other children encoded the situation in a surprising variety of ways. The most common notation, made by 15 children, used only the numerals. Prompts given to add more symbols led to further surprises.

---

2. The first-grade children were taken from two classrooms in separate low socioeconomic (Title I) Schools. The sample represents the total number of children with parental permission for participation in the interviews. The ethnic makeup of the group was white, black and Spanish-surnames. The Spanish surname children were judged to have sufficient English fluency to be placed in classrooms with English-only instruction. Each teacher used a different textbook.

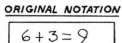

ORIGINAL NOTATION   *READ AS*
SIX PLUS THREE EQUALS 9.

$6+3=9$

Luther C.

Only 4 children wrote a complete number sentence.

ORIGINAL NOTATION   *READ AS*   **REVISED NOTATION**

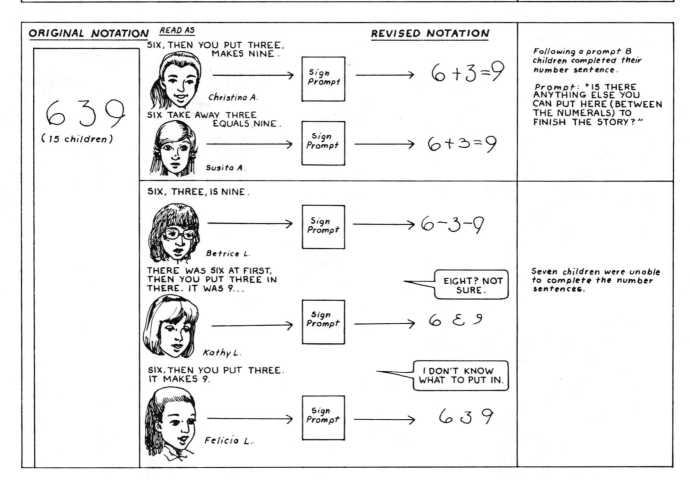

639
(15 children)

SIX, THEN YOU PUT THREE, MAKES NINE.
Christina A. → Sign Prompt → $6+3=9$

SIX TAKE AWAY THREE EQUALS NINE.
Susita A. → Sign Prompt → $6+3=9$

Following a prompt 8 children completed their number sentence.

Prompt: "IS THERE ANYTHING ELSE YOU CAN PUT HERE (BETWEEN THE NUMERALS) TO FINISH THE STORY?"

SIX, THREE, IS NINE.
Betrice L. → Sign Prompt → $6-3-9$

THERE WAS SIX AT FIRST, THEN YOU PUT THREE IN THERE. IT WAS 9...
Kathy L. → Sign Prompt → $6 \, \varepsilon \, 9$
EIGHT? NOT SURE.

SIX, THEN YOU PUT THREE. IT MAKES 9.
Felicia L. → Sign Prompt → 639
I DON'T KNOW WHAT TO PUT IN.

Seven children were unable to complete the number sentences.

SIX PLUS THREE IS NINE.

$6+39$

Pat M. → Sign Prompt → $6+39 \, \overset{\downarrow}{10}$

This child reads "is" into 6+3 9 but is unable to represent it in writing. When asked if "10" told about the balls he said, "NO."

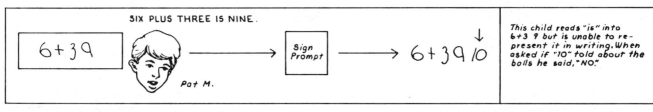

Following a non-numerical description and two prompts.

Steven writes.

0982

Prompt Given

HOW MANY DID WE START WITH?   AND THEN?

639

He writes symbols in response to prompts.

Final Product

$639 \; 63$

He pauses after writing each of the first two numerals. The others are written in rapid succession.

Read As:

SIXTY-THREE, NINETY-SIX, SIXTY-THREE.

Steven M.

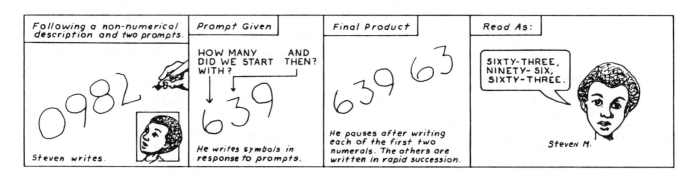

Examining the children's responses raises several questions. Is there any connection between the children's ability for oral description and their written notation? Do the children know their addition combinations? How do you explain these responses? These questions will be answered in the following task and in subsequent discussions.

Task B: Interpretation of Equality Sentence Notation—Addition

This task began with the child's being asked to complete an open equality sentence, just as had been done hundreds of times throughout the school year. The child was then asked to read the completed equality sentence and to model its meaning with blocks. In other words, the child was asked to decode the shorthand communication used by mathematicians after a year's exposure to this convention.

Completion of Sentence:	What number would finish this sentence (6 + 2 = ___)?
Reading of Sentence:	Read this number sentence.
Modeling with Blocks:	Do something with these blocks to show what this number sentence means.
	(A prompt was used if a child needed help getting started): Pretend that you are helping a kindergarten friend. What could you do with these blocks to help him get the answer and to understand this number sentence?
	(Following the modeling with blocks the child was asked):
	Show me the six in your blocks . . . the two. Show the eight (as each numeral was pointed out in the sentence).
	(The child was then given a doll and asked): Give the doll just as many blocks as the number sentence tells about.

Completion of Sentence. One of the questions raised earlier was whether these children knew their addition facts. You might reason that if they were able to answer 6 + 3 = ___, they would be able to write 6 + 3 = 9 as their notation for the demonstration in task A. In fact, given 6 + 2 = ___ in task B, all of the 33 children answered correctly. Although not having an immediate, memorized response, they did obtain satisfactory answers by counting on their fingers.

Reading of Completed Sentence. Nearly all the children read the completed sentence as expected. Five of the children ignored either the "=" sign or both "+" and "=" in their reading. Attempts to assist children in focusing on these signs met with mixed success.

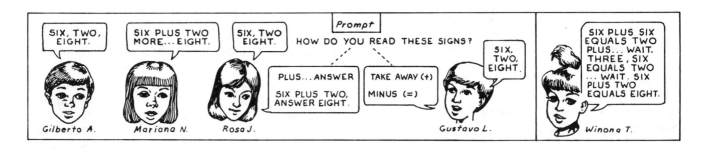

It is interesting to speculate on why children omit the names for signs in their reading. Consider the following possibilities:

They don't know what to call them.

They don't understand what the signs represent.

They simply forget to include them.

The signs are obvious to adult readers, so they needn't be included.

Modeling of Completed Sentences with Blocks. In this study, children who successfully completed open equality sentences such as 6 + 2 = ___ had earlier shown a surprising inability to encode a demonstrated physical combination as a conventional equality sentence. Now let's examine how they decoded or interpreted the completed equality sentence in terms of a physical situation.

In response to task introduction.

Christina shows groups of 6 and 2 blocks spontaneously joining them.

SHOW ME THE "TWO".
Child
6+2=8

SHOW THE "SIX"
6+2=8
After recounting 6 blocks.

SHOW THE "EIGHT."
She joins the blocks again, enveloping the blocks with her hands.
6+2=8

The request to identify the parts once combined into the whole, and to identify the whole once it had been decomposed again, was to provoke the children into thinking about the blocks' relationship. Half of the 33 children considered 8 as being composed of 6 and 2. Although most of these children identified this relation indirectly through their actions of combining and decomposing groups, others did so without this physical action. Are joining the two groups and thinking about them together equivalent ways of representing addition?

SHOW THE "EIGHT."
I HAVEN'T GOT EIGHT.
THIS IS THE EIGHT RIGHT HERE.
Gustavo rethinks his first response and joins the 2 groups.

SHOW ME THE "TWO"
ALL OF THEM.
Susita
First Susita makes a move to count another set of 8 blocks. Then she pushes the 2 groups together.

DO SOMETHING WITH THE BLOCKS TO SHOW WHAT THE SIGN MEANS.
Felicia constructs a plus sign with blocks.

SHOW THE "EIGHT."
Felicia
She points to both groups simultaneously.

What about the remaining half of the 33 first graders? Six of these 16 children initially showed sets of 6 and 2 objects, but when asked to show eight after the final group had been decomposed, they placed a set of 8 objects side by side with the other sets.

SHOW ME "EIGHT."
I DON'T HAVE EIGHT.
Rosa J.

LET'S SEE WHAT YOU'D DO TO GET EIGHT.
6+2
She counts out 8 more blocks.

The remaining 10 children initially modeled the completed equality sentence with sets of 6, 2, and 8 objects placed side by side. The questioning that followed provoked some of these children to rethink the need for a separate set of 8 objects, while others seemed satisfied with their model of the sentence. Does showing the second set of 8 objects model the sentence appropriately?

Stan's demonstration
Shown
1st 3rd
6+2=8
Stan
Do the blocks match the equality sentence?

Completed by joining all the sets.
The sets are joined spontaneously and a total of 16 is counted.
The interviewer reconstructs the child's original 3 sets for the next question.

GIVE THE DOLL JUST AS MANY BLOCKS AS THE NUMBER SENTENCE TELLS ABOUT.
6+2=8

HOW COME YOU GAVE THE DOLL 8 BLOCKS?
BECAUSE HE'S EIGHT YEARS OLD.
6+2=8

EXPLAINING CHILDREN'S DIFFICULTIES WITH NOTATION—TASKS A AND B

How can we explain children's difficulties in both encoding and decoding the mathematician's convention of the equality sentence? Prior to any detailed discussion, you are invited to participate in two related activities.

1. Look at this drawing and describe what you see.

2. Examine the following statements of explanation and select the most plausible ones.

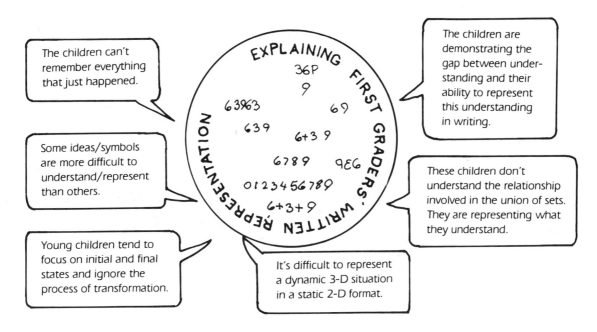

The children can't remember everything that just happened.

The children are demonstrating the gap between understanding and their ability to represent this understanding in writing.

Some ideas/symbols are more difficult to understand/represent than others.

These children don't understand the relationship involved in the union of sets. They are representing what they understand.

Young children tend to focus on initial and final states and ignore the process of transformation.

It's difficult to represent a dynamic 3-D situation in a static 2-D format.

Although all of the preceding statements of explanation have some merit, this section will discuss only two of them in detail.

The Whole-Parts Relation

To clarify children's understanding of the whole-parts relationship and its effect on their written nota-tion, we begin with a look at Piaget's class inclusion task. Prior knowledge of his systematic study of children on this task would have allowed us to predict some of the incomplete written notations of these six-and-a-half-year-old children.

Which color has more blocks? How many are blue? Yellow?	(Parts)
What are the blocks made of?	(Whole)
If you took all the blue blocks away, would there be any left?	(Parts)
If you took all the plastic blocks away, would there be any left?	(Whole)
Are there more blue blocks or more plastic blocks?	(Part/whole)

Examples of children's responses to the last question, following successful responses to the preceding ones, are illustrated in the following panel.

Although able to answer a long list of preliminary questions involving either the parts or the whole (where the information is readily available from the visible objects), most six-year-olds cannot consider the parts and the whole at the same time. If they think of the whole, the parts no longer exist; if they think of the parts, the whole no longer exists. Lacking the flexibility to reverse their thinking so that the parts and whole are considered almost simultaneously and as related, these children hear and answer the question in terms of what they understand. Most seven- to eight-year-olds have achieved this flexibility and, therefore, this logic in their thinking.

To appreciate the young child's experience, think back to the drawing we just had you describe (Thompson, 1982b). It can be seen in two ways—as a young beauty and as an old woman. In focusing on one, you lose sight of the other. Eventually you can look back and forth quickly enough so that the reversibility in focus gives the impression of seeing both women at the same time. It is also possible to relate the two women by identifying what focus it takes to shift from one to the other—for example, eye vs. ear. Try to relate this experience to that of a young child's dealings with the parts and the whole.

Children's textbooks emphasize an action model of addition in which two sets are combined physically. This dynamic sequence takes place over time and involves the union of sets where both original sets are encompassed in a larger superset. Although addition is an operation that relates the parts to the whole, this relationship is not easy for children to construct. In filling answer blanks in open addition sentences, a child can count the initial group, the second group, and the resultant larger group and record the numeral for the last counting word. Given more space, the child may make a notation of the counting process—for example, 1 2 3 4 5 6 7 8 9 or 6 7 8 9—rather than write the final number name. The first notation suggests a focus on individual entities, while the second indicates a clear focus on at least the first group (6) in counting-on 3 more to get 9. The first notation also indicates how a low level of functioning can still result in success in typical worksheet exercises and mask how the children think about addition. Because of the difficulty in considering the whole and its parts simultaneously, writing three numerals in a relationship is very difficult for first graders to handle. This difficulty is illustrated by other responses to Task A in which children focused on either the whole (9) or the parts only (6 3). In another study, C. Kamii (1981) interviewed a class of first graders earlier in the school year (spring) and found that 40 percent of the children wrote either of these incomplete notations. Another interesting response in my interview, 6 3 9 6 3, suggests a child's disequilibrium in considering the whole-parts relation.

Task B provided further evidence that the symbols of the open equality sentences that fill children's workbooks have no meaning to children in and of themselves. Although trained to respond accurately to open addition sentences, only half of the 33 children could show the whole-parts relationship in the completed equality sentence with physical models. Other children placed three groups of objects side by side with little apparent connection or connected by "+" and "=" symbols made of blocks. These responses suggest that the children were not trying to represent the meaning underlying the shorthand notation with blocks, but rather were attempting to represent the form of the sentence in another medium. The latter representation then misled a small number of children to join all three groups and give 16 blocks to the doll. Sometimes 8 blocks were given to the doll (out of the possible 16), but only one child could attach any meaning to this move. Observations similar to these have also been reported by C. Kamii (1980). Such observations may be accounted for by children's premature introduction to symbolism independent of physical contexts in which relations can be developed.

Increasing Abstraction of Written Symbols

Inflexibility in thinking about whole-parts relations can account for all three numerals not being recorded, but what of the omission of other signs? Fifteen of the 33 children who wrote all three numerals (6 3 9) omitted these signs. Prompts for signs brought only mixed success, so that children's omissions go beyond mere forgetting. Most of the other children wrote only one or two numerals without any signs. Only two children spontaneously wrote the complete equality sentence and two others did so following a sign prompt. In a similar study reported by C. Kamii (1980), 65 percent of the children used only numerals in their notation, 28 percent included an operational sign, and only 7 percent added the equality sign. In no case did a child write an equality sign without having written an operator sign. To Kamii these results suggested a developmental sequence. The representation of the operation (+) and the entire relationship (=) may be considered as successively more abstract than the representation of numbers. Even when reading a completed equality sentence, some children omit these signs, particularly equality signs, even though they may be identified independently. It's as if these children are unable to incorporate the signs into their thinking.

FIRST GRADERS' REPRESENTATION OF SUBTRACTION

I presented the group of 33 first-grade children with two subtraction tasks similar to those I gave them for addition. This section discusses tasks C and D and the children's responses to them.

Task C: Notation of Subtraction

In this task the children were asked to encode in written notation a demonstration of separating a group of objects into two subgroups. Prior to doing so, they were requested to describe the situation. Following their encoding, they were asked to read their own notation.

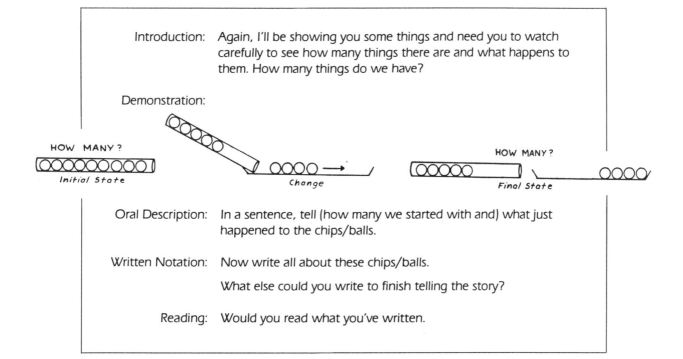

Introduction: Again, I'll be showing you some things and need you to watch carefully to see how many things there are and what happens to them. How many things do we have?

Demonstration:

HOW MANY ?
Initial State   Change   HOW MANY ?   Final State

Oral Description: In a sentence, tell (how many we started with and) what just happened to the chips/balls.

Written Notation: Now write all about these chips/balls.

What else could you write to finish telling the story?

Reading: Would you read what you've written.

Notice that the encoding task again has two versions based on different materials (chips/balls), and that an expansion of the directions for oral description is possible. Approximately half the children were exposed to each version.

You are invited to think ahead by considering the following:

1. *Analyze the encoding task in terms of the whole-parts relationship.*

2. *Predict the range of responses to this task by first graders.*

Oral Description.   Having previously examined children's written notation of an addition situation, we can anticipate their encoding of subtraction in a physical context. Prior to the written notation, only three children spontaneously offered detailed descriptions of the situation. Ten others were able to do so after a single prompt.

Some children gave incomplete descriptions following multiple prompts. The rewording of directions for the second version of the task seemed to facilitate more complete responses. Although the rewording evoked more awareness of the starting quantity (the whole), it did not eliminate the problem of incomplete description.

Actually, the rewording of the directions had a dual, counteracting effect. At the same time it seemed to help children to focus on the whole as well as the parts, it also triggered a number of responses of six—the starting quantity in the preceding addition task (task A).[3] Once the starting quantity was clarified as the one "just before we took some away," most of these children identified nine as the starting quantity. The description of children's written notation of subtraction that follows may shed light on the relationship between oral description and written notation abilities. Do children who give the most detailed descriptions

also write complete equality sentences? Are children who give incomplete oral descriptions capable of writing a complete equality sentence?

Writing and Reading Own Notation. It is interesting to note that two out of the three children giving complete descriptions were unable to write a complete equality sentence following a sign prompt. On the other hand, two children needing a prompt to identify more detail in their oral descriptions wrote complete equality sentences spontaneously.

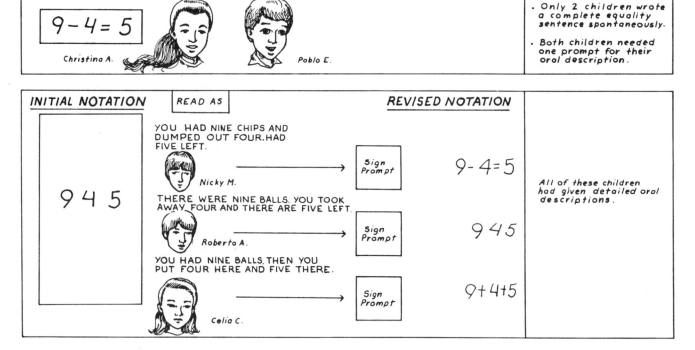

Can a child requiring multiple prompts for oral description write a complete equality sentence? This was the case for one child who encoded the action as addition initially. She and two other children who did the same were encouraged to reconsider the task from

another perspective by focusing on the starting quantity. The child in question was able to revise her equality sentence in terms of subtraction, while the other two children were unable to do so.

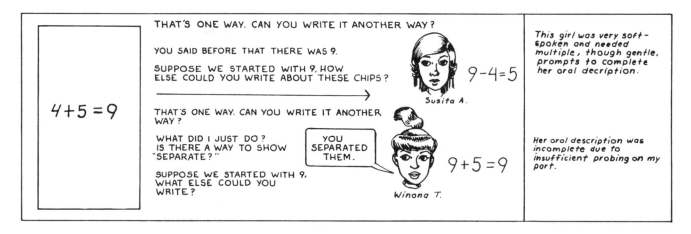

---

3. In the actual interview the encoding tasks for addition and subtraction preceded the decoding tasks.

First-grade children's resistance to inclusion of symbols other than numerals was amply demonstrated once again. Sign prompts seldom resulted in their inclusion. The most frequent response to "What else can you write here [pointing] to finish telling the story of the balls/chips?" was the insertion of numerals. Also, when children did write the appropriate three numerals, their notations did not preserve the order of the action. Illustrations of such responses follow.

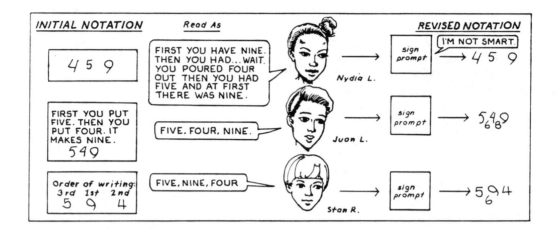

Subtraction is also an operation that relates the whole and its parts, requiring a simultaneous consideration of three numbers for a complete written notation of the operation. Children's inability to focus on the whole and its parts simultaneously is indicated clearly in the following responses. It is noteworthy that all of these notations followed difficulty in identifying the starting quantity during oral description. Also, in each instance the directions for the oral description had been abbreviated so that children had not been asked specifically to identify the starting quantity.

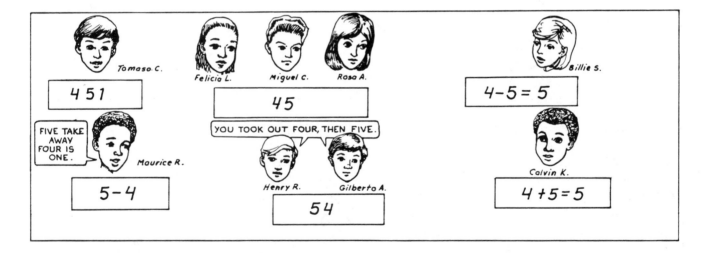

The 4 5 notation reflects an isolated focus on the parts as the children were unable to retain the whole mentally once it had been decomposed. A similar explanation can be given for the notations 4 5 1 and 5 – 4. In each case the whole could not be reconstructed, so a new whole was invented or a new arrangement devised. Like thinking can be attributed to the inappropriate sentence, 4 – 5 = 5, except that the 5 as the "visible answer" in the materials appears to be a strong influence here.[4]

---

4. One child's notation of 4 + 5 = 9 for the "take away" subtraction can also be interpreted in terms of the whole-parts relation. Once the whole had been decomposed, the child recombined it again as indicated in the notation. In doing so, he lost contact with the "take away" situation. The notation adequately represents the relationship of the numbers involved, but not the specific demonstration involving taking objects away.

Task D: Interpretation of Equality Sentence Notation—Subtraction

This task began with a familiar request to complete an open equality sentence for subtraction. The child then was asked to read the completed sentence and to model its meaning with blocks. In other words, the child was asked to decode the shorthand communication used by mathematicians following a year's exposure to this convention.

Completion of Sentence: What number would finish this number sentence
(7 – 3 = _____)?

Reading of Sentence: Read this number sentence.

Modeling with Blocks: Do something with these blocks to show what this number sentence means.

(Following the modeling with blocks the child is asked to show the quantity of blocks represented by each of the numerals in the equality sentence as pointed out individually.)

Think ahead by considering the following:

1. Analyze this decoding task in terms of the whole-parts relationship.

2. Predict the range of children's responses to this task.

Completion of Sentence.   Whereas the addition sentence 6 + 2 = ___ was completed accurately by all 33 children, 7 of the children responded incorrectly to 7 – 3 = ___. Six of these children interpreted it as an addition task. Once prompted to identify the operational sign they quickly readjusted their answers.

Reading of Completed Sentence.   Although the reading portion of this task was inadvertently omitted for more than half of the children, four children were observed with reading difficulties. Prompts to identify omitted signs prior to rereading the sentence were met with mixed success.

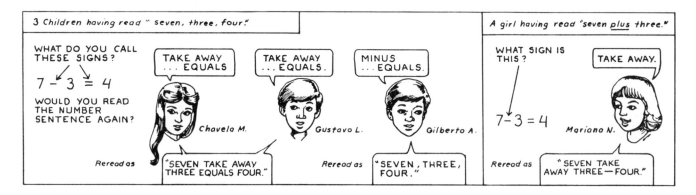

Modeling with Blocks. Half of the 33 children were able to model the equality sentence as a physical separation and to demonstrate the whole-parts relationship with blocks. Once showing seven blocks, some of the children hesitated while apparently considering whether three more blocks needed to be shown separately or whether they could be considered as a component part of the whole. Most of these children decided that seven blocks were sufficient for modeling the entire equality sentence.

Yet three of these children thought that more blocks were needed even after showing the separation. One of these children is illustrated below. When asked to show "seven" he added seven more blocks to the display rather than recombining the original blocks.

One-third of the children initially showed three sets of objects side by side, 7, 3, and 4, as indicated in the form of the equality sentence. Similar to the preceding example, some of the children concluded that the answer was eleven. The next example shows how one child arrived at an answer of eight.

This girl has some ideas about subtraction as separation and can count on her fingers to obtain an accurate answer. Yet she appears to be entrapped by the form of the equality sentence and cannot read a whole-parts meaning into it.[5]

EXPLAINING CHILDREN'S DIFFICULTIES WITH NOTATION—A CONTINUATION

It might be tempting to dismiss the surprising variety of notation and interpretations given by children in this interview study as a result of the size of sample or the low socioeconomic background of those studied. Yet, if we accept Piaget's principle that the intellectual development of all children passes through similar levels, at varying rates, the findings of this study are generalizable to other children. Furthermore, other studies conducted in a variety of locations by C. Kamii (1980), Allardice (1977), and Reyes and Suarez (1979) do report similar results following periods of extended exposure to the equality sentence. My own interviews with beginning second graders from an upper-middle-class community revealed several examples of such difficulties. A common example was the omission of the equality sign when encoding a physical combination or separation of groups. Another difficulty demonstrated by several children while decoding subtraction sentences is illustrated.

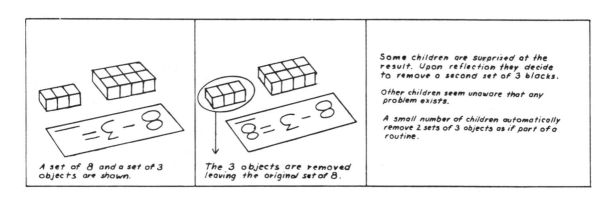

A set of 8 and a set of 3 objects are shown.

The 3 objects are removed leaving the original set of 8.

Some children are surprised at the result. Upon reflection they decide to remove a second set of 3 blocks.

Other children seem unaware that any problem exists.

A small number of children automatically remove 2 sets of 3 objects as if part of a routine.

The similarity of observations from diverse sources suggests that this difficulty with the formal convention of the equality sentence is widespread when children are asked to go beyond the typical workbook experience. The frequency and duration of these difficulties seems to vary with individuals and groups as some children construct the whole-parts relation sooner than others. According to Hendrickson (1980), appropriate teaching is also a contributing factor. He reports that some first graders encode physical situations of their own arrangement, such as $8 = 2 + 2 + 3 + 1$. The range of abilities observed earlier in children of the same age appears to exist in their written notation as well.

The Role of Materials in Encoding and Decoding

Contrary to a popular view, the use of materials does not automatically lead to correct answers and to meanings. This was illustrated by the preceding responses of children. The arbitrary imposition of the form of the equality sentence outside the context of familiar physical situations does not provide an opportunity for either developing relationships or for encoding. Rather it provides a form for presenting tasks for which answers are supplied in the blanks. At the same time, it eliminates the basis for decoding and hence elicits some bizarre responses, where the children are not trying to represent the meaning underlying the shorthand notation with blocks, but rather are attempting to represent the form of the sentence in another medium. Introducing the symbolism of the equality sentence as a shorthand notation for familiar physical situations prepares the children to encode other situations and also to decode the notation of others.[6]

---

5. In this child's display of objects, not only are each of the numbers represented as independent quantities, the minus sign is also represented.

6. In filling in blanks in open sentences, children focus on getting the answer, and can do so at a low level of ability. In encoding and decoding, they consider the total relationship, thus requiring a higher level of functioning.

The Role of Language in Encoding

One might expect the quality of children's oral descriptions of combining and separating demonstrations to relate directly to the quality of their equality sentence notations. Although this appeared to be the general trend in the interviews, there were also a number of interesting exceptions noted earlier. The observation that the best oral describers were not always the best encoders raises the following question: Does a complete oral description provide just a literal description or does it imply any arithmetic relationships beyond the quantities. Nesher, Greeno, and Riley (1981) have identified two distinct levels in children's language. In the context of everyday life children acquire a language of experience with objects—for example, using words such as *put, take away, makes* to indicate change. This ordinary language is distinct from the one developed in the context of numbers, operations on numbers, and relations among numbers, and is independent of physical contexts. Whereas ordinary language develops readily, the formal language of arithmetic—like *plus, minus,* and *equals* or *is the same as*—develops more gradually as meanings are abstracted/generalized from interaction with a variety of physical contexts. Once developed, meaningful arithmetic language can replace ordinary language and be directly related to the ability to encode equality sentences. Although this position appears plausible, it does not account for incomplete oral describers being able to encode. Nor does it account for the ability of oral describers using ordinary language to encode addition and subtraction situations. This dilemma prompts another question: Do children think in symbols other than language?

A related study by Hamrick (1979) tested the role of oral language in readiness for conventional written symbolization of addition and subtraction. In September, 38 first-grade children were placed into two groups according to their "verbal facility with words and sentences represented by arithmetic symbols." Each group was subdivided in turn for a 12 week period of instruction. Half of each group received regular instruction in which written symbolism was introduced simultaneously with addition and subtraction operations in the presence of materials. The second half of the first graders did not receive any instruction in written notation until the sixth week. During the earlier weeks, however, the focus was on concrete activities and verbal descriptions to develop both arithmetic relationships and vocabulary. The activities also included working from pictorial representations of combining and separating sets.

Following 12 weeks of total instruction, the post-

	Initial verbal facility	Initial lack of verbal facility
Early written symbols	GROUP A₁	GROUP B₁
Delayed written symbols	GROUP A₂	GROUP B₂

test revealed no difference in the encoding and decoding of equality sentences between groups of children having initial verbal facility (groups A_1 and A_2), despite a difference of five weeks of exposure to this form of written notation. Although there was no loss of ability resulting from the delay, some children were reported as being impatient or bored when held back. The children lacking verbal facility and provided five additional weeks to develop it together with the relations (group B_2) performed significantly better on decoding tasks than the "nonready" children (group B_1) who had been introduced to written notation five weeks earlier. In the encoding-task responses by the nonready children, those experiencing delayed symbolism scored consistently higher than those receiving immediate symbolism. Yet the scores of both groups (B_1 and B_2) were low compared to those of the ready groups (A_1 and A_2). Hamrick concluded that if a child isn't ready for introduction to written notation of addition and subtraction, the child's learning of the notation will be more meaningful if the notation is delayed until the child is ready. Her report indicates that verbal facility is an appropriate indicator of readiness and a basis for grouping children for instruction. Yet a precise definition of verbal facility is lacking in the report, so it is unclear to what extent ordinary language and arithmetic language were used as criteria.

Since language is being used as an indicator of a child's understanding, it is reasonable to ask whether children understand more than they are able to represent. On this question, Piaget (1977) wrote:

> . . . At all levels, including adolescence and in a systematic manner at the more elementary levels, the pupil will be far more capable of "doing" and "understanding in actions" than of expressing himself verbally . . . "Awareness" occurs long after the action. (p. 731).

At the same time, Piaget has stressed the importance of social interaction as a way of provoking awareness and rethinking. So although the relation between oral description and encoding is not crystal clear, a case can be made for the importance of children's social interaction in both physical situations for addition and subtraction and in their written notations.

The Role of Context in Notation

At the beginning of the chapter, Allardice's study (1977) highlighted children's notation of number ideas in the context of communication. She reported that many first graders spontaneously chose their own methods of informal notation of addition and subtraction in sending a message to Snoopy, despite a year's exposure to the equality sentence. This observation suggests children may resist formal notation because it has little or nothing to do with how they think about numbers and their combinations. It also questions the method of teaching and suggests that children may see the conventional format as purposeless within the context of filling in worksheet blanks. At the same time,

Allardice's work raises the possibility of respecting children's own representational abilities by dwelling and building on encoding and decoding activities at their level. She suggests that a child struggling to find a means of representing the dynamic aspects of addition as joining may accept the "+" sign more readily and have some appreciation of the economy of the formal convention. Similarly, a child dealing with decoding a variety of notations in the same classroom would eventually realize the advantage of a standard system for sender and receiver and would gain a greater appreciation of a standard, shorthand convention once it was introduced. In this approach, the children's range of representation would be explored and developed to the fullest rather than seduced by a formal convention that might only be empty symbolism to them. Here, the communication process, rather than its form, is given top priority.

IMPLICATIONS FOR TEACHING ARITHMETIC NOTATION

The research studies described in this chapter underscore the value of observing and interacting with children outside the context of structured workbook exercises. In doing so, we have uncovered both capacities and constraints that go unrecognized in most

classrooms. We are now able to contrast assumptions about learning and teaching written number notation held by textbooks with what we've learned from children. These contrasting assumptions are summarized in the following table.

CONTRASTING ASSUMPTIONS IN TEACHING ARITHMETIC NOTATION

TEXTBOOKS	WHAT WE'VE LEARNED FROM CHILDREN
Formal symbolism of equality sentences should be introduced to first graders in September simultaneously with addition and subtraction.	Formal symbolism of equality sentences should be delayed or introduced gradually. The full expression of $5 + 3 = 8$ can even be delayed until the second grade.
Children have no ideas of their own about written notation of number. They need to be given a ready-made system immediately.	Children have ideas of their own about representing number ideas on paper. These informal methods need to be developed in making the transition to a formal system having no apparent ties to reality.
Symbols have meaning in and of themselves, and can be readily explained by a quick demonstration with materials.	Symbols don't have meaning in and of themselves. Children can read into them only what they already understand about number relationships.
There's not much to talk about. It's all there in the symbols.	Talking about joining and separating situations (as well as other forms of addition and subtraction) can consolidate relations and develop math vocabulary. This spoken language and its underlying ideas can then be represented by the symbols.
The textbook represents an external system of knowledge that can be transmitted readily to a passive learner.	Mathematics is reconstructed in the mind of every child. It does not exist externally on a printed page. The child is an active learner constructing his mathematical understanding through interaction with physical objects, his existing ideas, and the ideas of others in adapting to new situations.
As long as children can fill in the answer blanks correctly in workbooks and on tests, they "know the material." The topic can be checked off as having been taught.	Equality sentences can be understood on different levels. Completion of open equality sentences involves lower levels of understanding. Encoding and decoding require higher levels of functioning that may not be "checked off" for some time. Using only workbooks interferes with determining children's level of functioning.

A second recurring message in the preceding research is the importance of avoiding introducing children prematurely to the formal symbolism of the equality sentence. Delay can be accomplished by first emphasizing children's construction of relations, own methods of representation, and oral language, and by introducing symbols gradually according to their level of difficulty.

1. *First emphasizing construction of relationships and problem solving in extended activities involving counting and derived strategies.* Many of the activities described in Chapters 3, 4, and 5 use concrete and pictorial referents and encourage development of meanings and language without emphasis on written representation. Initially, children's informal pictorial and written notation can be encouraged for keeping any records of interest to them. Eventually, the equality sentence can be introduced when the need for a shorthand recording of multiple relationships is realized. Since many number relationships will have been constructed at this point, the equality sentence can be learned readily as a shorthand representation of familiar ideas. With the focus on problem solving, the introduction to formal symbolism can be profitably delayed until the second grade. A similar focus has been advocated by C. Kamii (1982b).

2. *First exploring and developing to the fullest the potential of children's own informal notations in the context of communication.* The transition to the conventional equality sentence can be delayed until children's representations have been fully explored, their number relationships are under construction (suggested by disappearance of early forms of notation and acceptance of conventional use of numeral without any direct teaching), and the need for a standard convention is realized. Although we haven't learned how to explore this potential to the fullest, some starting points are provided on the following pages.

3. *First providing word problems in group settings to contribute to children's understanding of real-world contexts and the appropriate language to describe them.* The introduction of equality sentences can be delayed until they can serve as a shorthand representation of children's own words and experiences in familiar situations. Word problems also serve as an excellent vehicle for developing meanings. They can provide a variety of situations for addition and subtraction so that children can draw from a rich collection of experiences to generalize what these operations

are and when they occur. As further meanings are developed and refined, arithmetic language can be substituted for children's ordinary language. Some starting points suggested by Burns and Richardson (1981) are described on the following pages.

4. *Introducing the symbols gradually in the context of physical materials and oral discussion, according to their level of difficulty, prior to exposure to the entire form of the equality sentence.* For example, the numerals can be used first in written notation, followed by the operational signs, and finally the equality sentence. Classroom programs by Baratta-Lorton (1976, 1978) provide numerous examples of this approach.

Although the preceding variations have been outlined separately, they can be used productively in combination with one or two other approaches.

Introducing Notation in a Physical Context

The following sequence is suggested by Burns and Richardson (1981) for introducing children to equality sentences. It begins without using any numerical symbols and progresses toward the goal of enabling children to write appropriate equality sentences.

Using Children and Props. Make up problem situations that involve numbers for children to act out, varying the situations to include both addition and subtraction. For each situation, the procedure is the same: you tell the story, the children follow directions, then they count and determine the sums or differences. Use props you have in the room and use your students' names. After the children have acted out the situation, have them describe the story in their own words to further cement the experience.

Neal gives Ruth four books. Sara gets two more books and gives them to Ruth. How many books is Ruth holding? (Addition)

All the children in the third row line up. Miguel, Sonja, and Steven sit down. How many children are still standing? (Subtraction)

Introducing Symbols. Over time, as you continue to present experiences of this type, add numerical recording. You do the recording of the appropriate equation for each problem. The children need no previous experience with the plus and minus signs; by seeing the teacher write symbols that label the real situation, the children connect the symbols with their own actual experiences.

Luis lines up four chairs. Deane puts three more chairs in the line. How many chairs are in the line? (Addition)

There are seven chairs in a line. Neal, Josh, and Carla sit down in chairs. How many chairs are still empty? (Subtraction)

Think of the numerical recording as the mathematical counterpart of language experience. One way to write the story about Luis and Deane and the chairs is with words. Another way to write that story is $4 + 3 = 7$. Of course, $4 + 3 = 7$ can describe many stories, but in this case, it is an accurate description of the story presented. And it's a lot shorter to write than the version using words.

When writing the story in mathematical terms, don't focus on the sum. Keep the process of translating action into a numerical sentence uppermost. The "answer" in this activity is not just the numerical sum, but the entire sentence.

After you write an equation, read it back to the children, using the children's natural language first; that is, four and three is seven. After a brief period of time, introduce children to another way of reading the sentence: four plus three equals seven. After a while, the children will begin telling you what to write and will read the equations back to you.

Having the Children Record Equations. Continue to make up problems and act them out as before, but provide the children with individual chalkboards so that they can each record the equations as you do. This exercise is not to be a testing situation. Continue to model the correct equation so that the children still needing a check have one available to them. Deciding when to work at this level should depend on your particular children's abilities.

Not only is it important to introduce word problems using real situations, it is also important to vary their circumstances to ensure that children are not just learning to record equations in a narrow setting. Each of the situations can be experienced without numerals, with you doing the recording or with each child writing the appropriate equations. (pp. 29–30)

Other activities suggested include using blocks to represent cars, cookies, children, or whatever is referred to in the word problems. Again, children can carry out the activities at all three levels. They can be encouraged to tell the problem in their own words as well as to make up stories for others to show. Once the children have reached the level of numerical recording, some activities can begin with equality sentences and the children can be asked to decode them with blocks. They can also be asked to tell a story in their own words to fit the sentence or to draw a picture to represent it. Another possibility not mentioned by Burns and Richardson is to encourage children's informal notation prior to introducing them to the conventional form. A complete range of word problems is described in Chapter 8.

Gradually Introducing Conventional Symbolism

In her daily work with kindergarten and first-grade inner-city children, Mary Baratta-Lorton observed a gap between children's ability to act out operations on objects and their ability to symbolize these actions in a conventional form. She independently observed many of the difficulties described in this chapter. In adapting her teaching to the needs of her children, she devised a developmental sequence of learning activities for gradual introduction of the conventional symbolism in the context of physical situations and oral discussion. Following successful trial of this and similar sequences, she proceeded to have them published. It is important to note that this teacher observed what previously went unreported in research; and furthermore, she adapted to the needs of her children by going beyond existing textbooks and developing her own curriculum. She serves as a model of a *teacher as researcher and curriculum developer*. Two examples of her sequences are illustrated below. The first one is adapted from *Workjobs II: Number Activities for Early Childhood* (1978).

I	II	III
Pairs of children explore subtraction by taking turns making up situations and stories that demonstrate "take away." Colorful, intriguing boards and pieces depicting twenty different contexts such as watermelons, frogs and ponds, aquaria, airports, haunted houses, and so on encourage children to create, verbalize, and demonstrate situations.	Pairs of children repeat the activity of the first level. While one child is showing and telling, the other is using numeral cards to represent the quantities. Note that a record of the starting quantity (whole) is retained even though it has been decomposed. The display is now read by one of the children.	Individual children use a printed numeral card to indicate the starting quantity. They subtract as many objects as desired, making a display of their "operation." They then represent the situation by writing a conventional equality sentence on a separate strip of paper. Several such boards are displayed, providing the teacher with feedback in the absence of worksheets.

The second sequence is taken from Baratta-Lorton's *Mathematics Their Way* (1976). Notice that the final form is not always a horizontal sentence, but a vertical sentence. Although not technically an equality sentence, the vertical form is an alternative representation that avoids the need for an equality sign.

I	II	III
Children partition different combinations of a set of objects and orally describe that combination, "Three and two." The same set of objects is used repeatedly for the partitioning so the total quantity is known. There is no recording.	Arithmetic symbols are connected to familiar activities and objects. One child does the partitioning while the other records the process on a pictorial representation of the activity using printed numeral cards.	Written conventional notation is used to record the situation. One child again does the partitioning while the other does the recording. Worksheets with pictorial representations of the materials suggest the format of the notation (vertical or horizontal).

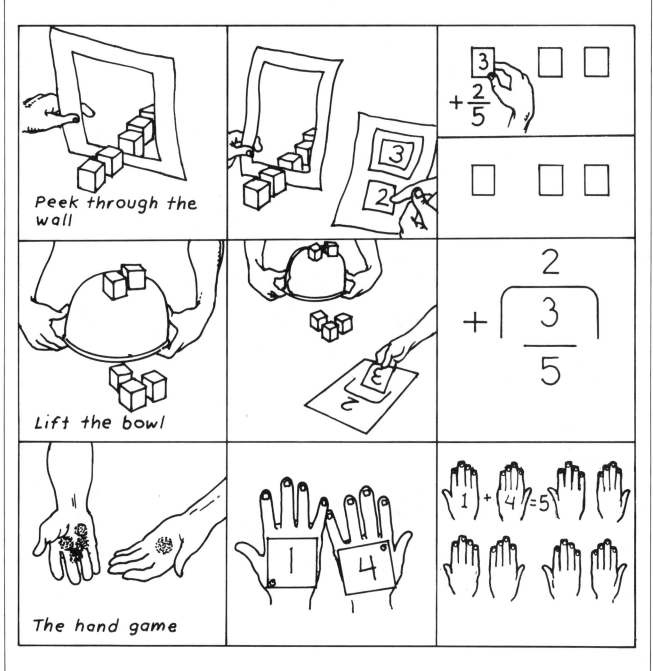

Peek through the wall

Lift the bowl

The hand game

Developing Relations and Communicating with Informal Notation

The next sequences offer starting points for teaching both addition relations and their informal representation. These activities convey to children the need for encoding to maintain a record and to communicate with others. They also illustrate how children can move back and forth between encoding and decoding in the context of communication. The organizing loops and split boards encourage children to think about wholes and parts and their relations. Also, children's informal representations of quantity are accepted. Progress from early forms of representation, facility in discussing the situations and considering whole and parts simultaneously, can be used as indicators of readiness for conventional notation. Sequences for two physical contexts for addition are illustrated.

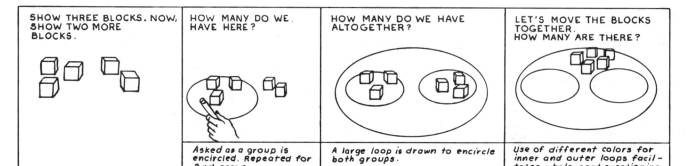

SHOW THREE BLOCKS. NOW, SHOW TWO MORE BLOCKS.	HOW MANY DO WE HAVE HERE?	HOW MANY DO WE HAVE ALTOGETHER?	LET'S MOVE THE BLOCKS TOGETHER. HOW MANY ARE THERE?
	Asked as a group is encircled. Repeated for 2nd group.	*A large loop is drawn to encircle both groups.*	*Use of different colors for inner and outer loops facilitates whole-part questioning.*

IF WE MOVE THESE BLOCKS UP HERE, HOW WILL WE KNOW HOW MANY WERE THERE (Small loops) BEFORE/FIRST?

SOMETIMES I FORGET. HOW CAN I KNOW TOMORROW?

CAN YOU LEAVE SOME MARKS ON PAPER THAT WILL LET ME KNOW TOMORROW:..?

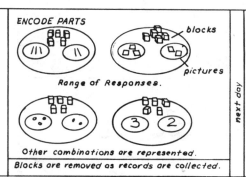

ENCODE PARTS

Range of Responses.

Other combinations are represented.

Blocks are removed as records are collected.

DECODE PARTS

HERE ARE SOME MESSAGES THAT YOU WROTE YESTERDAY. I'M NOT SURE WHAT ALL THE MESSAGES MEAN.

WOULD YOU PUT BLOCKS ON YOUR PAPER TO SHOW HOW MANY WE HAD FIRST AND THEN WHAT HAPPENED.

Children should receive their own records. They may also decode the records of others.

next day

ENCODE WHOLE/PARTS

SNOOPY WANTS TO KNOW HOW MANY THERE WERE AT FIRST, WHAT HAPPENED AND HOW MANY ALTOGETHER. WRITE SNOOPY A MESSAGE.

Sample written response.

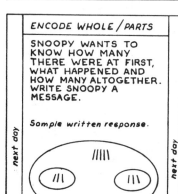

Other... combinations are encoded.

next day

DECODE/WHOLE PARTS

SNOOPY LIKES THESE MARKS BUT HE'S NOT SURE WHAT THEY ALL MEAN?

WITH THE CHIPS, SHOW SNOOPY

• HOW MANY AT FIRST.

• WHAT HAPPENED.

• HOW MANY ALTOGETHER.

TELL ABOUT THE CHIPS AND WHAT HAPPENED TO THEM.

next day

SHOW ME WHERE THE THREE CHIPS GO?

THE TWO CHIPS?

SHOW THE FIVE CHIPS.

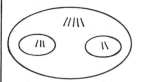

USING CHIPS, SHOW EVERYTHING THAT HAPPENED.

Have children discuss any disagreements in the group.

• *Children work in pairs. One child makes up a combination and demonstrates with blocks on an ⊙⊙ organizer. This child describes the actions orally.*

• *The partner makes a written representation on the same organizer card.*

• *Children decode the written representations with blocks on the next day, verbalizing as they do so. ("Plus" added to vocabulary.)*

• *These activities are repeated with different materials, quantities, and partners.*

several days or weeks as needed

The second sequence provides another perspective of addition and the resulting whole-parts relations. The sequence is extended to conventional symbols.

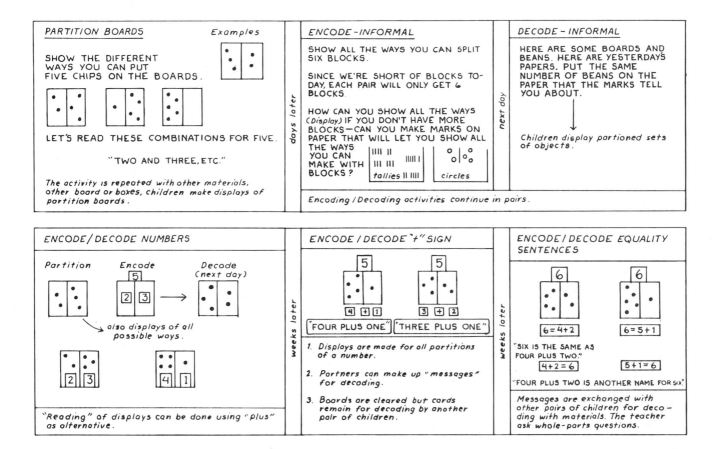

The partition boards may be covered with halves of different colored paper. They may also be plain, with partitioning achieved with a physical divider or by using two-colored objects such as lima beans spray-painted on one side, or Othello game pieces. Since the same quantity is being partitioned in different ways the whole-parts relation has the potential of being connected to conservation of number. The partition board, like the loops, allows reference to the quantity on the whole board and part of the board in discussing the relation.

Research Evidence for Alternative Teaching Methods

Research reported earlier in this chapter indicated that children have difficulty with encoding equality sentences even after a year's exposure to this conventional form in traditional classrooms. Hendrickson (1981) interviewed children selected randomly from an alternative teaching program combining methods described above—emphasizing the construction of relationships using materials, providing physical number story problem contexts, and delaying the introduction of conventional symbolism. Using the same encoding task described earlier in this chapter as Task A, he found that 38 of 47 children in their sixth month of the first grade wrote appropriate equality sentences. Three more children made appropriate corrections of their initial sentences. This finding suggests that the encoding of equality sentences need not present a major barrier for children. The use of teaching methods that are congruent with children's approaches to learning appear to be effective. However, the inclusion of a control group for comparison in the design of the study would have made the results more conclusive.

REFLECTIONS AND DIRECTIONS

A major classroom problem identified in this chapter is premature, arbitrary introduction of the conventional symbolism of arithmetic outside of familiar physical contexts. As children attempt to cope with the "arbitrary dance of the symbols" they may experience confusion and fear of failure (Casey, 1981).

They also may develop an insurmountable barrier to working with larger numbers (Ginsburg, 1977).

Furthermore, this may lead children to distrust their own thinking (C. Kamii, 1982a).

Four alternative methods of teaching arithmetic notation have been outlined for your examination and adaptation.

---

1. Reflect on the extent to which each of the alternative methods fits the following model.[7]

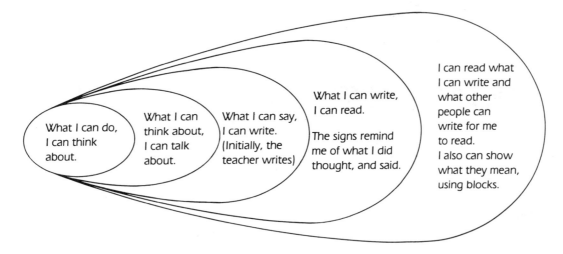

What I can do, I can think about.

What I can think about, I can talk about.

What I can say, I can write. (Initially, the teacher writes)

What I can write, I can read.

The signs remind me of what I did thought, and said.

I can read what I can write and what other people can write for me to read. I also can show what they mean, using blocks.

This chapter has dealt with both form and meaning.[8] Number meanings are gradually constructed and expressed in writing by children in ways of their own invention. The conventions of a standard form of written notation are based on the inventions of mathematicians. A certain amount of modeling is necessary to convey these conventions to children.

---

2. Reflect on the extent to which the four alternative methods facilitate the development of both a trust in one's own thinking and a feeling of control over one's learning by the relative emphasis given to **form** and **meaning**.

---

7. The model is adapted from Van Allen (1961).

8. Piaget distinguishes between meaning and form in his categories of knowledge—logicomathematical knowledge and social knowledge (discussed in Chapter 1).

7

CONSTRUCTION OF MATHEMATICAL EQUALITY

PREVIEW

The preceding chapter reported evidence of children's resistance to both the encoding and decoding of the "=" sign. Yet once a child adopts the use of the "=" sign and reads it as "equals," different levels of meaning may still be assigned to it. The concept of mathematical equality has many facets. Like the concept of number, notions of equality undergo continuous reconstruction at higher levels of mathematical relations. In other words, understanding the equality sign is not an all-or-nothing affair to be achieved instantaneously or to encompass the whole concept. While mathematicians read many logical properties into equality sentences, children read their own mean-

ings into these symbols. Since equality is central to the understanding of mathematics, further exploration of children's views of the different facets of mathematical equality is essential.

The first part of this chapter will examine second graders' understanding of equality as represented in a familiar form of equality sentence notation. It will also explore their interpretation of equality in unfamiliar contexts. These tasks will include the encoding of unfamiliar physical situations and children's acceptance or rejection of unfamiliar forms of the equality sentence.

Consider how second graders will respond to these tasks.[1]

Task A: Interpretation of "=" in familiar context

> (Following the completion of one or two open sentences,
> 4 + 2 = ___ or 7–3 = ___,
> the child is asked to read the completed sentence.)

What does this sign (=) mean?

Task B: Acceptability of unfamiliar forms of the equality sentence

> I'm going to write some number sentences. Read each one and tell me
> if it's correct or not.

7 = 5 + 2	7 = 4 + 3	
5 + 2 = 2 + 5	5 + 2 = 4 + 3	7 = 7

---

1. The questions are taken from a similar interview study by Behr, Erlwanger, and Nichols (1976).

> I'm going to read some number sentences. Tell me if they are correct (presented orally in the absence of written symbols).
>
> Seven equals five plus two.
>
> Seven is the same as four and three.
>
> etc.
>
> Task C: Encoding an unfamiliar situation
>
> (The child is shown two unequal groups of objects. After indicating that one group has more, he is asked to make the two groups the same.)
>
> Now write a number sentence that tells that the two groups are the same.

The children interviewed consisted of twenty-six of the original second graders described in Chapter 3, together with six additional second graders from a similar background. The exploration of children's views of equality served as a basis for designing a teaching/ learning experiment to facilitate second graders' construction of more sophisticated views.[2] The details of this teaching experiment will be presented in the second part of this chapter.

EXPLORING CHILDREN'S VIEWS OF EQUALITY

In preparation for the experiment, variations of different tasks were presented to the children. This served to allow me to not only learn from the children, but also to find out which tasks or questions had the most potential to uncover children's thinking or to provoke a rethinking of their view.

Task A: Interpretation of "=" Sign In a Familiar Context

All but one of the 32 second graders interviewed knew what to call the symbol "=" and to read it as "equals" in the context of the sentence. Yet the meanings they offered were limited. Lacking a strong basic notion of "sameness," they gave the symbol meaning in the best way they could. The great majority of children tended to view the equality sign from a problem-answer focus. For them, the equals sign either suggested the need to *do something to get the answer* or *to separate the answer from the problem.*

2. The children participating in the teaching/learning experiment will be identified with an asterisk.

Vince D.

TO EQUAL THE TWO NUMBERS UP... TO PUT THEM TO-GETHER TO TELL THE ANSWER... (P) THAT IT EQUALS BOTH OF THEM AT THE SAME TIME.

(P) Prompt: "HOW WOULD YOU EXPLAIN THIS TO A FRIEND IN KINDERGARTEN?"

Fay B.

EQUALS...I MEAN LIKE...IT LEAVES A LITTLE SPACE THERE SO YOU WON'T...IF YOU DIDN'T HAVE IT THERE IT WOULD BE 7-34.

Anne J.

IT EQUALS LIKE A NUMBER.

Vic J.

IT TELLS YOU THE PROBLEM IS OVER AND YOU HAVE TO WRITE AN ANSWER WHEN YOU THINK OF THE NUMBER.

Beth R.

TO EQUAL THE NUMBERS UP.

Nancy J.

MEANS WHEN YOU ADD THESE TWO THE SPACE IS WHERE YOU PUT THE ANSWER.

Perry J.

THE NUMBERS THAT YOU'RE ADDING HAVE TO EQUAL SOME-THING.

Nate D.

IT MAKES THE ANSWER.

Dom R.

IT TELLS THE ANSWER TO THE PROBLEM.

Rita C.

EQUALS.

No further response despite kindergarten prompt.

None of the 32 children gave a response that clearly focused on the "sameness" relation of the quantities on both sides of the expression. Although most children responded to the task as indicated, some of them had difficulty in discussing the meaning of the sign once they had labeled it. Encouraging them to explain it to a kindergartener in their own words had little effect. They were unable to "make any words up." Do some of the children understand more than they can verbalize?

Following a prompt to explain it to a kindergarten friend.

THAT'S AN EQUALS SIGN.

I JUST KNOW IT MYSELF. I...I CAN'T TELL ANY-BODY. I DON'T KNOW HOW TO EXPLAIN IT.

Mickey J.

(Shakes head)

Task B: Acceptability of Unfamiliar Forms of the Equality Sentence

If $5 + 2 = 7$, does $7 = 5 + 2$? If $4 + 3 = 7$, does $7 = 4 + 3$? Although mathematicians make no distinction between left and right sides of equality sentences, children often do.

Written:	$7 = 5 + 2$	Read the number sentence.
	$7 = 4 + 3$	Is the number sentence OK?

All but 3 of 20 children were able to read the equality sentences as given. The three children showed strong resistance to reading the equality sentences from left to right since the "answer" was now on the left-hand side of the sentence. They transposed the equality sentence and read what they understood.

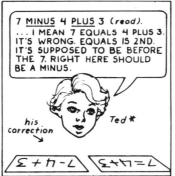

Although the remaining 17 children read with little difficulty, only 7 could accept the unfamiliar forms of equality sentences as correct. Most of the children felt the need to compare, or change it, to the more familiar format.

> Oral: I'm going to read some number sentences and need you to listen carefully so you can tell me if they're correct or not.
>
> Seven equals five plus two.
>
> Seven is the same as four and three.

When the same equality sentences were presented orally for reaction, the children demonstrated greater flexibility in accepting unfamiliar forms. The level of acceptance doubled (14 out of 20) in comparison to the written expressions. In justifying their acceptance, some children made reference to the sameness relation, while others appeared to recognize the reversibility of the relationship. Yet 7 children rejected both written and oral expressions.

If 5 + 2 = 7 and 2 + 5 = 7, does 5 + 2 = 2 + 5?

Written: 5 + 2 = 2 + 5   Read this number sentence.

Tell me if it's OK.

4 + 3 = 3 + 4   How do you know?

How would you correct it? (as needed)

Faced with another novel equality sentence, having an operational sign on both sides, 9 children out of 20 were able to make some sense of it.

Some children resisted this novel format for the equality sentence though seemingly dissatisfied with their responses. Others were unaware of any problems and showed impatience at being asked the obvious.

NO, 5+2 EQUALS 7, NOT 2+5.

AND 2+5?

SEVEN.

Nate D.

Although having the "basic facts" he is unable to coordinate them into a relationship.

IT'S SUPPOSED TO BE 7.

Ricky S.

He responds after crinkling his nose and shrugging.

THERE'S NO LAST LETTER TO EQUAL.

Vince D.

His correction shows 2 number sentences.

THAT'S 14. 5+2 EQUALS 2+5 (read). I CAN'T SAY REALLY.

Wendy K.

How did she get 14?

Oral: Listen to these number sentences and tell me if they're OK or not.

Five plus two is the same as two plus five.

Five plus two is the same as three plus four.

Again the oral statements appeared to give children greater flexibility in considering equality relations as the acceptance level nearly doubled compared to the written form. All but 2 of the 20 children accepted the oral expression. Four children's reactions to both forms of the task are illustrated side by side.

ORAL

BECAUSE YOU SWITCHED THE NUMBERS AWAY.

Mickey J.

WRITTEN

YOU FORGOT THE ANSWER (5+2 = 2+5).

BECAUSE BOTH EQUAL 7.

Perry J.

DOESN'T EQUAL 2+5, IT EQUALS 7.

BECAUSE 4+3 AND 5+2 ARE THE SAME.

Vince D.

ADD A 7 TO THIS SIDE.

4+3 IS THE SAME AS 5+2 BECAUSE 5 IS ONE NUMBER OVER 4 AND SO YOU HAVE TO USE 3 FOR THE 4 TO CATCH UP TO 5+2.

Ted*

He mentally coordinates/ compensates the variables in the relationships.

EQUALS HAS TO GO TO THE RIGHT. *He corrects as* 5+2 = 7, 4+3 = 7.

It is interesting to speculate on the effect of substituting "same as" for "equals" on children's openness to accepting the oral version of the novel sentence. Two versions of the oral expression would be needed to explore the question. However, the order of presentation of oral and written versions of the tasks was examined. During the interview the oral and written statements were grouped separately. Half the children received the oral statements prior to the written ones, with the order being reversed for the others. No noticeable effect on the children's responses was observed as a result of the order of presentation.

How do children view equality sentences showing no operation?

> Written: 7 = 7   Read this number sentence.
>
> Is it correct?
>
> How do you know?
>
> How would you correct it?

Out of the 20 children asked this question, 9 were able to accept this novel equality sentence implying no action.

The remaining 11 children, after rejecting the equality sentence, corrected it by including another numeral along with an operational sign. All but one of these children corrected the equality sentence according to the familiar format of the "answer" appearing on the right-hand side.

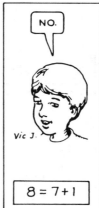

Task C: Encoding an Unfamiliar Situation

Earlier in the interview, all of the children were presented with variations of the same encoding task.

 Do we have the same number of blocks in both bunches or does one bunch have more?

What can you say about the bunches now (as one block is removed from larger set)?

Variation 1: Write a number sentence that shows that the two bunches are the same.

Variation 2: Pick one of these signs to put between the two bunches of show they're the same. Now, write a number sentence that tells they're the same.

Variation 3: The same task as variation 2, presented in the context of a balance.

Of the 14 children presented the first variation of the task, the entire group wrote an equality sentence containing an operational sign. Another group of 18 children was given the choice of symbol cards to place between the two groups in variation 2 of the task. Most of these children chose an operational sign. Four children chose the "=" sign following a lengthy elimination. However, with one exception, their written notations included operational signs.

Following a long period of contemplation in which she touched all symbol cards & picked up and replaced the "+" and ">" cards, she placed the "=" sign between the two groups.

| + | > | = |

YOU WEREN'T SURE BUT YOU THINK THIS IS THE BEST ONE.

BECAUSE 4... IF YOU PUT THE "+" SIGN IT'D BE 8... AND THE ">" SIGN... NONE OF THEM IS LESSER, AND 4 EQUALS 4 IS THE SAME THING.

Gloria*

She looks puzzled as she finishes her justification.

WRITE A NUMBER SENTENCE THAT SHOWS THAT THE GROUPS ARE THE SAME. $4 + 0 = 4$

THAT'S ONE WAY, IS THERE ANOTHER WAY TO SHOW THEY'RE THE SAME? $4 + 0 = 4$

SO YOU HAVE A SECOND WAY. LET'S TRY FOR A THIRD WAY.

NO, USE NUMBERS.

CAN I PUT 4 BLOCKS?

I WAS GOING TO PUT 4 BLOCKS EQUALS 4 BLOCKS.

On the 4th attempt.

4 EQUALS 4... THE SAME THING BECAUSE... (mumbled).

Notice the acceptance of responses above to maintain the flow of ideas.

Also notice the contrasting restriction of ideas when a pictorial representation was suggested.

The fourth child selected the "=" card with little hesitation, yet omitted the sign from his written notation (4 4). He resisted both adding a sign and reading his notation following prompts. The next panel illustrates how an inadvertent reversal in order of task presentation resulted in his notation 4 = 4.

Version 3 of the task was tried with a couple of children. The following panel illustrates one child's creative compromise solution: when unable to find an appropriate sign to place between the two groups, he found two signs to do the job.

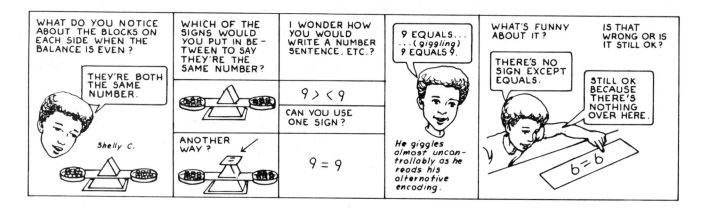

For this child, the cancellation of two inequality signs was a better indication of "sameness" than the equality sign.

A Summary

Children's interpretation of the equality sign is usually a unidirectional one. Reading from left to right in the equality sentence, children view the equality sign as either a "do something" signal or as a separator of the problem from the answer. By contrast, mathematicians view the equality sign as a bridge between quantitatively equivalent expressions. In other words, they view "=" as a relational symbol of "sameness." For them, the order of writing the sentence is unimportant since the relationship is bidirectional; also, the relationship does not imply action. Although some children gave such justifications in accepting novel forms of equality sentences, they lacked consistency across tasks. However, one child modified his view at the end of the sequence of tasks and spontaneously corrected the preceding ones to achieve this consistency. Other studies by Behr, Erlwanger, and Nichols (1976) and Van de Walle (1980) report similar limited views of the equality sign among elementary school children of varying ages.

Establishing Priorities and Searching for Explanations

Prior to reading ideas on the importance of the equality concept in the school curriculum and explanations of children's views of equality, reflect on the following pairs of statements.

1. a. As long as the children get the answers in school math it's not important if they understand the nuances of mathematical equality. Understanding will come as the children get older.
   b. It is critical that children develop a solid understanding of equality relations for success in elementary school mathematics.

2. a. Children's limited understanding of equality is the direct result of the limited focus of their textbooks and teachers. A broader understanding of equality will accompany their exposure to different equality sentence formats.
   b. Children have a limited capacity for dealing with logical thinking demands of equality as a bidirectional relationship.

The Importance of the Equality Relation. The equality relation lies at the heart of mathematics. Understanding of many areas of elementary school mathematics, including the following, requires the recognition of "=" as a relational symbol of sameness (Van de Walle, 1980):

1. Completion of missing addend sentences and similar expressions
2. Relations of new concepts or symbols to old (for example, $4 + 4 + 4 = 3 \times 4$)
3. Fractions, especially the concept of equivalent fractions ($3/4 = 9/12$)
4. Properties of the number system such as the distributive property or the commutative property
5. Ratio, proportion, and percentage concepts ($3/4 = .75 = 75\% =$ "three of four")
6. Expanded notation of place value ($243 = 200 + 40 + 3$)

Children studying in these areas without a bidirectional concept of equality are in danger of viewing mathematics as an arbitrary "dance of the symbols." The research of Behr, Erlwanger, and Nichols (1976) suggests that a more sophisticated concept of equality does not automatically develop with age. Yet a review of elementary school textbooks reveals that there is no systematic attempt to teach equality as a bidirectional relationship. Rather, bidirectionality seems to be viewed as a given—automatically built into the symbol itself. Instead of blindly accepting the school curriculum as "handed down from above," we need to examine all phases of it in terms of children's intellectual capacities and constraints.

The Influence of Current Teaching Methods. The second graders viewed both familiar and novel forms of equality sentences from a problem-answer focus. The equality sign was a signal to "do something" to get the answer. It also separated the problem from the answer. In the former view, the equality sign was indistinguishable from the operational sign, conveying the meaning "equals up to." Given the current status of classroom teaching, the children's views are quite understandable. Their general inflexibility in accepting novel written forms and their problem-answer focus can be attributed readily to a rigid written format used in workbook exercises that require "getting the answer." The blanks to be filled are almost invariably found to the right of the equality sign, and the sentence always includes two numerals separated by an operational sign to the left of the equals sign. "Getting the answer," if it is not already stored in memory, is usually accomplished by joining or separating actions in the context of blocks or fingers. In other words, children assimilate exercises such as $5 + 2 = $ ___ with what they already understand (C. Kamii, 1980b).

Take five	add two	makes seven
5	+ 2	= 7

In this context, the notation indicates the order of the action, with the result noted last (rightmost position). The workbook view of addition as joining action could also account for children's resistance to use of expressions such as "is the same as" or "is" as alternatives to "equals." Nichols (1976) suggests that children's view of equality may be better represented by a one-way arrow (→). Although second graders' view of equality in conventional written notation is understandable given current teaching methods, other methods have potential for helping them to expand their view.

The children's relative openness to oral statements of novel equality sentences suggests that they are open to the ideas presented as long as they don't clash with a rigid written form. Perhaps this rigidity can be avoided by exposing children to alternative forms of notation at the start. For example, a partition model of addition

indicates a static relationship where order of notation is no longer important. The equality sentence can be written as either 7 = 4 + 3 or 4 + 3 = 7. A child's encoding of this relationship can be thought of as the "answer" to the "problem" of encoding an arrangement of objects on the partition board. Rather than think of "=" in terms of doing something, it can be expressed naturally as "is the same as" or "is". Exposure to a variety of physical situations for encoding has the potential for not only increasing children's acceptance of more than one form of the equality sentence, but also for ex-

panding their view of equality as expressed in these notations. However, methods of teaching are not the only determinants of how children view mathematical equality.

The Influence of Calculators. Although not a major influence in this study, electronic calculators provide children with experiences that reinforce the prevailing operator view of the equality sign (Carpenter, 1981a). For example, the "problem," 5 + 2, is punched in first and then the "=" key is pushed to give the "answer," 7. Since the relational view of equality is important in children's understanding of mathematics, the counteracting influence of the calculator must be considered in designing an optimum curriculum for children.

The Influence of Children's Intellectual Development. Piaget's systematic study of children's thinking has identified characteristic ways of thought at different stages of intellectual development. By the end of the first grade (about age seven), many children are entering the stage of concrete operations, in which gradually emerging logical thinking abilities appear capable of supporting the construction of a notion of bidirectional equality. The following table describes some of these emerging thinking abilities in the context of a conservation of number task, juxtaposed with specific equality relations that require similar thinking.

THINKING ABOUT EQUALITY IN THE CONCRETE OPERATIONAL STAGE

EMERGING LOGICAL THINKING ABILITIES (SEVEN-EIGHT YEARS)	EQUALITY SENTENCES REQUIRING SIMILAR THINKING
Same number or more? ●●●●●●●● → ●●●●●●●● (green) ●●●●●●●● ●●●●●●●● (red) **Reversibility** Same because if you pushed them back together again they'd be the same. (mentally reversed)	Symmetry of 4 + 3 = 7 and 7 = 4 + 3. Consideration of whole and parts on both sides of equality sentence simultaneously.
Identity It's the same chips. All you did was spread some out. You didn't add any or take any away.	Identities: 8 = 8 They're the same number. 5 + 2 = 2 + 5 They're the same numbers. All you did was turn them around.
Compensation This line is longer but the spaces are bigger so it still has to be the same number.	Compensation: 5 + 2 = 4 + 3 They're the same. Five is 1 more than 4 and 2 is 1 less than 3.
Same number? ⊙⊙⊙⊙⊙⊙⊙⊙ (covered green) ●●●●●●●● (red) ▢▢▢▢▢▢▢▢ **Transitivity** Same, because the green and red chips were the same number. Now there's the same number of blocks as red chips. So, there has to be the same number of blocks as green chips.	Transitivity (relating equality sentences): If 5 + 2 = 7 and 4 + 3 = 7, then 5 + 2 and 4 + 3 are the same (5 + 2 = 4 + 3).

Such a comparison of the intellectual demands of considering mathematical equality with the thinking capacities of children provides a basis for judging the viability of the teaching methods suggested in the preceding chapter. Indeed, according to Hendrickson (1980), many children taught in the suggested manner, following some delay in introduction to symbolism, are capable of using a variety of forms of the equality sentence by the end of the first grade. However, Piaget's description of the gradual emergence of children's logical thinking abilities prompts me to wonder about the extent to which the children have constructed relational understanding beyond adapting to the varied forms of the equality sentence. Exploratory interviews are needed to pursue this concern.

AN EXPERIMENT EXPLORING LEARNING AND TEACHING

This part of the chapter shifts its focus from assessment of children's understanding of the equality sign to exploration of the effects of gentle provocation of children to rethink their existing notions and reconstruct them at a higher level of mathematical relations. With this shift in focus, my role will switch from interviewer to teacher. Analysis of the interaction with children may not only shed light on their learning processes, but may also offer insights into ways of facilitating their learning.

This *teaching/learning experiment*[3] was designed according to the procedures used by Piaget's colleagues, Inhelder, Sinclair, and Bovet (1974). It began with the *preliminary interviews* reported in the first part of this chapter. Since children's new constructions integrate existing ones, it is essential to first have an intimate knowledge of children's views of the equality sign. Another preliminary phase of the experiment was the *selection of tasks*. A pool of tasks likely to prompt a rethinking of existing notions of the equality sign was collected or devised, and included in various combinations near the end of the preliminary interviews. Tasks for the experiment were selected on the basis of the puzzlement generated in children, and resulting modifications in their thinking. *Preassessment and selection of the sample* made up the next phase of the experiment. Here, six end-of-the-year second graders in a traditional textbook program were given a preassessment interview to test both their understanding of the equality sign and their capacity for logical thinking. Four of these children were selected for participation in the teaching/learning experiment based on their unidirectional views of equality and their capacity for logical thinking on at least two Piagetian tasks, judged

to be sufficient to consider equality relations.[4] In the next phase—the *teaching/learning episodes*—twelve 45-minute sessions were scheduled over a three-week period. The interactional environment for these sessions was designed in accord with Piaget's theory of intellectual development (Inhelder, Sinclair, and Bovet, 1974). The intensive interactions of the sessions were documented on videotape and analyzed prior to planning for the next day. Selecting a small sample was critical to this careful documentation of teaching and learning. During the last phase of the experiment, the *postassessment*, the children were re-interviewed one month later, using both familiar and novel tasks. This delayed postassessment provided information on the stability of the children's learning.

In designing the teaching/learning interactions, I made the following constructivist assumptions:

1. Children construct their own mathematical relationships through interaction of already existing ideas and new input from the environment.
2. Young children learn best through mental activity using concrete materials, with opportunities for action and reflection.

I attempted to provide optimal possibilities for interaction as the children tested their existing notions of the equality sign in a range of novel physical contexts and were confronted by feedback from a variety of sources. The variety of object arrangements that served as springboards for interaction provided for continuous interplay between form and meaning. As a physical context prompted an encoding in a novel form of

---

3. Whereas Piaget's colleagues refer to this type of study as a learning experiment, mathematics educators have labeled it as a teaching experiment.

4. The children's preassessment responses were pooled with those of other second graders in the first part of the chapter.

In addition to the preassessment interviews, prior knowledge of the abilities of the two boys selected had been obtained in other exploratory interviews conducted several months earlier. Ted's task persistence and invention of a double count strategy have already been described in Chapter 4. Randy showed both an affinity for recognizing and extending number patterns, and a delight in his discoveries.

the equality sentence, children were provoked to reconsider the meanings of the symbols and to modify them accordingly. Children's testing, discussing, and evaluating ideas of form and meaning set the stage for modification of the ideas at higher levels of mathematical relations. The children were encouraged to make their own generalizations following experiences that had the potential for equipping them to do so.

My role as teacher was to facilitate further interaction by juxtaposing selected children's responses to sharpen the focus on any contradictions and to invite their reaction, as outlined in the following diagram:

Gentle confrontation of existing one-way notion of equality with

- feedback from its application in a novel physical context—for example, demonstration of equality with objects on a balance as a basis for accepting 4 = 4.
- conflicting notions expressed by the same child in another context—for example, accepting oral statements but rejecting written versions of same equality sentences.
- conflicting notions held by other children in the same group—for example, some children will reject an equality sentence while others accept it.
- conflicting notions held by a hypothetical child of the same age group—for example, "The other day a boy/girl told me that. . . . What do you think about his/her idea?" (countersuggestion)

A critical aspect of this gentle confrontation was to avoid imposing my teacher/adult authority on the children's discussions, while maintaining a friendly, neutral manner during the intense interactions. Since environmental feedback from objects and peers was maximized in these interactions, I attempted to accept each child's responses without any judgments of my own. I redirected authority for making evaluations back to the children by inviting their reactions. I made a further attempt to consider each response carefully as a potential indicator of the child's current thinking and as a source of hunches for the next question to ask or the next situation to devise.

Since the participating children had been exposed to a rigid form of the equality sentence in a problem-answer context on worksheets over a two-year period, any reconstruction of meaning had to not only overcome the resistance of existing notions, but also had to resolve the conflict with the authority of the textbook. The latter move would require considerable independence of thought on the part of a child.

Prior to describing the interactions of the teaching/learning episodes, it is important to highlight some of the differences between the teaching/learning experiment and the typical psychological experiment. Different approaches result from different theoretical perspectives.

DIFFERENCES IN EXPERIMENTAL APPROACHES

		THE TYPICAL EXPERIMENT	THE TEACHING/LEARNING EXPERIMENT
Sample	size: selection:	large (30+) random	small (1 to 8) based on predetermined criteria, such as specific capacities for logical thinking
Treatment		standardized; controlled	not standardized; loosely prepared and spontaneously modified for individuals
Data Collection	focus: kind: method: scheduling:	products of learning quantitative scores standardized, paper-and-pencil tests pre- and posttreatment	process and products of learning qualitative, detailed descriptions flexible interviews and observation of interactions pre- and posttreatment and continuous observation during teaching/learning episodes
Data Analysis	kind: result:	complex (inferential) statistics relationships between variables or group differences identified	simple (descriptive) statistics or categories processes of some kind that explain why things work in a particular way are identified
Generalizability of Findings		based on the assumption of a normal distribution in the population	based on the assumption that all children pass through the same levels of intellectual development

The teaching/learning experiment was characterized by the quality of the interaction between the teacher and the children and the tradeoff in the amount and kind of information gathered. Detailed descriptions of small groups of children replaced numerical information about large groups. Whereas in the typical experiment the testing and treatment are standard for each child, in the teaching/learning experiment the standard in both the flexible interviews and teaching/learning episodes was to make contact with each child's thinking. Rather than focusing only on the products of learning, the teaching/learning experiment was most interested in the process of learning itself. The intensive interaction of the episodes was documented on videotape with the intent of following children's thinking and capturing moments of reconstruction of ideas, and studying the quality of the interaction. Indicators

of the learning process in action included expressions of puzzlement or disorientation, vacillation between conflicting viewpoints, compromise solutions, and both spontaneous and continued use of novel forms. The small number of children, selected according to predetermined criteria, was justified by the assumption that there is a regularity in children's construction of knowledge—all children pass through the same levels of intellectual development (Easley, 1977).

The Preassessment

The following panels portray the four children participating in the teaching/learning experiment and their responses on the preassessment tasks. Try to anticipate who will make the most progress.

	CYNTHIA	GLORIA	RANDY	TED
Meaning of "=" symbol	"Equals the answer." "Both of them equals what it is."	"How much it is. You put the answer."	"That's a plus. I don't know what it means."	"It tells you to put the number down after."
Oral statements of equality: 7 equals 5 and 2. 7 is the same as 4 plus 3. 5 plus 2 is the same as 2 plus 5. 5 plus 2 is the same as 4 plus 3.	Accepts all oral statements except the first one.	Accepts all statements. "4 and 3 is the same as 2 + 5."	Accepts all statements. "4 + 3 is the same as 5 + 2 because one more way to make 7."	Accepts all but the first. "Equals doesn't go second." "5 is 1 over 4 and you use 4 with 3 to catch up 5 and 2."
Written equality sentences: $7 = 5 + 2$ $7 = 4 + 3$	"No, backwards." Corrects as 5 + 2 = 7 or 2 + 5 = 7.	"Wrong, backwards." Corrects as 2 + 5 = 7 or 5 + 2 = 7	"Wrong, backwards."	Read as 7 − 4 + 3" "Equals is second. It's supposed to be before 7." Corrects as 7 − 4 = 3.
$5+2 = 2 + 5$ $5+2 = 4+3$	"It should be a 7, not another problem with a plus."	"They do equals but there's no answer. No." (puzzled)	"Wrong. Whew! That means 7 and a 7 there (answer) not these two." Corrects as 5 + 2 = 7.	"Wrong. Equals has to go last. Well, maybe its right, but the equals has to go right here because 5 + 2 is 7 and 4 + 3 is 7."
$7 = 7$	"Well it does and it doesn't. It does equal 7 but it should be a plus or a take away." Correction: 0 + 7 = 7	"Seven equals 7 the same thing. Same amount of blocks."	"That means nothing."	

(continued)

165

(continued)

	CYNTHIA	GLORIA	RANDY	TED
Encoding of physical comparison	Chooses "+" card. Accepts "=" suggestion. Writes 4 + 4 = 8 8 − 4 = 4	Selects "=" by elimination. Writes 4 + 0 = 4 4 − 0 = 4 4 = 4	Selects ">< " Accepts "=" suggestion. Writes 4 = 4 = 0	Selects "+". Writes 4 + 4 = 8 8 − 4 = 4

Encoding of physical comparison (with two cookie-shaped groups each containing 4 cookies)

Conservation Task		CYNTHIA	GLORIA	RANDY	TED
	Reversibility: "If you push them back together they'll still be the same number." (Also assumed from successful class inclusion.)		✓		✓
	Identity: "It's the same because you didn't add any or take any away."	✓	✓	✓	✓
	Compensation: "It's longer but the spaces are bigger between them . . . so it's the same."				✓
	Transitivity: Same (A and C) because this one (hidden A) is the same as this one (B) and this one (C) is the same as this one (B)	✓		✓	✓

Episode 1: Group Comparisons and Their Written Representations (8 > 5, 9 = 9)

The episode began with children encoding the comparison of unequal groups. The unexpected appearance of equal-sized groups was expected to present a dilemma for resolution.

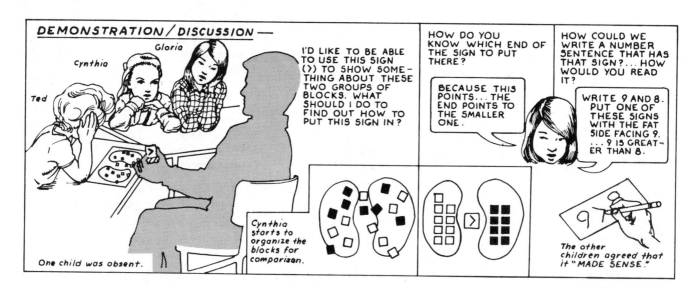

I'M GOING TO GIVE YOU SOME THINGS TO COMPARE AND TO WRITE NUMBER SENTENCES ABOUT.

FIRST IT'S GOING TO BE EASY. THEN THERE MIGHT BE A SURPRISE FOR YOU. I'LL BE INTERESTED IN HOW YOU FIGURED IT OUT.

The materials are distributed.

One child's materials are displayed as directions are given.

1. ARRANGE AND COMPARE.
2. WRITE A NUMBER SENTENCE.
3. ONE BAG OF OBJECTS IS COMPARED AT A TIME.
4. BAGS ARE STUDIED IN THE ORDER GIVEN.

The girls go through the activities for the first 4 bags in quick order. Ted works very slowly and needs encouragement to write number sentences. He writes $8>7, 9>8$ and $8>5$ once he got started.

Each bag contained quantities of objects of two different colors.

The Surprise: The last two bags contain groups of equal number.

THIS IS HARD.

Writes after deliberation.

THIS IS REALLY HARD... I DON'T KNOW.

She then held up her paper & called it "WEIRD."

CAN YOU WRITE IT ANOTHER WAY TO SHOW IT'S THE SAME?

I DON'T GET THIS... 9 AND 9 IS THE SAME. 9-9? 9+9?

Puzzled

Meanwhile...

OH, OH!...OH, NO!

She hits forehead with palm as she comes across the equal groups.

WHAT DID YOU TELL ME ABOUT THESE BLOCKS?

THEY'RE THE SAME NUMBER.

YEAH, THAT'S THE PROBLEM.

Girls look puzzled.

WHAT COULD YOU WRITE DOWN TO SHOW THAT THEY'RE SAME.

9 IS THE SAME AS 9?

9 EQUAL 9?

Gloria   Cynthia

---

9 EQUALS 9.   9 EQUALS 9.

The girls first look at each other in puzzlement, then in agreement.

They quickly write down two equality sentences.

$$9 = 9 \text{ and } 4 = 4$$

The girls begin to put away the materials

Since they were finished early, they were introduced to the game of "HANDFULS" a variation of the preceding activity.
Comparisons are generated by pairs of children as each picks up a handful of objects. Number sentences are recorded at each turn.

Meanwhile...

I DON'T UNDERSTAND THIS SAME KIND.

Ted

Ted comes upon the equal groupings.

HOW CAN YOU WRITE THAT FOUR AND FOUR ARE THE SAME?

4 PLUS 4 ARE THE SAME.

IS THERE ANOTHER WAY?... (<) motion

IS ONE GROUP LARGER?...   NO.

IS THERE A SIGN TO SHOW "THE SAME"?...

I DON'T KNOW.

COULD YOU GIRLS STOP FOR A MINUTE?... HERE'S WHAT ONE OF THE GIRLS PUT DOWN. HOW COME YOU WROTE IT THAT WAY?

TWO 4'S IN A SENTENCE IS 4 EQUALS 4.

4 EQUALS 4 - THE SAME AS 4 - BECAUSE 4 ISN'T A BIGGER NUMBER. 4 EQUALS THE SAME THING AS 4.

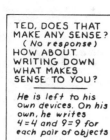

TED, DOES THAT MAKE ANY SENSE? (*No response*) HOW ABOUT WRITING DOWN WHAT MAKES SENSE TO YOU?

He is left to his own devices. On his own, he writes 4 = 4 and 9 = 9 for each pair of objects.

A summary discussion follows.

WHAT IS THE SAME ABOUT ALL OF THESE?... NOTICE ANYTHING THAT'S THE SAME ABOUT ALL OF THESE?

ALL OF THESE ARE THE SAME EXCEPT THIS ONE. ALL 3 HAVE GREATER...

THEY ALL HAVE SOMETHING IN THE MIDDLE.

Shakes head

| 8 < 6 | L < 8 | 6 = 6 | S < 8 |

(*The number sentences are displayed*)

SO, DO NUMBER SENTENCES NEED A "+" OR "−" SIGN? ...TED?

NOT THESE ONES.

Only the girls had their hands raised. The question was directed to Ted to bring him into the discussion.

THE LAST TIME WE TALKED (*interview*) ALL OF YOU WEREN'T SURE IF THIS KIND OF NUMBER SENTENCE WAS OK. WE'VE BEEN DOING SOMETHING WHERE THIS KIND OF NUMBER SENTENCE IS ALRIGHT.

Notice the following during this opening episode:

1. *The variable rate of operation among the children.* Each child began to grapple with the dilemma of encoding the comparison of equal-sized groups at different times.

2. *The value of social interaction in trying out new ideas.* The girls interacted spontaneously with each other, whereas the interaction with the boy was initiated by the teacher.

3. *The role of physical materials in confronting children's existing notions of equality and its representation.* The materials provided a novel physical context for extending their existing notions.

4. *The value of compromise solutions as indicators of mental activity.* The use of both "greater than" and "less than" signs together is a creative compromise for representing equality since they "cancel each other out." Although not a conventional form, it does reflect some underlying logic.

The children appear to have invented a novel form of the equality sentence reflecting a relational meaning. Will they use this form spontaneously in other contexts?

Episode 2: Partitioning Patterns and Their Written Representations (6 = 2 + 2 + 2, 6 = 5 + 1)

Children tend to view addition in terms of joining action. It is also possible to view addition as a partitioning of a number. When a row of objects is partitioned physically or mentally, no new objects are added; the row is merely subdivided. A row of six objects can be considered as groups of four and two, five and one, and so on. Rather than a joining action, addition is represented in static relationship. If rows of six objects are partitioned to show "different names for 6," perhaps children will be less inclined to see 6 as "the answer" and more inclined to start the equality sentence with it. Since partition into many subgroups is possible, perhaps children will accept equality sentences with more than one operational sign—for example, 6 = 2 + 2 + 2. The following patterns were used to induce such forms.

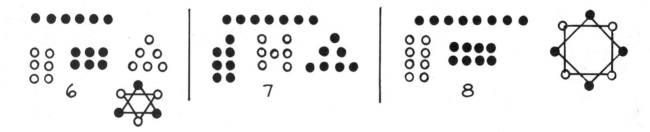

The illustrated episode that follows is actually composed of excerpts from activities that took place over several days.

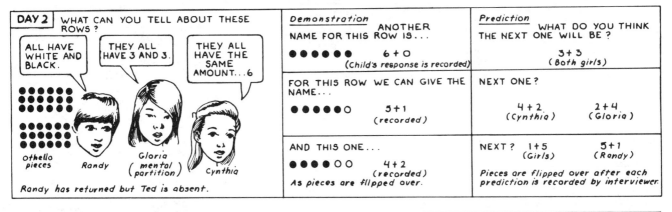

IS THERE ANOTHER WAY TO WRITE IT?

THIS SHOULD BE 3 + 3.

YOU COULD SCRAMBLE THE NUMBERS UP.

Having missed the first session, he objects to 1+2+3 or the number of pluses.   *In defense of 1+2+3 or 3+1+2 format.*

Interruption—Randy is called out.

ORAL STATEMENTS

Since the children had demonstrated some flexibility in accepting oral statements of equality in the pre-assessment, they were exposed to some at this point to see if a connection could be made between oral statement and physical context.

"6 IS THE SAME AS 4 + 2."
"6 IS ANOTHER NAME FOR 5+1."
"6 EQUALS 1 + 5."

All statements are accepted.

WRITTEN STATEMENTS

HOW ABOUT THIS ONE?

IT COULD BE RIGHT. YOU COULD READ IT AS 4 + 2 = 6.

IT'S BACKWARDS... THAT'S NOT THE WAY WE READ IT.

(objecting)

WHAT DID YOU TELL ME ABOUT EACH OF THESE ROWS BEFORE WE STARTED?

EVERY SINGLE ONE OF THEM HAD 6.

OKAY, WE ALREADY KNOW THAT EACH ONE HAS 6. SO, IS 6 THE ANSWER?

YEAH.

Cynthia   Gloria

WE KNOW THEY'RE ALL 6. COULD WE START OFF BY PUTTING A 6 HERE?

NO, IT'S BACKWARDS. THAT'S NOT THE WAY WE READ.

OH, I KNOW. LIKE IT'S . . . YOU DON'T HAVE TO READ IT THAT WAY IT'S SUPPOSED TO BE READ. YOU COULD READ IT THE OTHER WAY TOO IF YOU WANT. AND IF IT WAS THE WRONG WAY AND YOU READ IT THAT WAY . . . IF IT WAS THE SAME THING AS IF YOU READ IT THE OTHER WAY THEN IT WOULDN'T MATTER . . . THE OPPOSITE WAY OR THIS WAY.

Cynthia seems to work through some contradictions.

SO DO YOU THINK THEY ARE THE SAME THING?

YES.

'IT DOESN'T MATTER WHICH WAY YOU WRITE IT, 4 + 2 = 6, 2 + 4 = 6.

6 = 4 + 2   4 + 2 = 6

Notice that both girls agree to different questions.

DEMONSTRATION

"WATCH WHAT HAPPENS AND TELL ME IF IT'S OK."

This device was prepared in advance awaiting the critical moment for introduction. When it is rotated, the cards are free to maintain a vertical position.

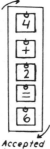

Accepted by all.

Rotation was continued

YES.   YES.   NO.

Cynthia   Gloria   Randy

Discussion continues—

SOMETIMES

YOU COULD IMAGINE THE EQUAL SIGN HERE.

Suggests that it can still be read from left to right.   *Supports Gloria's view.*

COULD I READ IT AS 6 = 2 + 4?

YEAH

YEAH, BECAUSE IF IT WAS IN YOUR MATH BOOK YOU CAN'T CHANGE YOUR MATH BOOK.

Silence

Gloria's response appears contradictory.

BUT IF IT WASN'T IN THAT MATH BOOK WOULD IT STILL BE OK?

I DON'T THINK IT'S THE RIGHT WAY.

I KNOW THAT IT'S NOT WRITTEN IN THE MATH BOOK BUT IT IS A WAY TO WRITE A NUMBER SENTENCE. *(Girls agreed)* I'D LIKE YOU TO PRACTICE.

IN THE READING BOOK YOU DON'T READ BACKWARDS.

Gloria immediately thinks of an objection (RAT ≠ TAR).

WOULD YOU WRITE NUMBER SENTENCES STARTING WITH 6 AND WRITE ABOUT ALL OF THESE (*Rows*)— JUST SO YOU REMEMBER IT IS <u>A</u> WAY OF DOING IT.

(The children clarified the task.)

The girls went to work immediately. Randy resisted getting started. Once he got started, he wrote 7 different number sentences beginning with 6.

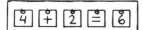

This format was imposed on the children when all were not ready to accept it. Will Randy and Gloria continue to use this format without encouragement?

READING NUMBER SENTENCES

TAKE TURNS READING YOUR NUMBER SENTENCES AND READ THEM IN A DIFFERENT WAY WITHOUT USING THE WORD "EQUALS"... MAKE UP SOME OF YOUR OWN WORDS FOR "EQUALS."

YOU MEAN 6...4+2 LIKE THAT?

Attempt to clarify

THAT NEEDS SOME WORDS IN BETWEEN.

6 IS THE SAME AS 3+2+1.

THAT'S ONE WAY TO SAY IT. SHE MADE UP HER OWN WORDS TO PUT IN THAT HAVE THE SAME MEANING AS EQUALS. CAN YOU THINK OF SOMETHING ELSE TO PUT IN?

6 EQUALS THE SAME AS 3+3. 6 IS...

6+0 IS THE SAME AS 0+6.

6 CAN BE 3+2+1.

This was not one of his written sentences.

After no further responses

HOW ABOUT 6 IS ANOTHER NAME FOR 4+2?... SO THERE ARE LOTS OF WAYS.

YOU SAID TO USE NUMBER SENTENCES AND THEY DON'T HAVE WORDS.

...WHEN YOU READ THEM YOU PUT THE WORDS IN.

Gloria points out the confusion in terms like READ and SENTENCE.

DAY 3   Ted returns, accepting the new format in context.

MAYBE ONLY FOR THOSE IF YOU'RE DOING IT THAT WAY.

6 = 4 + 2

3 children now accept the format. Randy still disagrees.

In reading a number sentence and substituting other words he reverses the order.

5+1 IS 6.

1 + 5 = 6

Although open to the possibility of 6=3+1, he is still not certain.

In the days following, the focus shifted to partitioning of other numbers, like 7 and 8, as described above.

DAY 7   NAMES FOR EIGHT

Demonstration.

An acetate sheet is placed over an outlined region. Eight objects are lined up in one half of that region. The row of 8 objects is partitioned as the clear acetate sheet beneath it is pulled over a region outlined on paper.

All children are present.

NOW THINK OF A NAME FOR 8.

8+1.   7+1.   BECAUSE THERE'S 7 HERE PLUS ONE AND ALL OF THESE ARE ARE 8. WHERE'S THE OTHER ONE?

Gloria   Cynthia

FIRST IT WAS IN THERE AND YOU MOVED IT.

DID I PUT ANY MORE IN?

NO.

WHAT WOULD YOU CALL IT, TED, 7+1 OR 8+1?

7+1

OH YEAH IT IS 7+1.

Gloria changes her view.

Ted

FROM NOW ON, WOULD YOU MAKE UP A NUMBER SENTENCE FOR 8 AND NOT USE THE WORD EQUALS.

I HAVE NO NAMES FOR THAT.

Randy

I THINK IT'S 6+2....8 IS THE SAME AS 6+2.

5+3 IS 8.

8 IS ANOTHER NAME FOR 3+5.

8 IS THE SAME AS 3+5 OR 8 IS THE SAME AS 5+3.

WHAT DO YOU WANT TO SAY ABOUT THAT LINE?

THERE'S A LINE SEPARATING...

SOMETIMES YOU COULD FORGET ABOUT THE LINE AND MAKE UP YOUR OWN PROBLEM.

171

SOMETHING ELSE, LIKE GLORIA SAID... YOU COULD JUST FORGET ABOUT THE LINE AND THINK THAT IT'S A PLUS. LIKE THERE'S A 4 ON THIS SIDE. IT'S A PLUS IN THE MIDDLE OF THE LINE.

SO, COULD WE PUT A PLUS THERE EVEN THOUGH WE'RE NOT PUTTING ANY MORE PIECES ON THERE?

YOU DON'T HAVE TO ADD ANY MORE NUMBERS (Pieces). WHEN YOU PUSH IT LIKE THAT (Demonstrating), YOU STILL HAVE 2 NUMBERS TO ADD TOGETHER.

YOU KNOW THIS IS CRAZY, EVERYTHING THEY SAY, I DON'T UNDERSTAND.

Randy protests the difficulty.

Having prior knowledge of Randy's abilities in mathematics, his protestations were not viewed with alarm. 5

DAY 6 — *NAMES FOR PARTS OF EIGHT*

NOW I'D LIKE TO SWITCH OUR FOCUS. LET'S LOOK AT NAMES FOR PARTS OF EIGHT.

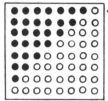

HERE, IF WE LOOK AT JUST BLACKS, WE HAVE 7 BLACKS. WHAT'S A NAME FOR 7 AS A PART OF 8? (No response)

7 = 8 - 1 (written)

Different names for 8 were studied in preceding activity.

BUT YOU SAID PARTS OF 8.

YES, 7 IS PART OF THE WHOLE ROW.

IF YOU TAKE AWAY FROM THE 8 IT WOULD BE THE SAME AS 7 OR IS THE ANSWER SUPPOSED TO BE 7 INSTEAD OF 8?

Attempting to clarify

WE'RE LOOKING AT THE BLACK PARTS NOW AND LOOKING FOR ANOTHER NAME FOR EACH. NOW IN THE ROW WE HAVE 6 BLACK ONES. AS PART OF THE WHOLE ROW OF 8, WHAT'S ANOTHER NAME FOR 6?

4 + 2 IS THE SAME AS 6.

Child's perspective

He renames 6 in another way.

THAT'S TRUE AND WE'RE LOOKING AT IT AS PART OF 8. SO COULD YOU ADD AN 8 TO YOUR NUMBER SENTENCE?

OK.   8 - 2 IS 6.

Ted.

SO ANOTHER NAME FOR 6 IS 8 - 2. (Written) NOW, HOW ABOUT THESE 5 AS PART OF THE 8?

I CAN'T THINK OF ONE TODAY.

OK, I KNOW ...AH, IT... WELL... 5 EQUALS 8 - 3.

Randy   Cynthia

HOW ABOUT A NAME FOR THE 4 BLACK ONES AS PART OF THE 8?

8... 4... 8 TAKE AWAY ...8 - 4?

Gloria

Ted was unable to respond earlier.

SO, 4 IS ANOTHER NAME FOR 8 - 4. (written).

NOW IT'S TIME FOR YOU TO WRITE. I'D LIKE YOU TO WRITE THE REST OF THE NAMES FOR 8 AND THE OTHER NAMES FOR PARTS OF 8.

Names for 8	Names for parts of 8
8 = 7 + 1	7 = 8 - 1
8 = 6 + 2	6 = 8 - 2
8 = 5 + 3	5 = 8 - 3
8 = 4 + 4	4 = 8 - 4

Other rows in the display are covered.

The boys ignore names for eight. With some help and encouragement they write 3 = 8 - 5. Randy then writes two more N.S. correctly. Ted writes 9 - 1 = 8.* Cynthia seems to avoid the difficulty by copying the number sentences discussed previously. She eventually does get started with encouragement. Gloria works systematically writing both names for eight (first) and names for parts of eight.

The remainder of the display of pieces was uncovered.

*Unrelated to context

Gloria's Worksheet

Names for 8	Names for parts of 8
8 = 3 + 5	3 = 8 - 5
8 = 2 + 6	2 = 8 - 6
8 = 1 + 7	1 = 8 - 7
8 = 0 + 8	0 = 8 - 8

Gloria also was aware that as some numerals in the series were "shrinking", others were "getting larger."

---

5. Several months earlier, during exploratory interviews, Randy showed an ability in recognition and extension of number patterns. Furthermore, he expressed sheer delight at his achievements.

The partition context had the potential for provoking the children's reconstruction of their notion of equality.

This lengthy episode shows not only the beginning of this process but also its gradual nature. Although we need change in order to grow, we resist new ideas. Both girls accepted the idea of the new form at one moment and even justified it, but then reverted to old notions at another moment. For example, Gloria seemed to believe the relationship to be true, yet she resisted the new form of the equality sentence because it contradicted the authority of her math textbook. At another time she showed an awareness of the static relationship of addition that would support a notion of bidirectional equality. She was vacillating between concerns for meaning and for a rigid form. The girls were aware of the need for change and their vacillations were a part of the process of change prior to reaching a stable construction. By contrast, Randy was unaware of the need for change. For him, there was no problem. The new form was obviously wrong. The novel physical contexts did not appear to provoke any rethinking on his part. At one point he wrote some number sentences in the new format to please me, but he resisted that same format consistently at other times. Although Ted seemed to accept the new number sentence format for the partition context, his absences didn't allow observation of his thinking from day to day.

Given the opportunity to invent other expressions for "equals," the children were quite productive. The ensuing usage by the children took on a pattern. The girls tried out a variety of expressions, whereas the boys preferred a particular expression. Gloria's invention of the redundant expression "equals the same as" suggests a compromise between two expressions that have different meanings for her. Although I had suggested "is another name for" as an alternative expression for "equals" and the girls tried it out, this expression did not gain general acceptance.

Episode 3: Equalizing Handfuls and Their Representations (8 = 5 + 3)

Following the comparison of two handfuls of objects the children were encouraged to equalize them and to represent the process of equalization in a sentence.

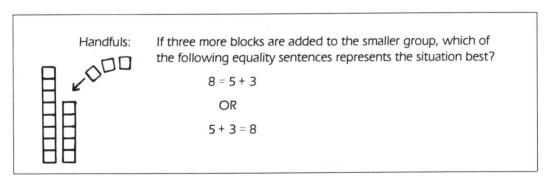

Handfuls:

If three more blocks are added to the smaller group, which of the following equality sentences represents the situation best?

$$8 = 5 + 3$$

OR

$$5 + 3 = 8$$

In presenting the task, the position of the larger set was varied to provoke the children to consider the possibility of a different notation—one that matched the spatial arrangement of the physical context. Once accepting this alternative, the children might realize that both notations were equivalent. Since children's ideas seldom proceed from being totally wrong to being totally right, the task was expected to provoke an intermediate perspective initially.

DAY 2 — EQUALIZING HANDFULS

WHO HAS MORE BLOCKS?

Gloria

Comparing handfuls

(Ted is absent.)

CYNTHIA, WOULD YOU WRITE A NUMBER SENTENCE THAT TELLS THAT SHE HAS MORE THAN I DO?

WHAT CAN WE DO TO MAKE BOTH GROUPS HAVE THE SAME NUMBER?

I KNOW. MAKE THEM THE SAME SIZE.

Randy

Six blocks are removed.

Cynthia writes.

$11 - 6 = 5$

PUT THE BLOCKS BACK. DO WE STILL HAVE 11?...WHAT'S ANOTHER WAY TO MAKE THEM THE SAME?

ADD SOME MORE.

Randy adds 6 blocks.

$11 = 6 + 11$

IS THAT RIGHT? WE HAVE 5 OVER HERE. WHAT DID WE ADD TO THE 5 TO MAKE THIS ROW THE SAME AS THAT?

WE ADDED 6, SO WRITE 5+6.

$5 + 6 = 11$

(Corrected)

Directions were given for work in pairs. (I teamed up with Randy.)

1. COMPARE HANDFULS. WRITE A NUMBER SENTENCE.

2. MAKE THE HANDFULS THE SAME. WRITE A NUMBER SENTENCE.

3. START OVER. FIND ANOTHER WAY TO MAKE THEM THE SAME. WRITE A NUMBER SENTENCE.

Randy had no difficulty in representing the equalizing situations. The girls avoided the activity by generating equal handfuls repeatedly.

DAY 3 — More Equalizing

Ted

Randy

while the girls are at a partition center.

The boys' representations

$5 < 8$ $8 > 5$

$8 - 5 = 3$ $8 - 3 = 5$

which N.S. represents the situation best?

LET'S LOOK AT WHAT YOU WROTE. HOW MANY BLOCKS DID YOU TAKE AWAY?

3, AND YOU THOUGHT IT WAS 5.

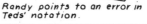

Randy points to an error in Teds' notation.

DOES THAT MAKE SENSE?...IF IT MAKES SENSE, WOULD YOU CHANGE IT?

Ted nods his head in agreement.

$8 - 3 = 5$

Ted changes his number sentence.

Another situation

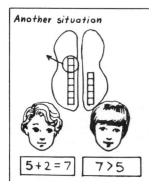

$5 + 2 = 7$ $7 > 5$

$5 + 5 = 10$ $7 - 2 = 5$

THAT'S NOT WHAT HAPPENED.

Randy protests.

COULD YOU TELL WHAT HAPPENED?

WELL...YOU TOOK 2 AWAY FROM THE 5.

Did Randy mean to say "AWAY FROM 7 LEAVING 5."?

YEAH, BUT I SEE IT AS... SEE, THIS IS 5 AND THIS IS 5. TOGETHER THEY MAKE 10.

WELL, NOT BOTH OF THEM. NOT BOTH OF THEM-10.

Ted clarifies his focus.

Teds' focus is on adding rather than comparing the groups and representing the prior equalizing.

An attempt to help Ted refocus on comparison.

TED, I'M WONDERING IF YOU COULD WRITE WHAT HAPPENED TO YOUR ROW SO THAT IT BE-CAME EQUAL TO HIS.

He rewrites his number sentence.

POINT TO THE BLOCKS THAT SHOW THE "7-2." (Pointing simultaneously to the equality sentence.)... POINT TO THE BLOCKS THAT SHOW THE FIVE.

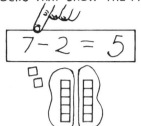

$$7 - 2 = 5$$

Randy points to different groups of 5 objects, coinciding with the equalization.

Ted points to the same group of 5 blocks as represented by both sides of equality sentence.

Other children have vacillated in indecision when asked to point.

Girls' Equality Sentences

$7 > 5$	These
$5 = 5$	sentences
	were in
$7 - 2 = 5$	agreement
$5 + 2 = 7$	with the context experienced.

Notice the spontaneous use of a new format for the equality sentence, $5 = 5$.

DAY 4

Continuing where they left off last day... They were asked what number sentence to write down as a representation of different phases of the equalizing process.

Comparing

$$7 > 5$$

Ted is absent.

Equalizing by Addition

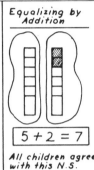

$$5 + 2 = 7$$

All children agree with this N.S.

The children decode N.S. by pointing to "7"   "5+2"

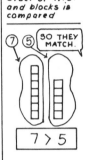

$$5 + 2 = 7$$

Order of N.S. and blocks is compared

⑦ ⑤ SO THEY MATCH.

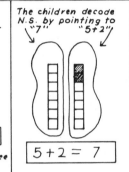

$$7 > 5$$

NOW, WHAT ABOUT THIS N.S.? YOU SAID 5+2 WAS AWAY OVER THERE ON THIS SIDE. CAN I TURN THIS N.S. AROUND AROUND?

YOU COULD PUT $7 = 7$.   I AGREE.

$$5 + 2 = 7$$

YES, AFTER WE MADE THEM EQUAL IT'S 7 AND 7.

The original question was repeated in an attempt to focus on the process of equalization rather than the final result.

7 EQUALS 2 + 5.

RANDY... IS THAT OK? ... HOW COME?

IT DOESN'T LOOK IT TO ME. IT'S... I JUST CAN'T TELL. IT'S A SECRET... WRITING IT BACKWARDS.

Another situation.

WHAT'S A N.S. FOR THIS?

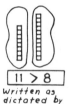

$$11 > 8$$

Written as dictated by children.

DOES THIS MATCH?

NO   8 IS LESS THAN 11.

I JUST CANT WORK ON IT. IT'S GETTING HARDER EVERY DAY

IT'S EASY.

RANDY, HOW WOULD YOU MAKE THESE EQUAL?

Randy removes 3 blocks

He is asked to leave them on the table in sight.

11-3 EQUALS 8, BUT IT HAS TO BE A DIFFERENT WAY BECAUSE 11 ISN'T HERE.

Gloria is concerned about matching the physical context. Cynthia is quiet during the discussion.

SO, HOW ELSE COULD YOU WRITE IT?

8 EQUALS 11 - 3.

$$8 = 11 - 3$$

Written as dictated.

NO, YOU DON'T HAVE TO MATCH THEM.

MAYBE YOU DON'T, BUT I WONDER IF IT WOULD BE EASIER FOR SOMEONE TO READ IF YOU MATCHED.

Three levels of interpretation were suggested by this activity, although only two of these levels had been observed directly thus far.

1. *Unidirectional—context independent.* There's only one way to write a number sentence, regardless of differing physical contexts.
2. *Bidirectional—context dependent.* There's more than one way to write a number sentence but it's the physical context that determines which way it needs to be written—the form of the sentence must match the spatial arrangement of the physical context.
3. *Bidirectional—context independent.* There's more than one way to write a number sentence and the physical context is not important. It's the relationship that counts regardless of the order of writing and the physical context.

Review this episode and determine which level of interpretation each child has given. With this information, predict whether any child will achieve the third level of interpretation. In reading the next episodes be aware of these levels.

Episode 4: Equalizing Handfuls on a Balance and its Representation (8 = 5 + 3 vs. 5 + 3 = 8)

Again, in the presentation of this task, the position of the larger set was varied to provoke the consideration of alternative forms of notation. The possibility of rotating the balance and its contents might incite a view of the equality sentence as a statement of relation regardless of context.

DAY 6 | *DEMONSTRATION*

I'VE SEEN YOU WRITE DOWN DIFFERENT NUMBER SENTENCES AND WONDER IF ONE IS BETTER THAN THE OTHER... ARE THEY BOTH JUST AS GOOD OR IS ONE BETTER?

7 > 5   5 < 7

ONE'S BETTER.

BECAUSE 7 IS MORE THAN 5.

I WOULD SAY BOTH OF THEM. IT'S THE SAME BECAUSE THE 7 — THE BIG PART'S FACING IT AND THE LITTLER PART IS FACING THE 5. AND HERE THE BIG PART, ETC.

(5 < 7) I THINK THIS ONE BECAUSE 5 (*blocks*) IS OVER HERE AND 7 (*blocks*) IS OVER THERE.

SO YOU'RE SAYING THAT THE 5 (*numeral*) MATCHES THE 5 BLOCKS HERE AND THEN 7 (*numeral*) MATCHES THE 7 BLOCKS HERE.

Reflecting Gloria's response.

I DISAGREE. I THINK THEY'RE BOTH RIGHT. CYNTHIA'S RIGHT. THEY'RE OPPOSITE, ONLY THE BIG SIDE IS TO THE 7 AND THE LITTLE SIDE IS TO THE 5.

Ted restates Cynthia's position.

Further Provocation

COULD BOTH N.S. FIT BOTH BALANCES OR DOES ONE N.S. FIT BETTER THAN ANOTHER?

5 < 7

7 > 5

Gloria repeats her "matching argument."

Randy's view:

THIS ONE GOES WITH THAT AND THAT.

THIS ONE GOES WITH THAT

7 > 5

Turns 7 > 5 / 2 < 7 *upside down.*

AND THAT.

Cynthia continues her argument.

I THINK THEY'RE BOTH THE SAME. BECAUSE IT DOESN'T MATTER IF BOTH ARE ON THE SAME SIDE. IT MATTERS IF IT'S THE SAME AMOUNT.

For Cynthia, the mathematical relationship overides the spatial relationship.

SO THERE'S MORE THAN ONE WAY TO LOOK AT THESE AND EVERYONE DOESN'T AGREE. BUT YOU ALL HAVE INTERESTING WAYS OF LOOKING AT THE BLOCKS AND NUMBERS.

Some children appear to be generalizing inequality relationships beyond the order of the physical context.

Will they be able to extend their thinking to equality situations?

EQUALIZATION

NOW I'M GOING TO WRITE THIS IN TWO DIFFERENT WAYS.

Two blocks are added.

5 + 2 = 7   7 = 5 + 2

THERE'S ONE MORE WAY. 7 EQUALS 7.

7 = 7

Written at Gloria's suggestion.

ARE ALL OF THESE N.S. TELLING ABOUT THE BALANCE?

YEAH.

Cynthia — 7 = 5 + 2 / 5 + 2 = 7   *Gloria* — 7 = 7

DOES ANYONE WANT TO SAY NO?

I GUESS I WILL. I DON'T LIKE IT WRITTEN BACKWARDS.

Randy — 7 = 5 + 2

WHAT DO YOU THINK, TED?

I THINK THAT 7 = 5 + 2 IS KIND OF BETTER THAN 7 = 7.

Ted

On further questioning Ted agrees that 7 = 7 is still OK.

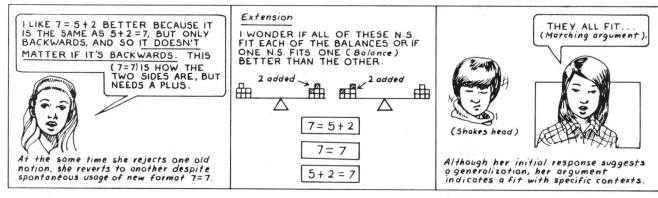

Cynthia appears to have achieved the highest level of understanding, in which equality relationships are generalized beyond the physical contexts. Gloria consistently accepts or writes number sentences in either direction but ties them to specific contexts. Ted appears to be in transition between two levels. He indicates a generalization beyond materials for inequalities but is tied to specific contexts for equalities. Randy demonstrates similar insight—generalization beyond the physical context for inequalities—yet he views equality sentences at the lowest level—one way, regardless of context.[6]

Episode 5: Confronting Authority

Two interesting confrontations with authority occurred on the seventh day. Earlier episodes have illustrated Randy's consistent resistance to an alternative format for equality sentences as well as Gloria's repeated concern about the absence of the novel formats in her math textbook. Some important questions were raised by the group's interaction on these issues.

6. Based on the performance of these children, a separate consideration of levels of understanding for inequalities and equalities is justified.

BUT SOMETIMES HE MAKES MISTAKES.

HE'S NOT PERFECT.

Shakes head

NOBODY'S PERFECT... YEAH,... GOD, NOBODY ELSE IS PERFECT.

Disturbed

The girls present a united front.

The last word —

THE KING OF AMERICA!

Randy is a recent arrival to America. He was born in a country having a "king".

At this point the discussion is refocused on partitioning.

B. THE TEXTBOOK ISSUE

The children were reacting to written statements of equality immediately following acceptance of similar oral statements.

TELL ME IF IT'S OK.

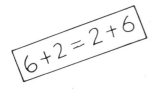
$$6 + 2 = 2 + 6$$

ARE YOU WRITING IT AS A PROBLEM? WHERE'D YOU PUT THE ANSWER DOWN?

IT'S THE SAME QUESTION AS BEFORE AND WE ALL THOUGHT IT WAS RIGHT. 6+2 IS THE SAME AS 2+6 BECAUSE IT'S JUST BACKWARDS... THAT THEY'RE TURNED AROUND.

Disturbed by format.

Connection is made between oral and written statements.

Encouraging Interaction on Different Views

WHAT DO YOU THINK GLORIA?

I AGREE, BUT I DON'T AGREE. IF IT WAS A PROBLEM IN THE MATH BOOK AND YOU PUT 6+2 IN ANSWER. THE ANSWER WOULD BE RIGHT BUT YOU WOULDN'T WRITE IT LIKE THAT.

Looks disturbed

Gloria accepts the relationships but still resists the novel form.

HOW COME IT COULD BE RIGHT?

BECAUSE THE ANSWER OF IT IS THE SAME THING. BECAUSE 6+2 IS 8, 2+6 IS 8. THE ANSWER IS THE SAME.

Other Opinions Solicited

I THINK I AGREE.

I DON'T.

Randy shrugs when asked to expand on his response.

Textbook Concern

YOU COULDN'T WRITE LIKE THAT IN THE MATH BOOK.

Randy's new watch diverts discussion.

HERE GOES ANOTHER...

WELL, I SAY IT'S RIGHT.

I AGREE.

BECAUSE 6 IS ONLY 1 AFTER 5 AND 3 IS ONE OVER 2... AND BOTH MAKE 8.

The question was shifted to the girls when I noticed Ted wasn't finished. Focusing back on Ted, we heard the completion of his thoughts.

JUST A SECOND, TED HAS SOMETHING ELSE.

ONLY THEY'RE DIFFERENT PROBLEMS FROM EACH OTHER BUT STILL CONTAIN 8.

THAT'S WHAT I WAS GOING TO SAY.

I DISAGREE.

Disgruntled

TELL US WHY.

IT'S NOT IN THE MATH BOOK.

IT DOESN'T MATTER IF IT WASN'T IN THE MATH BOOK.

Cynthia turns abruptly to react.

IF IT WASN'T A PROBLEM LIKE THAT, I WOULD AGREE.

SO YOU'RE REALLY NOT TOO SURE.

Randy agrees with Teds' explanation.

ONE LAST THING FROM CYNTHIA AND WE'LL MOVE TO SOMETHING ELSE.

IT DOESN'T MATTER IF IT'S IN THE MATH BOOK OR NOT, AS LONG AS IT'S RIGHT. IT DOESN'T MATTER IF IT'S IN THE MATH BOOK, JUST AS LONG AS WHAT YOU THINK IF IT'S RIGHT.

Cynthia argues for placing trust in your own thinking, and repeats herself for emphasis.

Gloria continues her argument with Cynthia as the activity is drawn to a close.

OK, WE'LL BE LOOKING AT MORE OF THESE.

The sequence of activities and subsequent discussions provoked a seven-and-a-half-year-old girl to not only generalize beyond the physical context, but also to confront the authority of the textbook. In this situation she achieved intellectual autonomy. Whereas Cynthia's rational thought superceded the authority of the her second-grade math book, Gloria's thinking was influenced by the book's limited content and Randy's thoughts were influenced by parental authority. Although Gloria could justify the correctness of the novel equality sentence, she was unable to liberate her thinking from concerns for authority, thus vacillating from one focus to the next.

What is the source of the children's demonstrated resistance to accepting novel forms of the equality sentence?

1. Lack of sufficient logical thinking abilities
2. Deference to the authority of the textbook/teacher/parent
3. A tendency to hold on to the familiar
4. Lack of confidence in one's logical thinking abilities
5. All of the above

Episode 6: Classifying, Coordinating, and Representing Variables (6 + 2 = 3 + 5)

The following task (Sinclair, 1973) provided the physical context for a static relationship with whole-parts considerations that could be encoded in a novel form of the equality sentence 6 + 2 = 3 + 5.

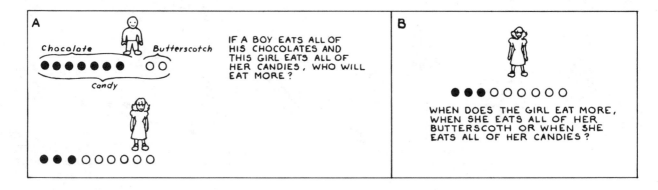

A

Chocolate   Butterscotch

Candy

IF A BOY EATS ALL OF HIS CHOCOLATES AND THIS GIRL EATS ALL OF HER CANDIES, WHO WILL EAT MORE?

B

WHEN DOES THE GIRL EAT MORE, WHEN SHE EATS ALL OF HER BUTTERSCOTH OR WHEN SHE EATS ALL OF HER CANDIES?

Encoding the distribution of chocolate and butterscotch candy between the two dolls requires the simultaneous coordination of relations and conservation of the whole. To accomplish this, each subset must be seen in relation to the other and both must be seen in relation to the sum.

Prior to the encoding phase of the task, the children were asked to deal with the whole-parts considerations indicated. They experienced difficulty in co-ordinating all the variables. The children often focused on the parts, having lost sight of the whole. Cynthia had a complex and lengthy argument for such a focus. In her persistence, she swayed the thinking of the other children. Vacillations typified the discussions taking place over three days. The illustrations that follow show the children's interactions on the next task requiring written representation of whole-part considerations.

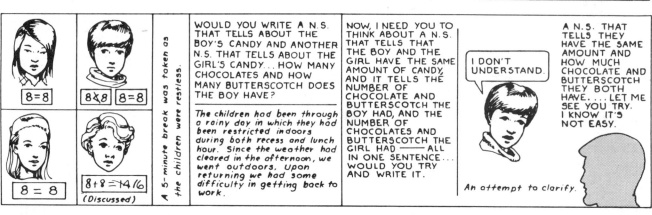

Row 1

The boys lose sight of the physical context.

I'M WONDERING IF YOU'RE WRITING ABOUT THE CANDIES. I SEE 4+4 OVER HERE. WHAT'S ON THE OTHER SIDE?

WOULD YOU WRITE ABOUT THAT?

2 PLUS 6.

Rewritten as
4+4 = 2+6

ARE YOU WRITING ABOUT THE CANDIES? ... WOULD YOU POINT TO THE 5 PLUS 3 IN THE CANDIES?

8=5+3=5×3=8

Notice his extension of format.

Why are the children having such a difficult time coordinating the variables and representing the physical context?

DAY 8

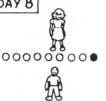

O O O O O O O O ●

● ● ● ● ● O O O O

RANDY, GIVE THE BOY THE SAME NUMBER OF CANDIES BUT GIVE HIM MORE CHOCOLATE AT THE SAME TIME. (*The result above was achieved after trial and error.*)

WOULD EVERYONE WRITE ONE N.S. THAT TELLS THAT THEY HAVE THE SAME AMOUNT OF CANDY.

9 = 9   8+1 = 1+8   8+1 = 9

9 = 1+8

HOW ARE YOUR NUMBER SENTENCES DIFFERENT?

IT TELLS THAT THEY HAVE THE SAME NUMBER OF CANDIES AND TELLS THAT 8 AND 1 IS THE SAME AS 1+8, ONLY BACKWARDS.

Cynthia appears to have lost sight of the physical context after focusing on one row.

An attempt to refocus.

WHAT DOES THE 8 STAND FOR, FOR THE GIRL? FOR THE BOY?

CARMELS — CHOCOLATES.

Row 3

DO YOU WANT TO SAY SOMETHING ABOUT YOURS, GLORIA? ... HOW DOES IT SHOW THAT THE DOLLS HAVE "THE SAME?"

TELLS THAT IT HAS THE SAME.

Following the girls' responses I decided to alter the activity.

Attempt to focus on context

I'M GOING TO WRITE SOME NUMBER SENTENCES AND ASK YOU TO EXPLAIN WHAT THE NUMBERS TELL YOU ABOUT THE CANDY.

IT'S ABOUT THEY BOTH HAVE THE SAME AMOUNT.

AND THAT'S THE EASIEST WAY TO TELL THAT.

9 = 9

WHAT DOES THAT (N.S.) TELL YOU?

IT TELLS YOU THAT IF YOU ADD 5 AND 4 TOGETHER YOU GET 9.

WHAT DOES IT TELL YOU ABOUT THE CANDY?

5 CHOCOLATES AND 4 MORE, THAT'S 9.

5+4 = 9

Row 4

WOULD YOU TALK ABOUT THIS N.S. AND WHAT IT TELLS ABOUT THE CANDIES.

IT TELLS NOTHING.

IT TELLS ABOUT THE N.S. BECAUSE THIS HAS 8 (*Pointing*) AND THIS HAS 1. IT DOESN'T TELL ME ABOUT THIS (*Boy's candy*) BECAUSE IT DOESN'T IT HAVE 8 AND 1, BUT THEY EQUAL THE SAME THING.

1+8 = 9

Another number sentence

IT'S THE HARDEST THING AROUND.

THEY BOTH HAVE THE SAME. THERE'S 9 FOR BOTH OF THEM. 5+4 IS THE SAME AS 1+8 SO YOU CAN SAY THEY BOTH HAVE 9.

1+8 = 5+4

Randy was encouraged to comment.

IT SHOWS IT A DIFFERENT WAY. IT SHOWS WHICH IS THIS WAY AND WHICH IS THAT WAY.

BUT IT DOESN'T HAVE TO BE, BUT IF YOU WANT IT TO BE LIKE IT IS ON THERE.

Concerned about matching context.

Changes arrangement of candies for an exact match.

LOOK AT EVERY ONE OF THESE SENTENCES AND SEE IF YOU CAN FIND SOME CONNECTION... HOW ARE ALL OF THESE RELATED OR CONNECTED?

$$9 = 9$$

$$1 + 8 = 9$$

$$5 + 4 = 9$$

$$1 + 8 = 5 + 4$$

THEY'RE ALL RELATED BECAUSE THEY ALL HAVE AN EQUALS SIGN. THEY ALL EQUAL 9. THEY'RE ALL IN THE SAME FAMILY BECAUSE THEY ALL EQUAL THE SAME.

"SAME AMOUNT—MORE CHOCOLATE."

Gloria miscounts as she gives the girl more chocolates, then self-corrects.

WRITE AS MANY N.S. AS YOU CAN ABOUT THESE CANDIES TO TELL HOW THEY'RE THE SAME.

Later —

CAN YOU WRITE IT IN ANOTHER WAY THAT SHOWS THEY HAVE THE SAME AMOUNT OF CANDY AND SHOW HOW MANY BUTTER-SCOTCH AND CHOCOLATES THEY HAVE?

DO ALL OF THESE EQUAL 9? ... 4 + 5, WHAT DOES THAT EQUAL TO (Pointing to candies)?

IS THERE SOMETHING ELSE YOU CAN PUT IN HERE?

9 AND 3

OH, I MADE A MISTAKE I GOT TO DO IT OVER.

His correction:
$$4 + 5 + 3 + 6 = 9$$

Randy's focus shifts from joining to relating, but he is unable to make all the necessary coordinations.

Gloria wrote ten number sentences equaling 9 but ignoring the context.

ARE THEY ALL ABOUT THE CANDIES?

I HAVEN'T DONE EXACTLY WHAT YOU WANTED.

Cynthia's Paper

$$6 + 3 = 5 + 4$$
$$3 + 6 = 4 + 5$$
$$5 + 4 = 4 + 5$$ —Unrelated to context.
$$9 = 4 + 5$$
$$9 = 5 + 4$$
$$9 = 3 + 6$$
$$9 = 6 + 3$$

Comparing Responses:

WOULD YOU LOOK AT THIS NUMBER SENTENCE THAT CYNTHIA HAS AND THE ONES THAT RANDY HAS.

$$6 + 3 = 5 + 4$$
$$4 + 5 + 3 + 6 = 18$$
$$4 + 5 + 3 + 6 = 9$$

I WONDER IF THEY'RE SAYING THE SAME THING. ...DO THEY BOTH TELL YOU THAT THE BOY AND THE GIRL HAVE THE SAME AMOUNT OF CANDY?

THIS DOESN'T TELL IT RIGHT BECAUSE HE SHOULD HAVE DONE... HE COULD HAVE LIKE THIS... 4 + 5 = 3 + 6. HE DIDN'T HAVE TO WRITE THAT (18) BECAUSE IT DOESN'T TELL ANYTHING ABOUT IT'S THE SAME THING, ETC.

Cynthia points to Randy's earlier attempt.

$$4 + 5 + 3 + 6 = 18$$

WHAT DO YOU THINK, RANDY?

I DON'T KNOW.

WHAT YOU HAVE TELLS ABOUT THE CANDIES THE BOY AND GIRL HAVE ALTOGETHER, BUT WE'RE WONDERING IF THAT TELLS THAT THEY HAD THE SAME AMOUNT?

Protesting

WE ALL DID THIS ONE WITH THE CANDIES.

SUPPOSE WE WRITE AN EQUALS SIGN HERE WHAT ELSE COULD WE WRITE NEXT?

9.

18

Ted's spontaneous written response (5 + 3 = 3 + 5 = 8) on DAY 7 prompted a spontaneous introduction of the preceding question at this moment.

In episode 6 of the task required simultaneous consideration of the relationship between two wholes (boy's candy and girl's candy) and the composition of their parts (subgroups of chocolate and butterscotch). Successful completion of the task required coordination of parts and wholes as represented in the equation $8 + 1 = 4 + 5$. A comparison of the boy's and girl's candies indicated that one of the subgroups was larger while the other subgroup was smaller. The two inequalities 8 (boy) > 4 (girl) and 1 (boy) < 5 (girl) compensated each other, with the increase in one subgroup being compensated by a decrease in the other. Children's written representations such as $8 = 2 + 6 + 4$ and $16 = 6 + 2 + 4 + 4$ illustrate the children's difficulty in dealing with the coordination demands of the task.

Episode 7: Encoding and Decoding Equality Sentences ($4 + 3 = 6 + 1$)

The following task using blocks on a balance presented less difficulty for the children. At the same time it further revealed how the children interpreted equality sentences of the format $4 + 3 = 6 + 1$.

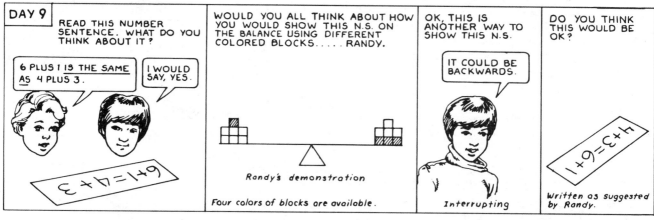

Consider why this task appears less demanding than the preceding one with candies.

Episode 8: Extending Equality Sentences

On day 12, with only the two girls present, I returned to a topic of controversy that had arisen spontaneously on days 8 and 9 but had remained unresolved.

EARLIER, YOU AGREED ON A N.S. LIKE THIS —

$$5 + 3 = 6 + 2.$$

THEN, WHEN I PUT AN EQUALS SIGN AT THE END, YOU DISAGREED. ONE OF YOU THOUGHT THIS WAS RIGHT,

$$5 + 3 = 6 + 2 = 8,$$

AND THE OTHER THOUGHT THIS WAS RIGHT,

$$5 + 3 = 6 + 2 = 16.$$

THIS IS TRUE, BUT $(5+3 = 6+2 = 16)$ IT'S NOT THE REAL KIND OF THING. THIS IS NOT TRUE $(5+3 = 6+2 = 8)$.

I AGREED ON THIS ONE $(5+3 = 6+2 = 8)$. IT ALREADY HAS AN EQUALS AND $6+2$ IS THE ANSWER ALREADY. BUT IT HAS ANOTHER EQUALS SO YOU HAVE TO GIVE THIS PROBLEM AN ANSWER.

Cynthia *Gloria*

BUT THEY'RE ALL TOGETHER. EVEN THOUGH IT'S LIKE THIS ALTOGETHER.

HOW COME THEY'RE TOGETHER?

BECAUSE 5 PLUS 3 EQUALS 6 PLUS 2 EQUALS 8...NO.

Cynthia's focus is a compromise of both relating and joining (conflicting views). Although Gloria extends the relationship, resulting in a novel form, she uses "problem-answer" in her argument.

IF YOU WANT TO DO THE 16, YOU TAKE OFF THE MIDDLE EQUALS SIGN.

WELL, I THINK IT'S RIGHT. IT'S TRUE, ALL OF THESE ALTOGETHER EQUAL 16.

EVEN WHEN THE EQUALS SIGN IS THERE *(Points)*?

UHHUH.

I DISAGREE.

BECAUSE ALL OF THESE TOGETHER ARE 16. ALL OF THESE TOGETHER AREN'T 8.

I DISAGREE.

YOU DISAGREE WITH ALL OF THEM.

The girls begin a contest of disagreement.

NOW ALL OF THESE THINGS WE LOOKED AT BEFORE ARE NAMES FOR 8. SO IF THIS IS A NAME FOR 8 AND THAT'S A NAME FOR 8. AREN'T THEY BOTH EQUAL TO 8?

NAMES FOR 8
$8 = 7 + 1$
$8 = 6 + 2$
$8 = 5 + 3$
$8 = 4 + 4$

WELL.

I AGREE AND YOU DISAGREE.

(Distracted by Gloria's response.)

IF THAT WAS TRUE WHAT ELSE COULD WE PUT HERE AND STILL HAVE IT BE TRUE?

$$5 + 3 = 6 + 2 = \boxed{}$$

After a long pause.

$4 + 4$.

I DISAGREE... 1 PLUS 7. I AGREE, BUT THERE COULD BE ALL KINDS OF THINGS. *(clarifies)*

WE'D HAVE TO DO THAT FOREVER!

Cynthia motions as if to write.

OK LET'S SEE THE LONGEST N.S. YOU CAN WRITE ABOUT 8.

OH GOD, I NEED MORE PAPER.

The girls were given 2 legal-sized sheets.

CAN YOU START WITH 8?

YEAH.

Cynthia points to the chart. The chart was removed once the girls got started.

Gloria's number sentence:

$$8 = 4+4 = 6+2 = 7+1 = 5+3 = 4+4 = 8+0 = 2+3+3 = 1+4+3 = 2+2+2+2 = 3+2+1+2$$

JUST ANOTHER MINUTE.

OH NO, JUST WHEN IT WHEN IT WAS ABOUT TO BE FUN.

Gloria's reaction to time limit.

I COULD MAKE IT LONGER.

For the first time, I was unable to contain my sheer delight with a child's response (the extended sentence) and went beyond my usual friendly, but neutral manner.

ISN'T THAT WILD!

Earlier, Gloria had vacillated between acceptance and rejection of the format $6 + 2 = 5 + 3$ owing to the pull of authority of the textbook. After experiencing physical contexts for the sentence format she was able to accept it when it precisely matched the context. In episode 8, in the absence of a physical context, she totally liberated her thinking from the confines of her math textbook with an extension of the number sentence based on the notion of different names for the same number. Despite showing this major reorganization in her thinking about equality sentences reflecting considerable mathematical sophistication, Gloria simultaneously demonstrated a limitation in her understanding of equality. When requested to substitute her own words for "equals" when reading the equality sentence, she was unable to do so despite prior facility. It's as if the coordination of one area of the equality

concept unsettles another area prior to the achievement of an overall coordination and a stable understanding. Following this experience, Gloria expressed no further concerns for the authority of the textbook.

Episode 9: Simultaneous Equalizing of Groups and Its Representation ($13 - 3 = 7 + 3$)

Another physical context (Piaget, 1965) provided the children with an opportunity to write yet another novel equality sentence—one containing two different operational signs. Now, not only must the quantities be coordinated, but also the inverse operations.

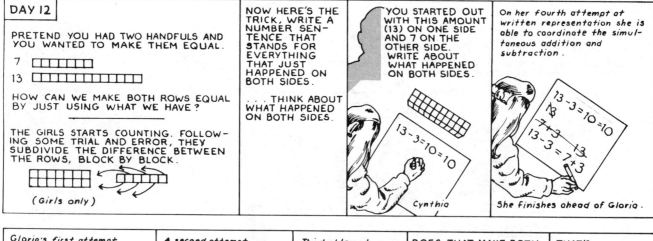

Both girls needed more than one attempt to coordinate both the quantities and the inverse operations.

Day 12 marked the last day of teaching/learning activities. The episodes included in this chapter describe the highlights of the children's interactions with novel contexts for testing their existing notions of equality and its representation. A couple of episodes have been left out:

1. Searching for patterns in a series of equality sentences (days 4 and 8).
2. Solving and representing number story problem formats revealing different facets of equality (days 9 to 12).

Both of these areas will be illustrated in the following chapter.

The Delayed Postassessment

Since we've kept track of the children's responses in different contexts, we're aware of changes in their notion of equality. To gather information on the stability of these changes, I returned for a final assessment interview after a one-month delay. In addition to including novel equality sentence formats for reaction, I provided some opportunities for spontaneous writing of such formats in variations of the original contexts. Instability in the children's reconstructed notions of equality would be indicated by reversion to earlier, more primitive, unidirectional notions.

Prior to examining the children's responses on the postassessment interview, you are invited to predict the stability of each child's reconstructed notions of the meaning of the equality sign. You may want to consider additional information such as Ted's absence from half of the sessions and his invention of a "double count system" several months earlier (Chapter 4, p. 79).

---

(1)

FINISH THIS N.S. 8+4=___. WRITE AS MANY N.S. AS YOU CAN, USING THE SAME THREE NUMBERS. YOU CAN USE A "+" OR A "—" SIGN.	8 + 4 = 12 4 + 8 = 12 12 - 4 = 8 12 - 8 = 4	8 + 4 = 12 4 + 8 = 12 12 - 4 = 8	8 + 4 = 12 4 + 8 = 12 12 - 8 = 4 12 - 4 = 8	8 + 4 = 12 4 + 4 = 8 + 4 4 + 8 = 12
PROMPT: CAN YOU WRITE ANY N.S. THAT ARE NOT IN THE TEXTBOOK BUT ARE STILL OK?	12 = 8 + 4 12 = 4 + 8	NO.	I DON'T SEE THEM *(in text).* HOW COULD I WRITE THEM IF I DON'T KNOW THEM. *(Figure them out)* I ALWAYS THINK BUT I CAN'T DO IT.	NO.

The children appeared sluggish at the start of the interview. Would they have demonstrated spontaneous usage of a novel format for the equality sentence if the task had been given later in the interview?

②				
$4 + 3 = 7$ READ THIS N.S. USING YOUR OWN WORDS FOR EQUALS.	IS IS THE SAME AS COULD BE IS ALWAYS	IS IS THE SAME AS CAN BE 4 + 3 EQUALS 7 BUT IT ISN'T THE SAME AS 7 BE-CAUSE THERE'S NOT ANOTHER PROBLEM. (AS IN 3 + 4 = 6 + 1)	ANSWER IS THE SAME AS	IS THE NEXT IS THE NUMBER THAT TELLS THE ANSWER.
COULD WE WRITE THIS N.S. ANOTHER WAY ? *(oral response)*	3 PLUS 4 EQUALS 7.	3 PLUS 4 EQUALS 7.	7 EQUALS 3 PLUS 4.	3 PLUS 4 EQUALS 7. 7 MINUS 3 EQUALS 4.

③				
READ THIS N.S. TELL IF IT'S OK OR NOT. $9 = 5 + 4$ $9 = 7 + 2$	(EQUALS). OK	9 DOES EQUAL	(IS THE SAME) RIGHT. BECAUSE EVEN IF YOU DO IT BACKWARDS, IT DOESN'T MATTER.	(EQUALS) IF YOU WANTED IT THAT WAY YOU HAVE TO PUT A MINUS — $9 - 5 = 4$
$5 + 4 = 4 + 5$ $5 + 4 = 7 + 2$	(EQUALS) RIGHT, BE-CAUSE 5 + 4 = 9. $4 + 5 = 9$ JUST ANOTHER WAY. (IS THE SAME AS) RIGHT $5 + 4 = 9$ $7 + 2 = 9$	(EQUALS) RIGHT, BE-CAUSE 5 + 4 = 9 AND 7 + 2 = 9	(IS THE SAME AS) RIGHT. BECAUSE BOTH IS 9. JUST BACKWARDS. ONE'S BACKWARDS.	(IS THE SAME AS) TRUE, BECAUSE IT'S JUST MIXED AROUND, BUT IT'S THE SAME, ETC.
$9 = 9$	RIGHT.	RIGHT.	(9 THE ANSWER IS 9.) (9 IS THE SAME AS 9.) RIGHT.	SAME, BECAUSE THERE'S AN EQUALS THAT SAYS THEY'RE THE SAME.

④				
PRETEND THESE ARE COOKIES. WHAT ARE YOUR FAVORITE COOKIES ?... PRETEND YOU HAD THIS MANY COOKIES TO EAT AT SCHOOL. ON MONDAY YOU ATE ALL OF THESE IN THE MORNING AND ONLY HAD ONE LEFT FOR THE AFTERNOON. ●●●●●●●● ● ON TUESDAY YOU HAVE JUST AS MANY COOKIES BUT YOU DECIDED TO SAVE SOME FOR THE AFTERNOON. → ●●●●● ●●●● DO YOU HAVE JUST AS MANY COOKIES TO EAT ON BOTH DAYS? *(Piaget, 1965)* *(First exposure for all children.)*	· MORE ON TUESDAY. · Question repeated. · She counted both groups as the same.	SAME	· Starts by comparing mornings on both days and afternoons separately without coordinating comparisons. · When asked to focus on cookies all day Monday and all day Tuesday → 9 ON BOTH DAYS.	· DIFFERENT NUMBER. · Counts → same.
WRITE A N.S. THAT TELLS HOW YOU HAD THE SAME NUMBER OF COOKIES ON BOTH DAYS AND THAT YOU ATE A DIFFERENT NUMBER OR COOKIES IN THE MORNINGS AND AFTERNOONS.	$8 + 1 = 4 + 5$ *(Instant response)*	$6 + 3 = 9$ *Directions repeated* $8 + 1 = 4 + 5$	$9 = 9$ A.M.-P.M. Prompt: $81 / 45 /$ Sign prompt: $81 = 45$ Sign prompt: $8/1 = 4/5$ *The partition was explained as :* IT COULD SEPARATE THEM. I DIDN'T EAT 18 ON ONE DAY AND 45 ON TUESDAY.	$9 + 9 = 18$ $9 - 1 = 8$

HOW COULD YOU MAKE THESE TWO ROWS EQUAL BY USING ONLY THE BLOCKS SHOWING ON THE TABLE? (Piaget, 1965)	✓	✓	✓	✓
WRITE A N.S. THAT TELLS THAT BOTH ROWS ARE EQUAL AND WHAT HAPPENED TO THE ROWS TO MAKE THE ROWS EQUAL. (Second exposure for girls) (First exposure for boys)	$11 - 3 = 5 + 3$	$11 - 6 = 5 + 3$ Focus on 6 removed prior to subdivision. DID YOU REALLY TAKE 6 AWAY? "NO" Changing to $11 - 3 = 5 + 3$ An explaining N.S. by action on blocks.	$11 - 3 + 5 = 8 = 8$ Randy proceeded to demonstrate its meaning with with blocks but was unable to coordinate the written representation.	$4 + 4 = 5 + 3$ HOW MANY DID YOU HAVE AT THE BEGINING? "11," CAN YOU SHOW WHAT HAPPENED TO THE 11 ON ONE SIDE? $11 - 3 = 8$ $5 + 3 = 8$ Arrow explained: I PUT THE (some) 3 RIGHT THERE. (His attempt to coordinate N.S.)

The thirty-minute interview also contained a variety of number story problems as time permitted.

The Process and Products of Learning

The preceding episodes confirm the contention that the meaning of the equality sign is not an all-or-nothing affair, to be achieved instantaneously and to encompass the whole concept. This teaching/learning experiment shows that second-grade children with a capacity for at least two kinds of logical thinking can construct and grapple with notions of bidirectional equality and its representation. The continuous interplay between form and meaning, prompted by the variety of physical contexts to encode, provoked the children to extend and to modify their views.

During the course of the episodes, the children not only demonstrated novel, more sophisticated responses, but also vacillated between sophisticated and unsophisticated responses in different contexts. This *vacillation* is an early indicator of progress since it is an acknowledgement of a problem to be resolved. This is a major step from the outright rejection of novel forms of the equality sentence. It is an essential step prior to the integration of conflicting viewpoints. *Partial constructions* or intermediate levels in the construction process were also evident, such as the need to match the spatial arrangement of the physical context in the written notation. Another intermediate level was the compromise of conflicting views to produce unconventional notations like $><$, or incorrect notations like $5 + 3 = 6 + 2 = 16$. Such examples of *compromise solutions* suggest considerable mental activity on the part of the child. Thus, the illustrated episodes captured many instances in the gradual process of the children's reconstructions of the meaning of the equality sign.

Both during the episodes and during the postassessment, the following indicators of understanding equality relations were shown by the children:

1. *Acceptance, justification, and continued usage of novel equality sentence formats that were not found in their mathematics textbooks*, like $4 = 4$, $7 = 5 + 2$, $4 + 5 = 6 + 3$. All four of the children appeared to have reached this level during the intervention, although at different times. In the postassessment, Ted rejected one of the above formats and both Randy and Ted experienced difficulties in writing equality sentences requiring considerable coordination of relations, such as $8 + 1 = 4 + 5$ and $11 - 3 = 5 + 3$.

2. *Invention and continued usage of relational synonyms for "equals"*, such as "is the same as," or "is." Although all four children adopted these terms, Gloria frequently referred to "equals the same as" without apparent awareness of the redundancy. My own suggestion of "is another name for" was tried on for size but never adopted.

3. *Generalization of relationships expressed in equality sentences beyond specific physical contexts.* Although all four children gave indications of progress in this area at different times during the

teaching/learning episodes, this progress was uneven. Once the generalization was observed for the first time, it usually did not appear consistently in different contexts. However, Cynthia did give strong indications of going beyond the spatial relations of the physical contexts to more general mathematical relations in different contexts. Although the children demonstrated considerable progress, they were unable to coordinate all the aspects of equality relations and their representation within the time limitations of the study.

Another indicator of progress would be the ability to discuss equality sentences as expressions of static relations without reference to the location of an "answer." Despite their growing awareness of "=" as a relational symbol, the children often made reference to different parts of the equality sentence in terms of problem and answer. This may indicate the coexistence of conflicting perspectives in children's minds. It also may indicate the lack of alternative terminology for making reference to left-and-right-hand sides of equality sentences, or to multiple extensions of equality sentences.

Although resistance to overcoming old notions is always a prominent factor in the learning process, in this study it was combined with resistance to confronting the authority of the textbook. Unfortunately, it is not possible to separate these two variables to find out the extent to which the children's progress was impeded by prior teaching. Yet reports of success (Hendrickson, 1980; Baroody and Ginsburg, 1982) in teaching children the flexible use of a variety of forms of the equality sentence in ways that avoid conflict with prior limited teaching suggests this impediment may have been considerable. At the same time that the lack of control over prior teaching confounded the evaluation of this experiment, it served to provide glimpses of the children's developing intellectual autonomy as they confronted the authority of the textbook.

Teaching as Facilitating Reconstruction

The teaching described in this chapter involved the design of various encounters with encoding novel physical contexts to provoke a rethinking of the meaning of the equality sign. These situations served as springboards for organized and spontaneous verbal encounters in which the children shared their views. As many as three different levels of interpretation of the same task led to some lively discussions and eventual reconstructions. The awareness that other children shared viewpoints different from their own played an important role in getting the children to rethink their ideas at more sophisticated levels of understanding. The teacher's role in the verbal encounters was to highlight the contradictions in viewpoints and to encourage the controversy so essential for intellectual development. This was accomplished by juxtaposing conflicting responses and inviting commentary and justification. Although the teacher continually challenged the children's thinking, time for both action and reflection was provided. Children were encouraged to risk trying out new ideas. Feedback on these ideas from the physical contexts and from their peers was maximized, at the same time that the teacher's adult authority and evaluative feedback was minimized. Praise and criticism were replaced largely by a friendly but neutral acceptance of all responses as indicators of a child's current level of understanding. Adult authority also was minimized by redirecting authority for making evaluations back to the children. In summary, the method of teaching facilitated reconstruction of ideas by optimizing possibilities for action, reflection, interaction, and self-evaluation. This approach to teaching is discussed in many parts of the book under the label *indirect teaching*.

The teaching/learning experiment provides many springboards for discussion in this and in subsequent chapters. The next chapter on children's interpretation of word problems features further episodes with Cynthia, Gloria, Randy, and Ted. Also, Chapter 9 discusses many aspects of classroom teaching with illustrative examples taken from the teaching/learning experiment.

REFLECTIONS AND DIRECTIONS

Discuss selected aspects of the teaching/learning experiment with at least one other person who has read the episodes. For example, you might consider:
- the learning process
- the indirect teaching process
- the implications for teaching equality sentences in the classroom

---

8

INTERPRETATION OF NUMBER STORY PROBLEMS

PREVIEW

Preceding chapters already have given indications of the importance of number story problems in the development of children's numerical thinking. Chapter 4 indicated the role of number story problems in stimulating children's construction of varied and more efficient counting strategies. Chapter 6 highlighted the role of number story problems in providing real-life physical contexts for introducing the formal notation of the equality sentence to children and for encouraging their continuing development of varied and deeper meanings to assign to its symbols. This chapter will expand on the latter role of number story problems.

The economy of mathematical notation results in different problem situations being encoded by identical equality sentences. For example, the following four problems can be represented as 5 + 3 = 8 despite differences in physical contexts.

---

1. Laurie has five comic books. Suzie has eight comic books. What could Laurie do to have as many comic books as Suzie?

2. Laurie is looking at her comic book collection. She has five old comic books. She also has three new comic books. How many comic books does she have?

3. Laurie had five comic books. She bought three more comic books. How many comic books does she have now?

4. Suzie has five comic books. Laurie has three more comic books than Suzie. How many comic books does Suzie have?

Physical models representing the different problem situations are shown on the following page to highlight the differences in context.

You are invited to match each number story with a physical model.

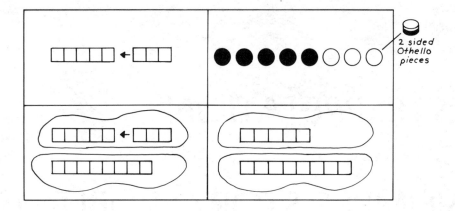

Here is another set of number story problems that can be encoded by the same equality sentence despite differences in physical context.

1. Ralph and Jenny are coming home from the library. Ralph is carrying eight books. Jenny has five books. What's the difference in the number of books they are carrying?

2. Gloria has eight record albums. She gives five record albums to Mary. How many albums does she have left?

3. There are eight girls and three boys in the reading group. How many girls have to leave the group for there to be the same number of boys and girls?

4. There are eight cans of paint. Five cans are red and the rest are blue. How many cans of blue paint are there?

5. George won eight prizes. His sister won five prizes. How many less prizes did his sister win?

Write an equality sentence that would represent all of the number stories in this set.

Match each number story with an illustrated model of the situation.

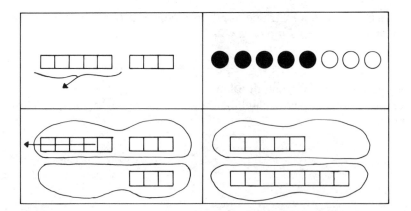

The preceding number story problems present different levels of difficulty for children. Certain problems are solved readily by kindergarten children while others may not be solved until children reach the second or third grade. Although a child may readily determine the answer to a problem, he may not be able to encode it in an equality sentence.

Further levels of difficulty result from varying the unknown in the preceding problem situations. For example, two such variations are illustrated in the following missing addend problems.

---

1. Jane had five books. She got more books from her teacher. Now she has eight books altogether. How many books did the teacher give her?

2. Johnny had some baseball cards. His friend gave him five more cards. Now Johnny has eight cards. How many cards did Johnny have to start?

Match each of the missing addend problems to one of the preceding physical models. Identify the problem situations you expect will present children with the most difficulty.

Number story problems have been a perennial stumbling block for children. In solving textbook problems, children are frequently heard to say, "Do I add or subtract?" Even when the correct operation is chosen, the answer is often surprising to most teachers. Even when the correct answer has been given, the equality sentence that follows may not contain this answer. Perhaps because of the prevalence of these difficulties the full range of possible number story problems (as sampled in the preceding activity) has been deemphasized in the classroom. Simple joining and separating action situations receive the greatest attention. Yet addition and subtraction are more than just joining and separating.

Interviews with second-grade children will illustrate both the power and the limitations of their problem-solving and encoding abilities as they tackle a variety of number story problems.[1] An analysis of their errors and an examination of related research may provide some insight into the perennial difficulties reported by teachers. A review of research on first grader's problem solving will seek to determine whether there is any basis for the traditional curriculum sequence where formal symbolism of the equality sentence precedes exposure to number story problems.

CATEGORIES OF NUMBER STORY PROBLEMS

Researchers in mathematics education have identified four categories of number story problems for both addition and subtraction based on characteristic differences in the situations they describe. A major dimension for categorizing number story problems is whether the problem situation suggests action or not. Action situations change the size of a set; static situations maintain the relationship among the components of a set. These *active* and *static* problem types have been labeled *moving* and *thinking* problems in discussions with children. A second dimension upon which number story problems can be classified is the number of sets in the outcome. The problem situation can indi-

ACTIVE	STATIC

One-Set Outcome		
Two-Set Outcome		

---

1. The second graders interviewed are those described in Chapters 4 and 7, including the four children from the teaching/learning experiment. All children were not given the same tasks, so the size of the groups discussed will vary: four, thirty, and thirty-four.

cate a *one-set outcome* or a *two-set outcome*. Earlier, when matching the number story problems to appropriate physical models, you were categorizing them along these dimensions. The four types of story problems fit into a matrix categorized along both dimensions.[2]

	ACTIVE	STATIC
One-Set Outcome	CHANGE	COMBINE
Two-Set Outcome	EQUALIZE	COMPARE

Examples of problems in each of the four categories for addition and subtraction are illustrated in the two following matrices.

ADDITION		
	ACTIVE (MOVING)	STATIC (THINKING)
One-Set Outcome	Laurie had five comic books. She bought three more comic books. How many comic books does she now have? CHANGE	Laurie is looking at her comic book collection. She has five old comic books. She also has three new comic books. How many comic books does she have? COMBINE
Two-Set Outcome	Laurie has five comic books. Suzie has eight comic books. What could Laurie do to have as many comic books as Suzie? EQUALIZE	Suzie has five comic books. Laurie has three more comic books than Suzie. How many comic books does Laurie have? COMPARE

---

2. These problem categories are adapted from the work of Carpenter and Moser (1979), Fuson (1979), and Riley, Greeno, and Heller (1982).

<table>
<tr>
<td colspan="3" align="center">SUBTRACTION</td>
</tr>
<tr>
<td></td>
<td align="center">ACTIVE (MOVING)</td>
<td align="center">STATIC (THINKING)</td>
</tr>
<tr>
<td>One-Set Outcome</td>
<td>Gloria has eight record albums. She gives five albums to Mary. How many albums does she have left?

 CHANGE </td>
<td>There are eight cans of paint. Five cans are red and the rest are blue. How many cans of blue paint are there?

 COMBINE </td>
</tr>
<tr>
<td>Two-Set Outcome</td>
<td>There are eight girls and three boys in the reading group. How many girls have to leave the group for there to be the same number of boys and girls?

 EQUALIZE </td>
<td>George won eight prizes. His sister won five prizes. How many less prizes did his sister win?

 COMPARE </td>
</tr>
</table>

For each of the preceding problem categories the interpretation of "+," "–," or "=" varies (Fuson, 1979). Thus "+" and "–" do not represent only joining and taking-away actions, but also other forms of the addition and subtraction operations. Similarly, the "=" sign can be interpreted in different ways. Action situations tend to show a unidirectional view of equality while static problem situations demonstrate a bidirectional equality relationship. Children's extended interaction with a variety of number story problems can, therefore, be an important avenue to deeper understanding of both operations and equality relations.

Although definite problem categories are evident to researchers in mathematics education, are they also evident to children? One perspective of these categories is that they exist "out there," independent of any problem-solver. In this case, we would expect them to be evident to children. A constructivist perspective is that such categories are invented by expert problem-solvers who read different meanings into different problem situations. Since second-grade children's understandings are in various levels of development, we can expect to see some apparent recognition of categories, some different interpretations, as well as some nonrecognition of categories (demonstrated by using similar procedures for different problem categories). Children can be expected to interpret problem situations in terms of existing ideas and available computational procedures. At the same time, specific problem situations may provoke rethinking of operations and equality relations. Evidence that children encounter different levels of difficulty in different problem situations, even those that can be encoded by the same equality sentence, suggests that certain problems require children to read more meaning into them to relate the quantities involved—both in solving the problems and in encoding them.

<table>
<tr>
<td></td>
<td align="center">ACTIVE</td>
<td align="center">STATIC</td>
</tr>
<tr>
<td>One-Set Outcome</td>
<td align="center">CHANGE</td>
<td align="center">COMBINE</td>
</tr>
<tr>
<td>Two-Set Outcome</td>
<td align="center">EQUALIZE</td>
<td align="center">COMPARE</td>
</tr>
</table>

CHANGE PROBLEMS: ADDITION AND SUBTRACTION

Change problems involve a joining or separating action that results in an increase or a decrease in the initial set to produce a final set. The action is the event that causes a change. Number story problems emphasized by textbooks are of the Change variety. Two types of Change problems involving joining and taking away are discussed here in terms of whole-parts relations.

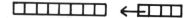

Before examining children's responses to these two types of Change problems, consider the interview procedure. The number story problems were presented verbally in the context of writing equality sentences for a variety of situations. Materials were made available but their use was optional.

This time I'll be reading you some number stories. You can use these blocks or your fingers to help you figure them out, if you want. First of all you need to listen to the story.
(Children who began writing before the problem was read in full were encouraged to slow down and listen to the entire problem. Also, children who seemed uncertain after several seconds of wait time were offered a rereading of the problem.)

Nearly all of the 34 second graders gave the correct answer verbally after using either blocks or fingers as models and counters, or using no visible strategy. In the latter case, a variety of mental strategies might be inferred. However, not all of the children were successful in representing the situation in an equality sentence. The responses of two such children are illustrated below.

Both Sammy and Wendy gave the correct sum verbally and then spontaneously recorded it. When asked to encode the situation in an equality sentence they appeared to lose sight of 12 (11) as the whole and began to think of it as a part that was then to be incremented by 8 or 4 to form a new whole. In the subtraction problem Wendy again had difficulty in thinking about the whole and its parts simultaneously.[3]

Following her removal of two blocks, Wendy considered the remaining part as the whole, which was to be decremented by 2 again. In both Change problems it is possible to obtain an answer without simultaneous consideration of the parts and the whole, yet encoding the situations may require such a consideration. For example, in Wendy's counting-all strategy in the first problem, it is likely that she lost sight of both the initial and the change set when they were joined and counted together. Yet it is possible to determine the final set without mentally holding both the initial and change sets. Although this inefficient counting procedure can serve children well for answering Change problems, it doesn't do so when the entire situation must be encoded.

Although this type of action problem has a one-set outcome, both Randy and Ted from the preceding chapter initially showed a two-set outcome in addition.

An interpretation of their modeling with blocks is that they were influenced by the form of an equality sentence they were envisioning.

Considerable research has been done on children's abilities to solve number story problems. The Change problems illustrated here (Change 1: addition and Change 2: subtraction) have been shown to offer children the least difficulty. Thus, reports of kindergarten children solving such problems involving smaller quantities are not uncommon. The difficulty of Change problems increases when the unknown is varied, as when the final set is known and either the change set or the initial set is unknown. The resulting situations are called missing addend problems. This type of problem will be discussed separately later in the chapter. Despite an abundance of research on children's abilities

---

3. For this problem, Sammy did not write the answer beforehand. In the process of encoding he indicated the wrong operation, writing 9 + 2 = 7.

to solve number story problems, there is a relative shortage of research in the area of encoding problem situations. Already, we have indications that the latter task is considerably more difficult than the former.

COMBINE PROBLEMS: ADDITION AND SUBTRACTION

Combine problems imply no action. Rather they involve distinct quantities that do not change. The relationship between these quantities is a static one in which a temporal order does not exist.

COMBINE 1: Addition—Combination Unknown
Diana has seven baby dolls and three grown-up dolls. How many dolls does Diana have?

◯ ◯ ◯ ◯ ◯ ◯ ● ● ● *(2 colored Othello pieces)*

A static relationship exists between the whole collection (10) and its component parts. The properties of the parts not only relate to the whole, but also allow the parts to be readily distinguished from each other.

$$10 = 7 + 3$$
$$7 + 3 = 10$$

COMBINE 2: Subtraction—Set Unknown
Larry has a collection of ten toy cars. Seven of them are red and the rest are silver. How many silver cars does he have?

◯ ◯ ◯ ◯ ◯ ◯ ◯ ● ● ●

The same static relationship exists between the whole and its component parts as for addition. Here a part is renamed in terms of the whole and the other part, with no action implied.

$$10 - 7 = 3$$
$$7 + \underline{3} = 10$$

Here, subtraction is finding the other part of the whole.

Most of the 34 second graders had no trouble in either giving the verbal answer to the addition problem or in encoding it in an equality sentence. Both Wendy and Sammy were successful in the total task. The strategies demonstrated by the second graders were similar to ones used in the Change 1 problem. One notable exception was a boy who demonstrated his solution using different colors to represent the parts in the whole. For the Change 1 problem he used only blocks of one color. This modeling suggests that Larry was distinguishing between two types of problems. Finding answers to Combine 1 problems presents little difficulty for children because the task can be interpreted in terms of existing count-all procedures. Keeping track of the initial and change sets while finding the answer is required to encode an equality sentence in which the "=" sign is interpreted as being unidirectional. A counting-on procedure increases the possibility of the information on initial and change sets being mentally held for recording. However, children may complete both aspects of the task successfully without reversible whole-part considerations. Since children tend to interpret Combine 1 problems just as Change 1 problems with no adverse consequences, Combine 1 problems present little difficulty for them.

Unfortunately, the subtraction task was inadvertently omitted from the interview tasks, leaving little direct information in this area of the study. However, a related task is described next as a possible source of information on children's interpretations of static situations.

I'm going to show you some blocks (while the blocks are being arranged behind a screen).

Write a number sentence to show your thinking about these blocks.

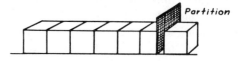

Half of the 30 children asked interpreted the situation as addition, lacking any obvious joining or taking away, and wrote the expected notation 6 + 1 = 7. On the other hand, the remaining half interpreted the situation as subtraction, but wrote unexpected notations. For example, ten children wrote 6 - 1 = 5 while two others wrote 6 - 1 = 6. Focusing on the parts, these children lost sight of the whole. Considerable hesitation was displayed by children appearing uncertain about their notation, but not knowing how to correct it. In his apparent disequilibrium, Ricky J. altered his notation from one incorrect form to another, 6 - 1 = 65. Only three children wrote the expected notation for subtraction, 7 - 1 = 6. Yet Larry did so only following some indecision and self-correction, 76 - 1 = 6. Another source of indecision was the choice of operation; the writing of correct notation following extended hesitation was interpreted as indicating this kind of indecision. As a follow-up, a handful of children were asked if they could write the notation in a different way. This extension revealed that at least one child, Sal S., could think of the same static situation as either an addition or a subtraction situation.

Second grader's interpretations of the preceding task corroborate the findings of other research on the relative difficulty of Combine 2 (static subtraction) problems in the absence of clarifying wording such as "of them" and "the rest."

> There are eight children on the playground.
>
> Five are girls.
>
> How many are boys?

> There are eight children on the playground.
>
> Five of them are girls.
>
> The rest are boys.
>
> How many boys are there?

Without such phrasing to help children focus on whole-parts relations, and without explicit indicators of action, children interpret Combine 2 problems in the best way they know how. For many children, this may involve putting out separate sets of ten objects and seven objects and counting-all. Whereas low-level strategies can produce correct answers for Combine 1 problems, they are not very helpful in solving Combine 2 problems. Here, whole-part relations are necessary to find the unknown part in terms of the whole and the other part. According to Riley, Greeno, and Heller (1982), Combine 2 problems present considerably more difficulty to children than other problem types considered so far. Here, subtraction is finding the other part of the whole by finding the *difference* rather than taking away.

		ACTIVE	STATIC
One-Set Outcome		CHANGE	COMBINE
Two-Set Outcome		EQUALIZE	COMPARE

EQUALIZE PROBLEMS: ADDITION AND SUBTRACTION

Equalize problems combine aspects of two other story problem types—Change and Comparison—with one-set and two-set outcomes. Equalizing is the process of changing one of two sets in order to make the two sets equal in quantity. Increasing or decreasing actions on one of the sets produces a one-set outcome. At the same time, a static equivalence relation is established between the two separate sets.

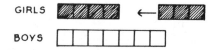

EQUALIZE 1:   Addition

There are four girls and seven boys on the class baseball team. How many more girls are needed so there will be the same number of boys and girls?

GIRLS

BOYS

The model includes two separate and distinct sets whose sizes are to be compared and equalized. The action represented indicates increasing the size of the smaller set to make it equal in size to the larger one.

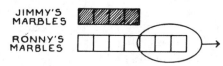

EQUALIZE 2:   Subtraction

Ronny has seven marbles and Jimmy has four marbles. How many marbles would Ronny need to give away so that he'd have the same number of marbles as Jimmy?

JIMMY'S MARBLES

RONNY'S MARBLES

In addition to the separate and distinct sets, the model represents the action of decreasing the larger of the two sets until both sets are equivalent.

Equalize problems were not included in the interviews with second graders. However, second graders' interactions with concrete equalizing situations independent of story problem contexts have been described in Chapter 7. No unusual difficulties were experienced by Cynthia, Gloria, and Randy in solving and encoding these tasks.[4] Although this category of number story problems has not been widely researched, one study by Carpenter, Hiebert, and Moser (1979) indicates that about 70 percent of a group of first graders were able to solve Equalize problems. Their performance on this task was roughly comparable to that in all the categories discussed thus far, with the exception of Combine 2 problems (46 percent). It is also interesting to note that several years ago an innovative elementary school mathematics curriculum project, *Developing Mathematical Processes* (Romberg, et al., 1974), used concrete equalizing situations to introduce first graders to the representation of addition and subtraction in equality sentences.

	ACTIVE	STATIC
One-Set Outcome	CHANGE	COMBINE
Two-Set Outcome	EQUALIZE	COMPARE

COMPARE PROBLEMS: ADDITION AND SUBTRACTION

Compare problems involve two distinct quantities that do not change and the difference that can be inferred from their comparison. The relationship between the quantities is a static one and therefore implies no action. Two types of Compare problems will be sampled and examined in detail: a Compare 3 problem with compared quantity unknown and a Compare 1 problem with difference unknown.

In examining these Compare problems you are invited to anticipate both the range of answers and equality sentences that second graders will generate.

---

4. Ted continued to write equality sentences that often had no connection to the action that had just taken place. He was obviously unaccustomed to working with objects.

COMPARE 3: Addition

Johnny has eight toy airplanes and Richard has three more toy airplanes than Johnny has. How many toy airplanes does Richard have?

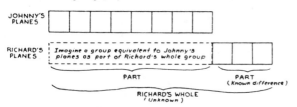

JOHNNY'S PLANES

RICHARD'S PLANES — Imagine a group equivalent to Johnny's planes as part of Richard's whole group

PART

PART (Known difference)

RICHARD'S WHOLE (Unknown)

8 + 3 = 11 represents the static relationship in Richard's whole containing the equivalent of Johnny's whole as one of its parts.

The problem must be modeled as a two-set outcome because the groups are separate and distinct. Neither one can be the subset of the other in the problem context. Only an imagined equivalent group can be included.

COMPARE 1: Subtraction

Bobby has eight record albums. Cindy has five record albums. How many more albums does Bobby have than Cindy?

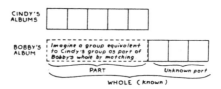

CINDY'S ALBUMS

BOBBY'S ALBUM — Imagine a group equivalent to Cindy's group as part of Bobby's whole by matching

PART

Unknown part

WHOLE (known)

8 – 5 = 3. The unknown part is represented in a relationship between the whole and its parts in which a group equivalent to the other comparison group is now part of the whole.

Although the comparison groups exist side by side physically, they are mentally related.

Here, subtraction is viewed as the difference between quantities.

As you have already anticipated, these comparison problems were quite challenging for the second graders. This portion of the interview was marked by lengthy pauses and requests for rereading of the problems. After a long pause I also asked whether they would like the problem read over. Most children accepted the offer. Yet nearly all of the 30 children gave expected answers. It was the encoding of the comparison situations that presented the greatest challenge for these second graders. Only about half the children wrote the equality sentences expected.

Examples of children's responses to the first problem follow.

Spontaneous verbolizing

OK, JOHNNY HAS 8 AND RICHARD HAS 3 MORE.

Fay B.

Notice her interpretation of "3 MORE" in the problem context. Also notice the potential value of having children repeat the problem and model it with blocks.

LET ME READ THIS OVER (Reread)

NOW, SO THEY HAVE 8 AND THE OTHER FRIEND HAS 3 MORE. HMM,...8 ...8. SO RICHARD HAS 11 AIRPLANES.

Written as : 8 + 3 = 11 with no further reference to the blocks. She imagined equivalent groups before considering the difference.

11 OR 12.

Written as problem is read.

Suzie H. began to write prior to giving her answer. No materials are used.

WRITE THIS AS A NUMBER SENTENCE.

① Writes the given quantities in relation.
② Counts-on 3 more fingers from 8.
③ Recounts 11.

Her counting produces a check on an earlier procedure.

Nearly all of the 30 children gave answers of 11 or reasonable approximations but only 18 wrote 8 + 3 = 11 following a variety of strategies sampled in the illustrations. Six additional children misapplied their strategies and recorded answers of 10 or 12. Not one child demonstrated matched rows of blocks. The only child using a matching strategy, Fay, misapplied it by matching rows of eight and three blocks before switching to a successful mental matching strategy. The other children using blocks modeled the problem as a simple joining action. The long pauses, the need for encouragement, and the rash of computational errors all suggest that the children were uncertain about their responses and disoriented by this uncertainty. Nancy,

the last child illustrated, showed us in Chapter 4 that she was an efficient and accurate counter. Despite an obvious ability to compute the answer to 8 + 3, she experienced enough difficulty in conceptualizing the comparison relationship that she was unable to either give the expected answer or write an appropriate notation. She fit the problem into an existing procedure for finding "How many altogether?" In this perspective, Richard would have only three airplanes. Although she was dissatisfied with her perspective, she was unable to alter it. The interactions of Nancy and other children with a Compare 1 problem, given in another interview, follow.

Nearly all of the 30 children gave correct answers of "three", yet only half of them also wrote acceptable equality sentences: 8 – 5 = 3, 8 – 3 = 5, and 5 + 3 = 8. At the same time that each sentence is an alternate expression of the same relationship, it suggests how the children view the Compare problem. Only two children, Sal and Shelly, wrote 8 – 5 = 3, the clearest expression of the difference. The more common notations of 8 – 3 = 5 and 5 + 3 = 8 are more suggestive of equalizing the two sets. Also, 5 + 3 = 8 may suggest fitting the problem into an existing procedure for finding "how many more" having a one-set outcome (as in the Change 3 problem to be discussed). Children's use of materials and verbalizations usually provides a solid basis for further interpretation of their responses. Yet all the children did not choose to use the blocks and some children showed no visible strategy. Also, the distinction between how children are seeing the problem and how they are computing the answer is not always clear. Furthermore, children's apparent redundant actions—for example, Marsha's use of her fingers to compute a quantity that is already given in the reading of the problem and modeled in her display of matched sets—confuse our attempts at interpretation. The children writing acceptable notations used the following models/procedures:

Matched sets of blocks and fingers 1

Unmatched sets of blocks (8 and 5) 3

Sets of 5 and 3 (different color) combined 1

Sets of 5 and 3 (same color) combined
or displayed 4

Fingers only 2

No visible model/strategy 4

Additional data suggest that the most common interpretation leading to appropriate notation was that of Change rather than Compare.

The inability to arrive at more precise interpretations of children's perspectives of Compare problems suggests that the interview tasks should be extended to gather additional information. All children could be asked to represent the problems with materials, drawings, or open sentences prior to computing an answer. Furthermore, they could be asked to circle their answer in the completed sentence and to decode the notation in terms of the problem. In addition to providing further information on a child's thinking for the interviewer, the additional requirements could also extend a child's thinking about the problem situation.

Although nearly all the children gave correct answers to the problem, four of the children gave one of the known quantities (8 or 5). Two of these children corrected their answers when the problem was reread. These responses are typical of younger children. For example, the Compare 1 problem may be interpreted as a Change 3 problem—"Got five. How many more to get eight?" In the process of incrementing from the set of five blocks to read the goal of eight, the child is likely to lose track of both the starting set and the incrementation. Hence the goal becomes the child's answer. The other set is ignored.

As in the preceding Compare task, the interview was marked by long pauses, puzzled looks, requests for rereading of the problem, and the need for encouragement.

Here, Unie's responses suggest the size of the gap between the ability to model the problem, give a verbal answer, and represent the whole-parts relationship inferred from separate and distinct sets.

The second graders' inability to appropriately encode problems they had already solved was punctuated by notations that failed to include their original answers.

Similar notations were given by 11 of the 15 remaining children. Nine of these children had given "three" as their answer but ignored it in their encoding.

How can you explain this discrepancy between the children's oral answers and their written representations?

Rather than comparing the given quantities, 11 children combined them in writing 8 + 5 = 13, as if fitting

the problem into an existing procedure to produce "more." Yet these children must have had a different perspective of the problem to arrive at an answer of "three" earlier. It is interesting to speculate on what may have occurred. Here's one conjecture that assumes the physical comparison model in explaining children's inability to write equality sentences for problems they've already solved: It's much easier to abstract "three more" from a model of matched sets than it is to abstract the total whole-part relationship. In the physi-

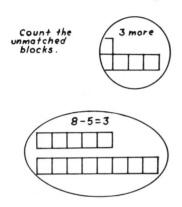

cal model the matching makes the abstraction of "three more" possible without the need to even consider the absolute size of the given quantities. On the other hand, these quantities must be related to write $8 - 5 = 3$ by considering a whole (5) as part of the larger whole (8). Since this is no easy task, a child may construct a comparison model, yet not be able to adequately represent it in arithmetic notation.

Other equality sentences written following oral answers of "three" included $5 - 2 = 3$ and $8 + 3 = 11$. Both of these responses suggest a serious attempt to sort out the whole-part relationship. In both cases one of the known quantities and the unknown difference are placed into a relationship with a third quantity. However, in each case, one of the wholes is overlooked in the process. A child's notation of $8 + 3 = 11$ may be considered as a compromise solution in his struggle between comparing and joining the sets.

Most children seemed unaware of the discrepancy between oral and written responses.

Would the children respond differently if they were made aware of this discrepancy, with questions like the following?

Earlier when I read the problem, how many did you tell me?

How come I don't see 3 in your number sentence now?

How would children who modeled the problem with blocks as a one-set outcome respond to the following situation?

Are these Bobby's albums? (Yes.)
(The blocks are then covered.)
Now show me Cindy's albums.

In the preassessment interviews of the four children in the preceding chapter, they were encouraged beforehand to demonstrate with blocks as an aid in helping me to understand their thinking. When showing a one-set outcome they were exposed to a friendly confrontation with the reality within the problem context. Similarly, the contrasting oral and written re-

sponses of a child were juxtaposed when the occasion arose to highlight the discrepancy and to encourage rethinking and integration of the responses. On the following page we join Cynthia, Gloria, Randy, and Ted as they interact with their second Compare problem in a series of three.

Betty has nine record albums and Gloria has six record albums. How many more albums does Betty have than Gloria?

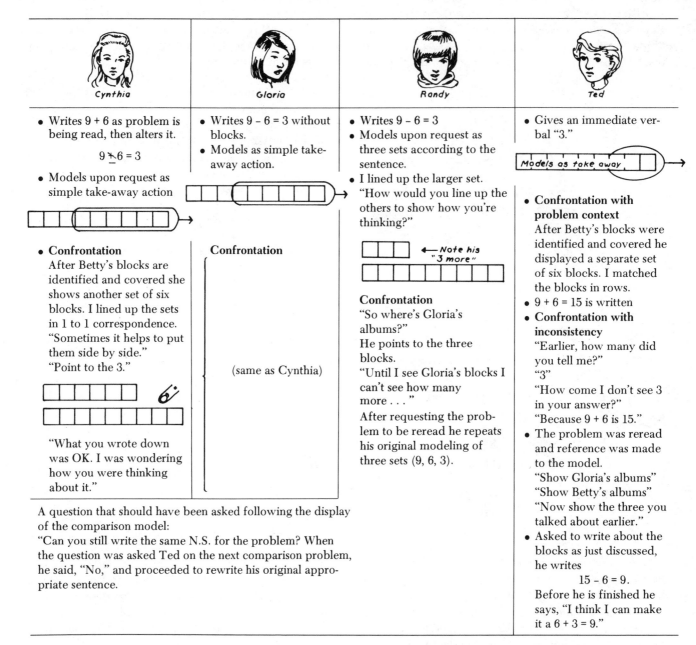

Cynthia	Gloria	Randy	Ted

Cynthia
- Writes 9 + 6 as problem is being read, then alters it.

  9 ⊼ 6 = 3

- Models upon request as simple take-away action

- **Confrontation**
  After Betty's blocks are identified and covered she shows another set of six blocks. I lined up the sets in 1 to 1 correspondence. "Sometimes it helps to put them side by side." "Point to the 3."

  "What you wrote down was OK. I was wondering how you were thinking about it."

A question that should have been asked following the display of the comparison model: "Can you still write the same N.S. for the problem? When the question was asked Ted on the next comparison problem, he said, "No," and proceeded to rewrite his original appropriate sentence.

Gloria
- Writes 9 – 6 = 3 without blocks.
- Models as simple take-away action.

Confrontation

(same as Cynthia)

Randy
- Writes 9 – 6 = 3
- Models upon request as three sets according to the sentence.
- I lined up the larger set. "How would you line up the others to show how you're thinking?"

  ← Note his "3 more"

Confrontation
"So where's Gloria's albums?"
He points to the three blocks.
"Until I see Gloria's blocks I can't see how many more . . ."
After requesting the problem to be reread he repeats his original modeling of three sets (9, 6, 3).

Ted
- Gives an immediate verbal "3."

  Models as take away

- **Confrontation with problem context**
  After Betty's blocks were identified and covered he displayed a separate set of six blocks. I matched the blocks in rows.
- 9 + 6 = 15 is written
- **Confrontation with inconsistency**
  "Earlier, how many did you tell me?"
  "3"
  "How come I don't see 3 in your answer?"
  "Because 9 + 6 is 15."
- The problem was reread and reference was made to the model.
  "Show Gloria's albums"
  "Show Betty's albums"
  "Now show the three you talked about earlier."
- Asked to write about the blocks as just discussed, he writes

  15 – 6 = 9.

  Before he is finished he says, "I think I can make it a 6 + 3 = 9."

This method of making children aware of a discrepancy in either their modeling or in their contrasting answers to the same problem was used in two out of the three comparison problems in the preassessment. A couple of weeks later, during the teaching/learning session, when the children returned to Compare story problems, the girls consistently modeled the problems with matching sets. Gloria used a different color for the unmatched blocks "to help a first grader understand." Her embellishment suggests that she may have been viewing the quantities in a static relationship. Ted continued to have difficulty representing two comparison sets in an equality sentence. Randy adapted his three-set model to a compromise Equalizing situation, thus incorporating the third set. Since the

boys and girls had contrasting interpretations of the Compare problem, a boy and girl were paired to stimulate discussion and possible rethinking of opposing views. The four sessions provided didn't allow enough time for the boys to reorganize their thinking.

Compare problems offer considerable challenge to children. Their degree of difficulty is roughly comparable to Combine 2 problems (Riley, et al., 1982).

However, variations of Compare problems offer different degrees of difficulty as well. An example of the most difficult Compare problem—Compare 6—follows.

> Bobby has three record albums.
> He has five albums less than Cindy.
> How many record albums does Cindy have?

Although most research on children's number story problem solving does not include their encoding of the problem situations, Riley and Robinson (1980) report on children's inability to write appropriate equality sentences for Compare problems already solved.

	ACTIVE	STATIC
One-Set Outcome	CHANGE	COMBINE
Two-Set Outcome	EQUALIZE	COMPARE

CHANGE PROBLEMS (cont'd): MISSING ADDENDS

No discussion of story problems would be complete without the inclusion of missing addend problems. These problems are variations of Change problems in which rather than the result's being unknown as in Change 1 and 2 problems, the change or the initial set is unknown.[5] This section will discuss two such variations of Change problems.

CHANGE 3:   Change Unknown

Pretend you have six pennies and you want to buy a cookie that costs ten cents.
How many more pennies do you need?

In this implied action situation the child must reconstruct the action in the problem situation represented by $6 + \underline{\hspace{1em}} = 10$. The action suggested is joining or counting-on four more pennies. However, mathematicians view this as an equation to be solved by subtraction, $10 - 6 = \underline{\hspace{1em}}$, calling it the additive form of subtraction. This view of the problem superimposes a whole-parts relation in which the unknown part is seen as part of the whole. The reversibility of this relationship allows an operation that is not identical to the

implied action of the problem. It will be interesting to see how the children view the problem.

Of the 34 children asked the problem, 25 wrote an appropriate equality sentence for this missing addend problem. All but two of these children interpreted the operation as addition, with counting-on being the predominant visible strategy. The children incremented the known set until the known result was reached, keeping track of the change. Despite the high success level, the problem challenged many children.

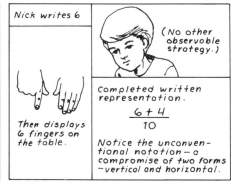

---

5. Furthermore, the change can be an increase or a decrease. Here, both problems indicate an increase. In the generally accepted system of categorizing story problems Change 1, 3, and 5 problems involve an increase while Change 2, 4 and 6 problems involve a decrease. A complete listing of the different types of problems is found in Appendix E.

207

Two other children attempting to interpret the situation as subtraction reversed the order of the quantities: 6 – 10 = 4. Other children responded orally with one of the known quantities (10 or 6) rather than the unknown —responses typical of younger children. The remaining children combined the quantities given, treating them as parts to be combined rather than considering them in an inclusive relationship.

Sammy reacts to the surface features of the problem. He displays and joins (counts-on) quantities mentioned in a procedure producing "more." The mention of the unknown change appears to be ignored since he lacks a whole-parts perspective to incorporate this information. Whereas low-level counting strategies allow children to obtain answers to Change 1 and 2 problems, more efficient counting strategies and whole-parts relations are needed for solving Change 3 problems.

In this problem situation, encoded as ___ + 3 = 9, some means of representing the unknown set to reconstruct the joining action must be found or a whole-parts perspective must be applied.[6] This higher-level perspective permits an unknown part to be inferred from the known whole and other part without regard for matching the specific action implied in the prob-

lem. In other words, the reversible relation permits interpreting the problem as 9 – 3 = ___.

Sixteen of the 30 children wrote an appropriate number sentence showing addition, while only one child did so in terms of subtraction. However, the latter form was an unexpected one.

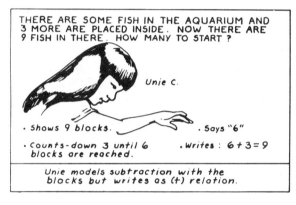

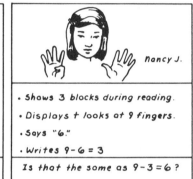

Vic's responses reflect the challenge presented by this Change 5 (missing addend) problem.

The difficulty of this Change 5 problem was so indicated by 6 of the 30 children giving wrong oral answers.

Earlier, 4 of these children gave an oral response of one of the knowns (3 or 9) with one of the children switching from one of the known quantities to the other. Two other children gave verbal responses of "twelve". In encoding the problem situation, 6 children lost sight of the context and considered the whole

6. It is possible to estimate the size of the initial set and count-on 3 more. If "nine" is not reached or it is bypassed once 3 are counted-on, an adjustment is made in the estimate.

and its known part as just parts to be combined. Having no clear whole-parts perspective for incorporating information about the unknown set, they reacted to superficial word clues in the problem. The following panel shows Vince's disequilibrium prior to settling for the notation 3 + 9 = 12.

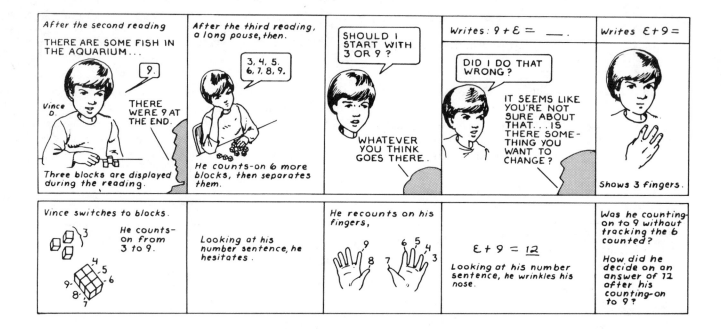

Vince's hesitation and wrinkled nose suggests that he was aware that something was wrong, yet he was unable to resolve his difficulty. Although he was observed to count-on from 3 to 9, he did not audibly verbalize "six" as the quantity of blocks counted-on. The inconsistency between his equality sentence and his repeated counting action further suggests a conflict of ideas. The ease of arbitrarily combining known quantities eventually overcame the difficulty of relating them within the problem context and finding the other part of the whole.

RELATED RESEARCH AND THEORY

Although researchers in mathematics education have identified convenient categories for discussing number story problems, children may or may not distinguish between these categories, depending on the knowledge that they bring to the problems. For example, without a whole-parts perspective for organizing problem information and inferring the unknown, the child can only pay attention to isolated verbal clues. Whereas attention to only the surface features of a problem is a useful strategy for some problems, it is misleading for others and produces predictable errors. A detailed study of how children approach these number story problems has led researchers to propose at least three levels of problem-solving ability. The following description of these levels has been adapted from Riley, Greeno, and Heller (1982) and Nesher, Greeno, and Riley (1982). Examples of these levels have been illustrated in the preceding interviews with second graders.

Level One	The child represents the quantities mentioned in the problem with blocks and shows the change suggested by isolated verbal clues. In joining and counting-all the blocks, the total is identified but the initial and change sets become indistinguishable. Since this information is not held mentally, representation of the problem is limited to the external display of blocks. Success at this level is limited to problems such as Change 1 and 2 and Combine 1 problems. However, difficulties are experienced in encoding these problem situations.
Level Two	Again the child represents the quantities mentioned in the problem with objects or fingers and shows the change suggested by isolated verbal clues. In counting-on and keeping mental track of the process, information on the initial set and the amount of change is retained, together with the total. If counting-back abilities are available, they are not yet related to counting-on, so that addition and subtraction are distinct procedures. Thus the counting procedure selected follows the direction of the change suggested in the problem. This level of ability supports the solution of a greater range of problems, including Change 3 problems.
Level Three	A whole-parts perspective provides a basis for noticing relevant features of the problem, organizing this information, and inferring the unknown. This mental representation overrides distracting features of the problem situation so that the operation can be subtraction even though the verbal clue is "more." Addition and subtraction are related. Equality relations are bidirectional. This level can account for success on the most difficult problems as well as in encoding the situations.

Problem-solving abilities increase as the child mentally represents problem situations within an expanding relational framework in ways that lead to appropriate computational procedures and equality sentences.[7]

The problems categorized by researchers vary in difficulty even though they may require the same operation for solution. This suggests that problem solving extends beyond knowing how to find the combination of two numbers. The following table, adapted from Riley et al. (1982), indicates the relative difficulty of different problem categories for children at four grade levels. The table has been reduced to include only categories discussed in this chapter. A complete listing of the different types of problems is found in Appendix E.

RELATIVE DIFFICULTY OF DIFFERENT PROBLEM CATEGORIES

	PROPORTIONS OF CHILDREN WHO PERFORMED CORRECTLY USING OBJECTS			
	GRADE			
PROBLEM TYPE	K	1	2	3
Change 1	87%	100%	100%	100%
Change 2	100%	100%	100%	100%
Change 3	61%	56%	100%	100%
Change 5	9%	28%	80%	95%
Combine 1	100%	100%	100%	100%
Combine 2	22%	39%	70%	100%
Compare 1	17%	28%	85%	100%
Compare 3	13%	17%	80%	100%
Compare 6	0%	6%	35%	75%

The preceding data indicate only the relative difficulty of giving answers to the different problems and not on appropriate encoding of equality sentences. Based on what we've learned from children's construction of numerical abilities it should not be surprising that young children demonstrate some problem-solving abilities even prior to any teaching. Although classroom teaching is restricted mainly to Change 1 and Change 2 problems, children gradually show abilities to solve other, more difficult problems.

However, children's early introduction to the equality sentence may influence how they approach these problems. The research of Carpenter and Moser (1979) involved interviewing a group of first graders in February on a variety of number story problems prior to any formal instruction in addition and subtraction and its notation in equality sentences. The researchers report that these children used different models/strategies for different situations involving addition and subtraction. They tended to match their physical models and computational strategies to fit the implied action or relationship. For example, out of 43 first graders, 17 of them spontaneously modeled comparison problems with matched sets of blocks and even fingers (when comparing 5 and 3). This performance is in marked contrast with both the second graders described in this chapter and those studied by Carpenter and Moser, who tended to model comparison situations as having one-set outcomes. These first graders demonstrated greater flexibility in representing the problem situations than did older children. Older children have a greater tendency to model the problems as joining or taking away actions regardless of the problem situation, and even to choose the wrong operation.

Although research on children's solving of number story problems is extensive, their encoding of problem situations and the influence of classroom

---

7. In Chapter 5 we noticed that children invent other computational strategies besides counting. Further attention needs to be paid to the role played by derived strategies in children's problem solving.

teaching on their problem solving need further study. For a thorough review of existing research and discussion of theoretical issues, consult Riley, Greeno, and Heller (1982).

TEACHING METHODS EMPHASIZING NUMBER STORY PROBLEMS

An explanation of the greater flexibility of representation by first-grade children in Carpenter and Moser's study (1979) suggests itself in an examination of the typical sequence of classroom instruction. In this traditional sequence, the emphasis is on the mastery of number combinations presented in equality sentences within a limited view of operations (joining or taking away) and of equality (unidirectional). Number story problems are presented last in the sequence. Once children reach this part of the sequence they may not know when to use addition or subtraction because of lack of experience and knowledge of the range of real-world situations giving rise to these operations. Early introduction to the formal notation of the equality sentence, and its restricted use in the worksheet format, also encourages an attitude of "getting an answer" with little or no reflection. Instead of paying attention to the relevant features of the problem, the children tend to focus on its surface features. For some problems, they get the right answers for the wrong reasons; for others they make predictable errors.

Number story problems offer a viable alternative for resequencing instruction since children already demonstrate natural abilities to represent and solve problems. Rather than being an appendage in the instructional sequence, number story problems can lead the way in providing an appropriate avenue for the introduction of addition and subtraction and for the development of a variety of counting strategies (Carpenter and Moser, 1979). Furthermore, number story problems offer opportunities for broader interpretations of addition and subtraction, and for the gradual construction of critical whole-parts and equality relations. By building on children's natural problem-solving abilities and by providing sufficient time for these conceptions to develop prior to the introduction of equality sentences, instruction can encourage children to bring richer meaning to the symbols. At some point, the encoding process itself can encourage reflection on the meaning of the symbols. At first the encoding of varied problem situations by the same equality sentence may be a source of disequilibrium. However, continuous interplay between problem situations and notation can further enrich the meaning read into the symbols.

It may be argued that it's too early to pursue a major revision of the teaching sequence based on limited classroom research. On the other hand, one might ask whether the traditional instructional sequence is based on any research of children's natural abilities or whether it has been imposed arbitrarily on children and teachers by textbook publishers.

A Continuing Program

In Chapter 6, methods have been outlined for using number story problems to introduce addition and subtraction operations and the formal notation of the equality sentence. There, the introduction of equality sentences was delayed until meanings were under construction. A classroom program incorporating such methods has been reported by Hendrickson and Thompson (1982). This program includes all categories of story problems and, therefore, has been expanded to include grades one, two, and three. The program's effectiveness in helping children become good solvers of verbal problems is attributed to three main factors:

1. Extended use of countable materials for modeling problem situations

2. Delayed introduction of equality sentences with use of numeral and sign cards for encoding prior to actual writing of sentences

3. The practice of having children encircle their answers in equality sentences to encourage reflection on their encoding

The latter practice also provides teachers with a basis for interpreting children's thinking about particular problem situations. Another indicator for teachers of the level of children's thinking in this program is a child's alternative use of addition and subtraction sentences for problems in which either is appropriate. This behavior can be used as a benchmark of children's successful understanding of whole-parts relations.

Other Promising Practices

In addition to the promising practices identified in the preceding program for grades one to three, three other practices have potential for helping children develop problem-solving ability. These are:

1. Extending opportunities for problem representation prior to computation
2. Providing children with opportunities for writing their own problems
3. Providing a variety of real-world situations giving rise to addition and subtraction

These are offered as suggestions for use in any combination that makes the most sense to you.

Extending Opportunities for Problem Representation Prior to Computation. The importance of representation of relevant aspects of problem situations has been highlighted throughout the chapter. By solving problems without analysis, some second graders have arrived at right answers to problems for the wrong reasons or obtained predictably wrong answers by paying attention to only the surface features of the problem situations and guessing at which operations to use. A critical aspect of encouraging children's representation is the provision of adequate time for reflection. The prevailing classroom emphasis on speed is not compatible with problem solving with any understanding.

Possible methods for problem representation are numerous, and include the following (Lindvall, 1982):

1. Acting out the stories using props
2. Acting out the stories using puppets, dolls, and so on.
3. Using blocks or other countables to model the story

4. Using diagrams to model the problem
5. Writing an open number sentence to represent the problem situation, like 4 + ___ = 7.

Most of these methods have been discussed earlier in Chapter 6.

Rather than circumventing thinking about relevant aspects of different problems, activities can actively involve children in interaction on similarities and differences between problem categories and their representations. An episode from the teaching/learning experiment will illustrate one such interaction. Prior to the episode, the children had been introduced to models of different problem categories composed of both objects and drawings. These models were arranged in

an unmarked matrix by dimensions of action and number of sets in the outcome. The teacher led a discussion of similarities and differences between each of the models. As the episode begins, Gloria and Cynthia have just been given five number story problem cards to match up with the four models. The four models can be encoded by the same number sentence, 5 + 3 = 8. Notice that although the children are not involved in computing answers to the problems, they are active in identifying relevant features of different problems and their representation, and in joint decision making.

Providing Children with Opportunities to Write Their Own Problems. Situations that create such opportunities include completing unfinished problems by providing the necessary information or question, or composing a complete story problem. Children can also write number story problems for themselves and solve them. Some examples follow.

Suppose you have only three different numbers and a question mark. How many different problems can you make from this story outline?	Dora had ___ marbles and Jim had ___ marbles. Jim has ___ more marbles than Dora.
Fill in the blanks in the story outline with reasonable numbers, making statements you can defend.[8] What are the smallest numbers you can use in the problem?	The Dooley twins organized a club. It had ___ members. After the twins made some rules, ___ of the members quit. The next week ___ more kids joined. Now the club has ___ members.

(continued)

8. Wirtz (1982a).

Make up your own question to finish the story problem.	Jimmy has eight toy airplanes.
How many different problems can you make up by changing the question?	Steve has five toy airplanes.
Exchange problems with a friend.	
Work out your friend's problems.	
Write a number story problem in your own words that can be shown by this number sentence.	$6 + \underline{} = 9$
Write a number story in your own words.	No equality sentence is given.
On another piece of paper write the number sentence for the story.	
Exchange problems in your group and work them out.	
Compare your number sentences for each problem and decide if there is a best way to show each one.[9]	

Each of the preceding situations provides a spring board for further interaction between children in small-group and whole-class discussion. For example, bringing together children with the following perspectives of the same equality sentence, $6 + \underline{} = 9$, has potential for some lively and fruitful interaction.

Second Graders' Story Problems for $6 + \underline{} = 9$ (Hendrickson, 1981)

Joetta

Kari has six boxes. Shelly gives her sum boxes win she was dun she had nine boxes How meny boxes did Shelly give Kari?

Sharron

Jim has six cars and Tommy came and gave him three how many does Jim have now?

Douglas

Pilp has 6 cars Nicky gave him 9 more haw meny dose Pilp have? 15

Steve

$6 + \square = 9$ Kari has six marbles Mark and had nine how many more does Mark have.

Writing a number story problem for an equality sentence is a form of decoding. In reading each other's problems and comparing their verbal interpretations with concrete models, the children become aware that other children have viewpoints different from their own. This awareness can play an important role in stimulating rethinking of their ideas at more sophisticated levels of understanding.

---

9. Meyer in Burns and Richardson (1981).

Providing a Variety of Real-World Situations Giving Rise to Addition and Subtraction. Should the problems described in this chapter always appear in repetitious blocks of a particular category, the stories soon would become mere exercises for the mechanical application of a procedure. They would no longer present problems for the children. Kilpatrick (1982) cautions that this approach can create an unrealistic expectation that "problems of a feather flock together" in real life, complete with accompanying instructions. As an alternative to this typical approach, Kilpatrick (1982) offers the following guidelines.

> If and when problem solving becomes the heart of instruction in school mathematics, problems will be seen to occupy several roles. Some problems will be used to introduce a mathematical topic, providing motivation to learn more about the topic as students attempt to understand the problem. Some problems, perhaps most, will be used essentially as exercise material through which to practice skills and techniques. Judicious selection and organization in assigning problems can keep the "one-rule-under-your-nose" problems to a minimum. Some problems will be used to synthesize what students have learned, providing them with an opportunity to use a variety of skills and techniques, either working alone or together with other students to develop skill in communicating

mathematical ideas. Some problems will be used to develop in students a greater sense of what mathematics is and how it has been created. (p. 7–8)

In the context of the teaching methods described in this chapter, number story problems can make a contribution to all of these areas.

A recent innovation that allows children to encounter a variety of real-life quantitative situations is the *Thinking Story Book* (Willoughby et al., 1981). This series of books contains extended number stories in which a variety of problems are embedded. The same characters running through the series of interesting stories have peculiarities that children come to know and to anticipate. Mr. Sleeby, for example, is extremely forgetful; Ferdie readily jumps to conclusions; Mr. Breezy always diverts by giving extraneous information and so makes easy problems appear difficult. As each story unfolds the children are asked to spot what is wrong with a character's actions, to anticipate the consequences of these actions, to identify relevant numerical information, and to select the appropriate operation. Each thinking story is followed by a set of similar but shorter number story problems based on the story context. The following problem is taken from the set immediately following the story "Mrs. Mudanza's Computer."[10]

Manolita's mother wrote a program for the computer that said, "1. Add. Add." To try it out, Manolita typed in the numbers 10 and 4. The computer typed out 15. Manolita typed in 5 and 2. The computer typed out 8. Manolita typed in 2 and 2. The computer typed out 5.

Can you figure out what the program is telling the computer to do?

If Manolita types in 4 and 2, what will the computer type out?

The *Thinking Story Book* frees classroom number story problems from a rigid, predictable format, yet encourages flexible problem solving in story contexts that are interesting to children.

Although the problem categories described in this chapter and in Appendix E seem exhaustive, all story problems need not follow this rigid form. For example, a problem may contain both addition and subtraction operations, as illustrated earlier in the problem about the Dooley twins' club with member-

ship increases and decreases (p. 214). This combination of operations opens the way for multiple problem-solving strategies and multiple ways of encoding. Manolita's computer problem also breaks the mold of the problem categories discussed in this chapter. This number story requires a child to infer a pattern from a series of actions producing pairs of numbers. A number story problem I devised requiring a more extended investigation is given on the facing page.

---

10. *Bargains Galore: Thinking Book—Level Three*, p. 66.

There once was a Square-Block Man that kept growing and growing. He kept gobbling up other square blocks as he grew bigger and bigger. He didn't grow a little at a time, he grew by leaps and bounds! Each time he grew bigger, he gobbled up more and more blocks.

Here's the Square-Block Man:

In order to grow to his next size and still be a square, how many more blocks does he need?

What is the Square-Block Man's new size in blocks?

To watch the Square-Block Man grow, keep a record of his starting size, how much he grows and his new size. Here's one way: $1 + 3 = 4$

Notice how the Square-Block Man grows two or three more times, using your blocks.

Keep a record of how he grows.

Do you notice a pattern in your record of how he grows?

Without building the next size of Square-Block Man, look at the numbers and figure out

      (a) how many more blocks will be gobbled up, and
      (b) the new size of Square-Block Man.

Now check your thinking by using the blocks.

What do you know about the next size of Square-Block Man?

How do other blocks in Blockland grow?

Notice how the story context provides both an overview and motivation for involvement. Here problem solving involves construction, computation, organization of data, searching for patterns, prediction, and testing. Thus the form of the story problem is not as important as its ability to involve a child in stretching his thinking.

The square growth problem was given to Gloria, Cynthia, Ted, and Randy without the story context. The girls quickly became involved in constructing larger squares and recording their data in a series of equality sentences.[11]

---

11. The form of these equality sentences was influenced by the context of the teaching/learning experiment. It is also possible to determine a pattern from the following series: $1 + 3 = 4$
$$4 + 5 = 9$$
$$9 + 7 = 16$$
$$16 + 9 = 25$$

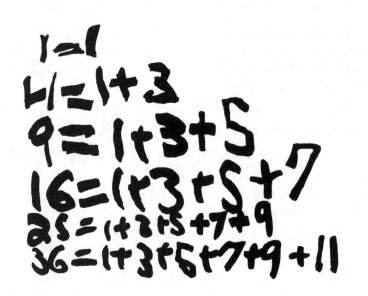

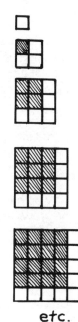

etc.

After recording the fifth stage of growth, they noticed a pattern in the number of additional blocks required for each succeeding stage. On the basis of this pattern they predicted the size of the next square to be forty-nine blocks. Gloria and Cynthia were experiencing the power of mathematics and enjoying it. Meanwhile, Randy and Ted made some larger squares and watched the girls' activity with little apparent interest. They did not accept the problem, illustrating the point that what may be a problem for one child need not be a problem for another.[12] Children screen out potential problems on the basis of a lack of understanding or lack of motivation. The Square-Block Man story context was devised later as a way of providing both overview information for understanding the problem and a source of motivation.

In providing children with a variety of problem situations it is essential that teachers free themselves from a central role as the sole presenter of problems. In the early grades all story problems must be read to the children. This can be done by other adults in the classroom, cross-age tutors, or even by good readers among the children. An alternative method of presentation, over which the children exercise more control, is an electronic card reader.[13] Each problem is recorded

---

12. Based on interviews conducted earlier in the year, I was aware that Ted was capable of considerable perseverance once he accepted the problem as his own (Chapter 4, p. 79) and that Randy was capable of recognizing patterns in numbers.

13. The card reader used in an unillustrated episode of the teaching/learning experiment was an Audiotronics Tutorette Model 822 operating at half speed on fourteen-inch cards. The device also has a capability of recording your own "talking cards." It is available from Audiotronics Inc., North Hollywood, CA. A headset is also available for use with the card reader.

both in print and on a strip of audio tape attached to the card.

PROBLEM CARD READER

The problem can be replayed at a touch of the button, to capture any missed details. Another button plays a second tape track containing reminders such as, "Is it an action or a thinking problem?" Following a brief demonstration of the operation of the card reader, individuals or small groups of children can work it at a center. The teacher need not be a central figure in the activity. For example, during the teaching/learning experiment some of the children's categorization of story problems took place in a center activity with the aid of the card reader. Although these second graders could read the problems themselves, the card reader was available for providing reminders in the absence of the teacher. Still, the children were in the driver's seat—they had control over receiving assistance. For some problems they were satisfied that they had made the right decision and chose not to press the button for reminders. The interaction between the children at the center continued at a high level as they listened and reacted to each other's viewpoints, sometimes integrating perspectives.

BEYOND NUMBER STORY PROBLEMS

Once children have begun to develop meanings and are "friendly with numbers," it is possible to present investigations and number puzzles without story contexts and without concrete materials.[14] This is illustrated by the following open-ended problem adapted from Wirtz (1981).

---

2   3   4   6   8   9   11   12   15   16   18   20

Using this list of numbers, how many different number sentences can you write?

Although an open-ended problem can be stated in a single general question, additional questions may be needed to help children open up other possibilities in their first exposure to this kind of problem solving. Additional questions can include:

How many different number sentences can you write with two or more "+" or "−" signs?

How many different number sentences can you write containing both "+" and "−" signs?

What is the longest number sentence you can write?

What is the longest sentence you can write using only ___ signs?

Also, more specific starters can be used such as a blank sentence form: ___ + ___ + ___ = ___. However, once children have compared their interpretations in group discussions, they are capable of initiating progressively longer investigations on different and longer lists of numbers. From the preceding list, the following forms of equality sentences can be generated:

$$3 + 6 = 9 \qquad 9 = 3 + 6 \qquad 11 - 9 = 2 \qquad 2 = 11 - 9$$

$$4 + 6 + 8 = 18 \qquad\qquad 3 + 6 = 11 - 2$$

$$2 + 3 + 4 + 6 + 9 + 11 + 12 + 15 = 8 + 16 + 18 + 20$$

$$20 - 18 = 18 - 16 = 8 - 6 = 6 - 4 = 4 - 2 = 11 - 9$$

Although computational skills are required to solve these problems, pursuing such investigations both require and develop other critical attitudes and abilities, such as reflection, ingenuity, sustained effort, searching for patterns, and perseverance in the face of difficulty (Wirtz, 1981). Whereas the discussion of categories of number story problems in this chapter has emphasized the development of rich mathematical meanings, the preceding open-ended problem has brought critical problem-solving skills and attitudes into focus.

---

14. Other sources of number puzzles and open-ended number investigations are found in Wirtz's *Banking on Problem Solving* (1976) and O'Brien's *Puzzle Tables* (1980). Other invitations for studying the number patterns in the growth of geometric configurations, such as the Square Block-Man, are presented graphically in Roper and Harvey's *The Pattern Factory: Elementary Problem Solving Through Patterning* (1980). The configurations are selected carefully so that children not only construct the shapes and predict the size of a shape without constructing it, but also identify a general rule for growth.

REFLECTIONS AND DIRECTIONS

In the first half of this book, a picture has been emerging of how school mathematics might be different for children. Rather than being an appendage to the primary mathematics curriculum, in which rote number combinations and worksheets are emphasized, problem solving can be the very heart of the curriculum.

Children invent varied and more efficient counting strategies in adapting to new situations. Number story problems provide physical contexts for such construction.

Children have difficulties with arbitrary premature introduction to the symbolism of the equality sentence—reading limited meanings into these symbols. Number story problems in the presence of materials provide physical contexts that initiate children's development of these ideas. They also provide interesting things to do while delaying the introduction of formal symbolism.

The meanings assigned to symbols by mathematicians are context-free. Number story problems provoke children to extend their understandings beyond specific physical contexts—beyond materials, space, and time.

---

1. Consider the implications of these findings for your classroom. Discuss your ideas with at least one other person.

Although this chapter has focused on children's development of mathematical meanings through problem solving, it has also highlighted their development of critical abilities and attitudes.

---

2. Consider how the following statement by Wirtz (1981) can apply to the number story problems discussed throughout the chapter, as well as to open-ended investigations such as the number list problem.

> The goals of public education must be to help all children find the most satisfying ways to approach all problem-solving situations—and all real-life uncertainty—with a confidence that they are not trapped. They must know that there are always alternatives from which they can choose—and that if they mobilize their own resources they can find an optimal solution. (p. 6)

9

TEACHING DIRECTIONS, DECISIONS, AND DILEMMAS

PREVIEW

Earlier chapters in this book have focused mainly on the state of children's numerical thinking as revealed by individual interviews. Later chapters have been interspersed with accounts of teaching episodes in which the interactions have had the potential for provoking a child to reorganize his thinking. Sometimes there is a fine line between interviewing and teaching, with the interviewer and the teacher demonstrating similar questioning, listening, and responding behaviors. This chapter will examine teaching behaviors as displayed in the teaching/learning episodes of Chapters 7 and 8, and will explore directions for change in the classroom. The discussion will extend beyond the promise of these methods to some of the dilemmas experienced by teachers attempting this kind of teaching in overcrowded classrooms. The problems I experienced in teaching only four children will be used as springboards for discussion of the realities of teaching a group of as many as thirty-five children.

First, you are invited to consider Duckworth's statement on classroom reality (1975):

> In most classrooms, it is the quick right answer that is appreciated. Knowledge of the answer ahead of time is, on the whole, valued more than the process of figuring it out. (p. 64)

Discuss Duckworth's statement in terms of teacher behaviors such as
    questioning
    listening
    responding

and other variables—
    class size
    group size

with at least one other person.

TEACHING BY QUESTIONING

The art of effective questioning is recognized universally as a crucial aspect of teaching. One way to focus on your questioning process is to tape-record a lesson and to examine the kinds of questions you are asking. There are several systems available for categorizing questions. In the following table, sample questions from the teaching/learning episodes are classified within a system devised by Lowery (1975). An examination of the four different types of questions illustrates how the teacher's construction of the question can encourage children to become engaged in specified types of mental activity. You are invited to examine the system in the following table before reading on.

A SYSTEM FOR CLASSIFYING QUESTIONS

NARROW →

QUESTION TYPES	EXAMPLES	DISCUSSION
Direct information: Recall or recognize information that is already known or readily available.	–What did you just tell me about the blocks? –How many blocks did you take away? –What did you tell me about each of the rows before we started? –How would you read this number sentence?	Direct information questions ask the child to recognize information readily available or to recall information that is already known. A review of information can serve as a basis for reconstructing a relationship.
Focusing: Transform information by comparing, analyzing, constructing relationships, procedures, or explanations.	–What is the same about all of these number sentences? –So, could we put a "+" there even though we're not putting on any more pieces? How come? –If that were true (5 + 3 = 6 + 2), what else could we put here (5 + 3 = 6 + 2 = ____) and still have it be true? How come? –Write a number sentence that tells us that the boy and girl have the same amount of candy and that they have a different number of butterscotch and chocolate candies. –Earlier you told me 3 was the answer. How come I don't see a 3 in your number sentence?	Focusing questions guide the child in a given direction or help him to focus on the relevant aspects of a problem, but the child constructs that response independently. Arriving at the response requires a comparison or analysis of the information or a reorganization of the information to construct a relationship, generalization, or explanation. Any probes or challenges that encourage the child to reconsider and restate a response to a refocusing question or to provide evidence for it can also be included in this category. Although it is narrow in focus, this type of question encourages a high level of thinking.
Valuing: Determine the goodness, appropriateness, or effectiveness of a product, position, or procedure.	–Could both number sentences fit both balances or does one number sentence fit better than the other? How come? –Would you look at the number sentence that Randy has and the number sentence that Cynthia has. I wonder if they're saying the same thing? (comparison leading to evaluation) –I'm wondering if it made sense to you or you believe it because your Dad told you? –But if it wasn't in the math book, would it be alright? How come?	Valuing questions encourage children to judge the excellence, appropriateness, or effectiveness of a product, position, or procedure. They are often accompanied by a request to explain the criteria for this judgment. Valuing questions are considered as broad questions because more than one defensible position is possible. A thoughtful response requires a very high level of mental activity that incorporates the processes involved in the above categories.
Open-ended: Broaden an area of study by generating multiple possibilities, interpretations, or solutions.	–Take turns reading your number sentences and read them in a different way without using the word *equals*. –Write the longest number sentence that you can write about '8'. –Write as many number sentences as you can about the candies.	Open-ended questions can encourage children to generate possibilities that broaden a new area of study. The possibilities can be generated through direct activity or through discussion that follows activity with materials. Sometimes the discussion can have the spontaneity of a brainstorming session, while at other times it can be paced to encourage thoughtful alternatives.

← BROAD

The teacher's distribution of the different question types is a strong determiner of the level of thinking involved in a discussion. The children's reconstruction of ideas and the high level of interaction in the teaching/learning episodes of Chapter 7 are a reflection of the number of focusing and valuing questions asked. Classroom research consistently indicates that although teachers ask many questions, most of them are of the direct-information variety. Those teachers who demonstrate a greater variety of questioning generate a higher level of discussion and greater student achievement. Research has identified two other classroom variables that affect the quality of children's thinking. These variables will be discussed in the following sections. First let's examine the need for sensitivity and flexibility in questioning.

Although questioning is certainly an important teaching skill, it must be employed with both sensitivity and flexibility. Extended opportunities to work with concrete objects allow young children both the opportunity and the time to develop intuitive ideas. This experience must precede attempts to reorganize understanding at a level of action to one at a higher plane. Asking focusing questions prematurely in an attempt to initiate this reorganization can interfere with the gradual process. Making an educated, sensitive guess at the appropriate timing of such a question is part of the art of teaching. Although the categorization of questions is a useful exercise in learning to phrase questions that encourage specific kinds of thinking, it must be applied with flexibility. The developmental level of the child and the timing of the question may interact to elicit unexpected responses. The child's available organization of ideas allow him to see only what he is prepared to see. There are times when children disagree with what they observe, illustrating that even the outcome of a direct information question is not always predictable. Similarly, focusing questions are directed towards a specific response. Yet, if a focusing question is directed at a group of children who are at various levels of understanding in a given area, widely different responses are possible at different levels of complexity and completeness. Each child will interpret the "narrow question" in terms of his own understanding.

There have been ample examples of multiple interpretations of focusing questions in preceding chapters. These examples underscore the need to consider the question categories flexibly and to ask questions with sensitivity. They also stress the importance of listening to the content of a child's response rather than listening for a category of response.

TEACHING BY LISTENING

Listening is another critical tool for raising the level of children's thinking during discussion. Easley (1980a) identifies some potential outcomes of "teaching by listening":

> Just listening closely to a pupil's erroneous explanation often seems to stimulate the student's incubation of richer, better understandings. . . . The student often straightens things out a day or so after he/she has tried hard to explain things to a good listener. (p. 37)

This section will examine what makes good listening as contrasted to what usually passes for listening in the classroom.

Describing good listening is much easier to do than to carry it out in practice. I made a conscious effort to be a good listener during the teaching/learning experiment (Chapter 7). The children, reconstructing existing notions of the equality sign, needed time to reflect on their ideas prior to responding to novel and discrepant situations. To facilitate the interaction, I also needed time to reflect on the meaning of their responses. Thus, armed with prior knowledge of Rowe's research on classroom listening (1973), I was consciously aware of the silent time I allowed to pass before and after a child responded to a question. After asking a question I tried to pause for at least three seconds prior to calling on a particular child to respond. Thus all children were encouraged to reflect on the question. Sometimes the length of the pause was consciously varied with the different types of questions asked, with direct information questions given the shortest pause. I also attempted to wait at least three seconds after a child's response before reacting to the response, redirecting the question, or asking a new question. The second wait-time allows for the possibility that a child may want to expand on the initial response. Rowe indicates that to be able to follow a child's reasoning, a teacher must match the length of the second silent period (wait time #2) with the length of the pauses that mark children's speech patterns during the spontaneous construction of novel responses.

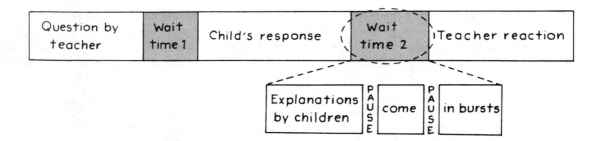

In one of the teaching/learning episodes (Chapter 7, p. 224), the length of my second wait time was shorter than the pauses in Ted's response. After four seconds I redirected the question to the girls at the same instant that Ted began to continue his response. Noticing this, I was able to return to Ted and catch the completion of his thoughts. In replaying the videotape of the episode, the amount of body movement accompanying his verbal response came into focus for me. With it came a greater appreciation of the effort that must have gone into the construction of his response.

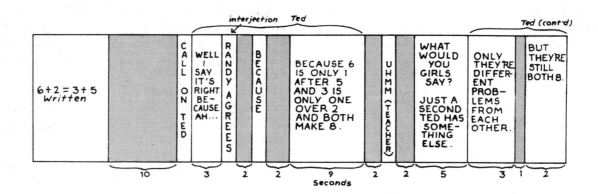

Since a child's explanation or other thoughtful response comes in bursts, an extended pause before reacting increases the possibility that either the child will expand the response or that other children will add to it.

In every group of children there are those who arrive at their responses quickly and are anxious to be first in giving their answer, as evidenced by "ooh" and "I know!" accompanying exuberant arm-waving. Unless you take the initiative to control the silent periods, you may find yourself conducting a dialogue with only a handful of children and assuming the others think alike. However, this control of pacing can only be attained with the children's cooperation. Methods that I used for informing children of the rationale behind wait time and for reminding them to wait are described below.

Introduction:
I've been talking to children since the beginning of the school year. What I've learned is that everyone takes time to think, but everyone doesn't take the same amount of time to think and everyone doesn't think the same way. I'm interested in finding out how everyone thinks and that's not easy because there're four of you. So when I ask a question I'm not so much interested in whether you raise your hand. I might call on you even though your hand isn't raised, to find out how everyone thinks. So that everyone has time to think I'm going to ask you to wait and not call out your answer. This way everyone will have a chance to figure it out and then I'll have a chance to listen to your different ideas. You'll find that you'll be learning from each other because everybody won't have the same answer.

Reminder 1 (the following day after the children had started to call out):
I know the answer. I need to be able to call on different people to find out how they think about it. (The answer is called out again.) I'm not sure anybody's listening because I still get people calling out. I'm trying to give you each enough time to figure it out. If someone calls out the answer and you're still thinking about it, you'll stop thinking when you hear the answer. Does that make sense? (Ted nods.) I'm going to need your help.

Reminder 2:
The next one's a hard one. Take time to think about it.

Nonverbal Reminders:
The hand-waving and calling out were reduced by conscious ignoring of the first child's wave, extending the pause and calling on another child. This was combined with signals indicating stop 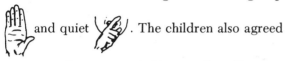 and quiet . The children also agreed to raise a finger as an indicator of readiness to respond in place of hand-waving. Once the children understood the purpose of the waiting and several reminders established the procedure, the pace of the discussions became more relaxed and reflective. The frequent "I disagree" indicated that the children were listening to each other.

Research by Rowe (1973) in two hundred classrooms indicated that average teacher wait times were about one second, and that the quality of classroom thinking was limited. Some teachers asked up to ten questions per minute, leaving no time for child or teacher reflection. However, working with groups of various sizes (four to thirty), Rowe found that if teachers with short wait times were trained to extend them to at least an average of three seconds, profound changes took place in the classroom interaction. From an analysis of over nine hundred audio tapes of classroom discussion made by these teachers, there is evidence of the following changes in children's behavior:

1. Children gave longer responses and the contributions of "slow" children increased.

2. Children initiated more responses that were appropriate and also asked more questions.

3. Children gave more alternative explanations, demonstrating speculative thinking.

4. Children made more and better connections among observations and inferences.

5. Children appeared to be more confident in their answers as they decreased the number of inflected responses.

Important changes in teacher behavior were also demonstrated in the wait-time tapes.

1. The teachers' questions decreased in number but showed greater variety and quality.

2. The teachers' responses were more flexible, reflecting a willingness to listen to diverse answers and reflect on their plausibility.

3. The teachers' expectations of "slow" children improved as they began to listen to them and to recognize their capacities.

4. The number of management moves by teachers decreased.

The last item, indicating control by teacher silence, seems paradoxical. Teachers, uncomfortable with silence, fall into the trap of speeding up the interaction to gain apparent control of the situation. However, the resultant barrage of questioning turns off the children and is accompanied by a loss of control. I am aware of this dilemma, having felt a momentary loss of control of a discussion when children appeared unwilling to wait. My first reaction was to regain control by calling on the child prior to asking the question. This is another trap that reduces participation. It is important to be wary of these potential traps when experiencing the discomfort of initiating the process. Furthermore, keeping in mind the powerful outcomes of the process may inject some needed perseverance during this uncertain period of transition.

Although Rowe's wait-time data were collected in the context of classroom science discussions, it is reasonable to generalize the conclusions to other subject areas that value thinking. In both science and mathematics, children learn best through mental activity in physical contexts with opportunities for action, reflection, and discussion.

Listening can increase the quality of interaction and the level of thinking in the classroom, but the impact of the listening technique is tied closely to ways in which teachers respond to children's diverse answers.

Praise is productive.

Praise is destructive.

Decide whether each statement is true or false.

Discuss your views with at least one other person.

TEACHING BY ACCEPTANCE AND ENCOURAGEMENT OF SELF-EVALUATION

During the late 1960s a number of innovative elementary science programs were designed to provoke inquiry by children based on firsthand experience with materials in interactions among small groups of children. The outcome of this inquiry was to be the construction of scientific concepts and generalizations. A follow-up study of teachers who had received inservice training to implement these programs indicated little inquiry in progress, despite admirable goals and an abundance of interesting materials and activities. Rather than inquiry, in which the process of figuring out was emphasized, there was an inquisition characterized by a barrage of questions and evaluative reactions by the teachers. Rather than the children deciding on the worth of a particular idea, judgments repeatedly came from the teacher. As much as 20 percent of teacher talk involved evaluative reactions such as "Right," "Good," and so on. Out of two hundred teachers studied, only a handful were facilitating inquiry in which teachers and children worked together in developing and refining ideas. The teaching style that supported this inquiry included not only the ability to listen, as indicated by extended wait times, but also the ability to refrain from use of either praise or blaming statements.

Although there is considerable research evidence of the effectiveness of praise on attainment of low-level tasks, such as memorization of addition "facts," we have only begun to look at its effect on more intellectually demanding tasks. Rowe's research indicates that praise has a destructive effect on the exploration of new ideas and the development of relationships. In classrooms where verbal rewards by the teacher were high, the following observations were made by Rowe (1973, and in Martin, 1977):

1. The children demonstrated low self-confidence as indicated by
   a. inflected responses having a self-doubting tone,
   b. low task persistence,
   c. eye-checking the teacher.
2. The children's answers were abbreviated, with incomplete explanations. Alternative explanations were seldom observed.
3. The children seldom compared answers or listened to or built on each other's ideas. The interaction was marked by hand-waving and competing for the teacher's attention.

Students who are praised heavily become conditioned to the "quick payoff" and are likely to avoid challenging tasks requiring persistence and complex reasoning. On the other hand, the teacher, by avoiding praising and blaming statements, can provide an atmosphere where it is safe to risk undertaking challenges. In this atmosphere of psychological safety inquiry is likely to flourish. Children receive alternative rewards such as inner satisfaction of self-discovery, self-evaluation, and working in cooperation with others.

Piaget's writings on education, although limited in number, emphasize the need for classroom intellectual freedom that gives the child the right to be wrong. He points out that in a developmental perspective children's ideas cannot be viewed simply as right or wrong for quick sanction by the teacher (1973):

In order to understand certain basic phenomena through the combination of deductive reasoning and the data of experience, the child must pass through a certain number of ideas which will later be judged as erroneous but which appear necessary in order to reach the final correct solution. (p. 27)

In this framework it is important to consider children's wrong answers as being appropriate to their current level of development. Teachers who insist on "right" answers are demanding something that young children are incapable of giving, or demanding something trivial. When teachers regard their incomplete answers as "wrong," children soon learn to distrust their own abilities and believe that the correct answer exists only in the teacher's head (Kamii, 1973). They begin to watch the teacher's eyes as if asking, "Is this what you want?" When children do not produce adult responses, teachers tend to impose ready-made answers on them.

Piaget views a child's errors not as responses to be eradicated, but rather as things to be examined and appreciated as windows to his thinking and essentials to his construction of knowledge (Piaget and Duckworth, 1973):

> Yes, I think children learn from trying to work out their own ways of doing things—even if it does not end up as we might expect. But children's errors are so instructive for teachers. Above all, teachers should be able to see the reasons behind the errors. Very often a child's errors are valuable clues to his thinking. . . . A child always answers his own questions correctly; the cause of an apparent error is that he did not ask himself the same question you asked him. (p. 24)

Thus, rather than view children's responses as things to be judged right or wrong, it is possible to accept them as potential indicators of children's current thinking and as a resource for asking the next question or devising the next task.

The work of both Piaget and Rowe guided my interactions with children in the preceding teaching/ learning episodes. I made an attempt to provide optimal possibilities for interaction as the children tested their existing notions in a range of novel physical contexts and were confronted by feedback from a variety of sources. Testing, discussing, and evaluating ideas made in physical contexts was the setting for children to reconstruct equality concepts at higher levels of mathematical relations. To encourage children to risk trying out new ideas, feedback on these ideas from the physical contexts and from their peers was maximized at the same time that my adult authority and evaluative feedback was minimized. Praise and criticism usually were replaced by a friendly but neutral acceptance of all responses as valuable and as indicators of a child's current level of understanding. Instead of making repeated statements of praise, blame, or "Yes, but . . . ," I provided a general atmosphere of acceptance. Some statements of acceptance I used in the teaching/learning episodes, and their apparent effect on children, are indicated in the following table.

TYPE OF ACCEPTANCE	EXAMPLE	CHILD'S INTERPRETATION
Silence		I'm doing different from Gloria, but the teacher didn't tell Gloria that hers was good and he didn't say mine was bad. I must be doing as well as Susie..
Acknowledgment	Uhmm, I see, OK, (nod)	He heard me.
Summary (after hearing responses from all the children)	So there's more than one way to look at these. Everybody doesn't agree, but you all have interesting ways to look at the blocks and numbers.	It's OK if I don't think like everybody else.

Acceptance is not agreement. Rather, it indicates that the child's response has been heard without judgment. Children are free to build on each other's views or reconstruct their own views.

Although the teacher may not agree with the children's ideas, his own ideas are not imposed upon them. Children may consider another child's view as just another opinion, but give the teacher's view more weight and accept it as fact without any true understanding. Rather than judge the children's responses, the teacher can redirect the authority for making evaluations back to the children by inviting their reactions. Examples of such redirection of authority used in the teaching/learning experiment follow.

WAYS TO REDIRECT AUTHORITY

TYPE OF REDIRECTION	EXAMPLE	CHILD'S INTERPRETATION
Redirection of question to the other children	Uhmm . . . any other ideas? or What do you think, Randy?	Well, he always either asks you to think more about your answer or asks for other people's ideas whether you're right or wrong, so my idea can still be OK. He's interested in what I think.
Redirection of a child's response to another for an opinion	Cynthia, what do you think about Ted's idea?	
Redirection of a response to a hypothetical peer (countersuggestion)	The other day a boy/girl said that . . . What do you think about his/her idea?	
Redirection of response for checking against a physical model	Which of those are Johnny's airplanes? Now, show me Scott's airplanes (comparison number story).	

In one-trial learning, the teacher imposes a judgment on the first response or may redirect the question to another child only when the first response is wrong. In the latter case, the judgment is given after the second response. In this kind of interaction, the redirection of a question can be a traumatic experience for the first child. On the other hand, when children's responses are accepted by the teacher, at the same time that feedback from other sources is maximized, the process of "figuring out" can flourish in an atmosphere of psychological safety. As the source of authority is redirected from the teacher to the children, both the children's competence in constructing meaning and their confidence are free to develop. Cynthia's confrontation of the authority of the math textbook with the authority of her own logic epitomizes the autonomy of thought possible by children.

IT DOESN'T MATTER IF IT'S IN THE MATH BOOK OR NOT, AS LONG AS YOU THINK IT'S RIGHT.

Cynthia

At this point you may have the impression that there is no room for praise in the classroom when higher-level thinking is involved. Both global praise and praise of the child are potentially harmful, as suggested in the next two examples.

While both global praise and praise of the child have the potential of inhibiting a child's performance, other forms of praise may encourage children to extend their performance. Two such forms of praise are descriptive praise of a child's response and a spontaneous and authentic expression of appreciation of a child's work.

The teacher's description of the child's response brings his process into focus and encourages the child to build on this increased awareness. Both descriptive and appreciative responses can be nonevaluative, allowing children to evaluate themselves. Any time the teacher's response is evaluative, the authority for making judgments is being redirected back to the teacher. One of the dangers of praise, however carefully worded, is that it can create a dependence on the teacher.

Although the preceding alternatives to familiar forms of praise can facilitate thinking in some cases, they may also inhibit thinking in others. Earlier we described Cynthia's preference for her own logical thinking over the authority of the math textbook. This demonstration of confidence and autonomy occurred in the total absence of praise. In the delayed postassessment (one month later) I inadvertently slipped into a praise response pattern instead of the usual neutral acceptance. This change was accompanied by a simultaneous change in Cynthia's response pattern, suggesting a shift in her confidence level.

TASK AND CYNTHIA'S RESPONSE	MY RESPONSE	CYNTHIA'S REACTION
Novel task in physical context:		
She wrote 8 + 1 = 4 + 5	Wow! That was a lot to think about in one sentence and you did it.	Was I right?
	Does is look right?	Yes.
Simultaneous equalizing of unequal groups:		
She wrote 11 − 3 = 5 + 3.	Hmm, you did it again!	
Number story problem with single joining action:		
She demonstrated the physical model.	No response.	Am I right?
	Do you have to ask me or do you know?	I know.

Cynthia's need to know whether her answers were right had not been observed in prior interviews or teaching/learning sessions. How do you account for this sudden shift in behavior? For me, the preceding incident demonstrates the fragile nature of children's intellectual autonomy in the presence of adult power of evaluation.

Although we have been discussing listening and responding separately, they are obviously intertwined. Easley and Zwoyer (1975) relate the two in discussing teaching by listening and identify them as critical to the art of teaching:

If you can both listen to children and accept their answers not as things to just be judged right or wrong but as pieces of information which may reveal what the child is thinking you will have taken a giant step toward becoming a master teacher rather than merely a disseminator of information. (p. 25)

They view this change as a giant step because of the demands of listening. Not only is listening exhausting, it also reveals unexpected challenges in children's responses.

ON WORKING IN GROUPS

The urgency of organizing the classroom for small group or individual work is brought home to teachers having the experience of interviewing several different children individually on the same task. The range of approaches required to make contact with the children's thinking provides a convincing demonstration of the futility in expecting true communication with a whole class of children at any given time. Rather, the classroom can be organized for both teacher-directed work with individuals and groups and for self-directed group work, with the latter grouping making the former possible. The following section offers guidelines for working in groups and also discusses dilemmas faced by teachers in group work. It begins with a discussion of my experience with group work in the teaching/learning experiment.

Group Mix

In working with the four children in the teaching/learning experiment I found them to have sufficiently diverse views of the equality sign to stimulate group interaction leading to a refinement and reorganization of original ideas by individual children. Although I chanced on this combination, there are times in the classroom when group membership is purposefully structured to maximize interaction. Despite the rich mix of ideas generated by the variety of tasks in the teaching/learning experiment, there were other problems in the group's interaction. A discussion of these difficulties may suggest other reasons for paying attention to group membership.

Despite prior knowledge of these children from individual interviews, I found it very difficult to predict how they would react to each other and to specific tasks in a group situation. All of the children had been interviewed twice in the preassessment. In addition, the two boys had been interviewed on other tasks several months earlier. During the interviews all of the children demonstrated an affinity for reasoning about numbers. The girls were very quiet during the interviews while the boys were more talkative. The teaching/learning experiment brought together two girls from one classroom and two boys from another. In the group situation, the girls became very talkative, interacting with the ideas and with each other. The once-talkative boys seemed to be overshadowed by the girls in thinking and verbalizing their ideas. The girls were also more mature physically and when they first met, the girls thought that Ted was a first grader. One of the problems I experienced was keeping the boys involved. I attempted to solve this by distributing participation in discussion as much as possible. Undoubtedly, knowledge of the state of each child's numerical thinking is crucial in teaching a class, but the complexities of classroom dynamics make it clear that other issues must be considered in predicting the performance of individual children. A discussion of group mix in the context of self-directed groups is found in Rowe (1973).

Guidelines for Group Process

At times I paired up children with contrasting views of the equality sign for an intensive interaction. Although classroom teachers are asked to involve thirty-five children simultaneously in the process of learning, there were times when two children seemed too many to me for following the thinking of an individual child and for provoking any rethinking. Teachers must be aware of many simultaneous happenings in their classroom and must handle these multiple tasks without being sidetracked. There is an obvious conflict between thinking about a child's reasoning and thinking about what might be going wrong with the rest of the class. This dilemma may underlie the superficiality of many classroom interactions. Some directions for resolving this dilemma and gaining the freedom for intensive interactions with small groups involve a reliance on the responsibility of children outside the teacher's current group or on adult assistance.

In a classroom where a variety of interesting problems and materials are available, children can not only become involved, but also raise their own questions and pursue multiple solutions. Rowe's research (1973) reveals that in classrooms with a low frequency of praise the children are less dependent on the teacher and tend to resolve their differences through discussions and comparison with other children. Burns (1980) offers some practical guidelines for working groups of four children that encourage both cooperative learning and less reliance on the teacher as a resource person:

---

1. Everyone in the group is responsible for himself.

2. Everyone in the group should be willing to give help to other group members when asked.

3. You may not ask the teacher for help until everyone in the group has the same question.

As was the case with initiating listening (wait-time) behaviors, it is important to have children review these group guidelines prior to the activity. Furthermore, Burns suggests that following the activity, the children be asked to evaluate their group performance based on the guidelines. Some teachers have achieved freedom for concentrated work with groups of children through assistance of other adults such as paid aides, volunteers, or a second teacher with whom they team.

A second group-learning situation involves cooperative learning by all children working in groups on problems of interest. In this situation, the teacher circulates around the room, observing and participating as needed. If the group is seen to be doing well, the teacher avoids any intrusion. Sometimes the group can be joined and observed without intrusion by kneeling or sitting down at their level and avoiding eye contact. Burns (1980) suggests tuning into the group process and ascertaining the missing ingredient when a group is bogging down. Since successful group interaction is usually dependent on group members assuming four roles—the doer, the questioner, the prober, the summarizer—absence of one or more of these roles can account for the interaction breaking down. By interjecting at this point and assuming the missing role, the teacher can revive the process and then leave. Joining a group without intrusion and participating only as needed is another example of redirecting authority back to the children.

There is no single best way to arrange learning groups. Actually there is an advantage to children experiencing learning in a balanced variety of groupings—independent work, cooperative groups of four children, teacher-led small groups, whole-class group—selected for their appropriateness to the situation.

Class Size

As a counterpoint to the preceding discussion of an ideal towards which teachers can strive, I provide the following honest glimpse of classroom realities as experienced by many teachers. In an excerpt from an interview with two British teachers, we share in their dilemmas in teaching thirty to forty children, their struggles, their guilt at failing to reach all of them, and their current method of coping. Their comments on teaching also include an expression of frustration and anger at people like myself who write books about teaching methods.

INSIDE CLASSROOMS[1]

Q   What do you think out of the classroom then?

T₁   I think what a miserable failure I am.

Q   Why?

T₁   Because I could do so much better in other circumstances if I had more knowledge, if I had more know-how, if I had more time, if there was more of me, if there were less kids, if I had more apparatus and knew how to use them, all those kind of things.

T₂   Yes, I think that teaching from the time that you are a student teacher onwards is set up by people who are not teachers, who write books and who work in institutions other than the sort of schools we teach in, in such a way that you're bound to fail to do the job as they set it out. You cannot possibly teach successfully in the way that it's laid down that you should do—that is, unless you totally ignore the amount that the children are learning and just concentrate on your speech delivery and the kind of things you do as a teacher. And I think you can see something happening to teachers after a year or two, they suddenly notice that they are dashing round frantically from child to child in a desperate attempt—I did this myself—you catch yourself out dashing around trying to do the impossible, trying to split into thirty different people, to cope with thirty or forty different children, and once you've done this you have to face the fact that you're failing that way, and that perhaps any other way you can think of acting you will continue to fail to meet the needs that you can see you could meet if there were thirty of you. You can always see many more times the amount of work you could do than you can actually get done in a day. But you have to learn to relax and B and I have both been doing the job long enough to have passed through this stage of frenzy.

Q   But you say you catch yourself doing the impossible, or trying to rush round to meet the needs of thirty children. One solution to that would appear to be to teach all thirty as a unit.

T₁   Yes, that is one solution, but that's an even bigger failure.

Q   Why?

T₁   Because there are thirty different people not one person. There are thirty different mentalities, intellects, abilities, range of interests, ways in which they learn and stages and speeds at which they learn.

Q   But J says it's impossible trying to cope with thirty as individuals and you're saying it's impossible to cope with the thirty as a unit. Why is the one preferable to the other? You both seem to have chosen the thirty as individuals.

T₂   I look at it now this way: whatever I do I'm going to fail most of the children most of the time. I'm going to fail to be with them, I'm going to fail to be part of their learning experience in school. So I can do various things about this; I can choose the way of failing which I as an individual find least unsatisfactory or most satisfying. One of the ways I like best is to try and work with a group of children until they are happy with their place, their room in school, and happy with one another enough to come in and get on with a variety of activities there which don't require my presence as a policeman or referee or helper of any kind. Then I can turn my back on them and sit in a corner with one, two, three or four or however many I choose of the class, and I can be confident that for half an hour, or an hour or a whole morning if necessary, or even a whole day, I can basically ignore ninety per cent of the children who are in my class and who are my responsibility. This is just one way of coping with failure. It seems to me that all the people who don't teach in this situation, starting with non-teaching Heads, do go in for the most incredible fantasies about what is actually happening in classrooms and what would help in classrooms. I think that this makes a lot of the advice that is given by non-teaching Heads, by advisers and teacher-trainers, and a lot of the advice written in books, come from a different world and that's why it often seems, to people who are actually teaching, very unhelpful. I am not prepared to say that reality is all on our side and fantasies all on their side, but their visions of what's going on, and what is possible, are just different visions, and there is possibly less communication between these two kinds of vision of what is going on in the school and the classroom than is normally admitted.

As reflected in the preceding commentary on teaching, a frustration experienced by most teachers is the size of the classes they are assigned. A 1978 research survey by Glass and Smith of eighty different studies confirmed this pervading concern of teachers. The study concluded that:

---

1. Association of Teachers of Mathematics. *Notes on Mathematics for Children.* London: Cambridge University Press, 1977, pp. 201-02.

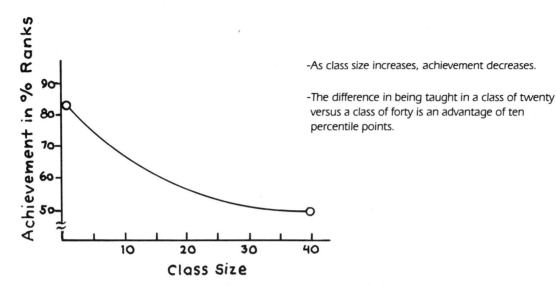

-As class size increases, achievement decreases.

-The difference in being taught in a class of twenty versus a class of forty is an advantage of ten percentile points.

The researchers commented on the advantage of smaller class size by stating, "Few resources at the command of educators will reliably produce effects of that magnitude." In a follow-up study (Smith and Glass, 1979), the researchers examined fifty-nine studies published since 1928, and found further evidence in support of smaller class size. The report contains the following conclusions:

Class size affects the quality of the classroom environment. In a smaller class there are many opportunities to adapt learning programs to the needs of individuals. Many teachers avail themselves of these opportunities; others would need training to do so. Chances are good that the climate is friendlier and more conducive to learning. Students are more directly and personally involved in learning.

Class size affects pupils' attitudes, either as a function of better performance or contributing to it. In smaller classes, pupils have more interest in learning. Perhaps there is less distraction. There seems to be less apathy, friction, frustration.

Class size affects teachers. In smaller classes their morale is better; they like their pupils better, have time to plan, diversify; are more satisfied with their performance.

Since research in this area is difficult both to design and to conduct, available studies are open to different interpretations. Thus the conclusions on the positive effect of reduced class size on achievement, pupil affect, and on the teaching process are not accepted universally in the education community. And although the results of the above research indicate a general trend, a reduction in class size is not automatically accompanied by improved classroom interaction and achievement. In other words, smaller is not always better. To maximize the impact of reduction in class size, teachers must conceptualize teaching and learning differently. In doing so, the importance of paying attention to the quality of questioning, listening, and responding comes into focus.

ON MANAGING MATERIALS

Materials are critical to children's development of mathematical relationships, but managing them is the bane of many teachers. Several potential problem areas are the following (Labinowicz, 1980):

1. *Acquiring sufficient materials:* Sharing intriguing materials that are in short supply can lead to conflicts among children.

2. *Distributing materials:* Some children may complete an introductory activity before others have received the materials.

3. *Collecting materials:* Some materials are just never returned.

4. *Scheduling and situating discussions:* Distribution of intriguing materials prior to giving directions can sabotage your efforts with large groups.

5. *Terminating the activity:* Children can become extremely frustrated by interruption of their work.

6. *Maintaining a quiet environment:* Activities with materials are much noisier than those with workbooks.

7. *Scheduling activities:* Activities with materials take longer than paper-and-pencil activities.

All of these potential problems can be minimized through teacher awareness and cooperative planning with children. To expand your awareness, identify the problem then brainstorm a range of solutions for each problem area with two other people or with your class. Following the brainstorming, compare the advantages and disadvantages of the different solutions for a particular situation as the basis for your classroom decision making. There are no universal solutions for classroom problems, but intelligent choices are possible if you are aware of the trade-offs.

In my own work with children on equality, I was aware of conscious decision making in the first episode (Chapter 7, p. 167). Here, I examined the trade-offs involved in two different approaches to the comparison of sets. I made a conscious choice in favor of increasing the impact of the encounter, despite the need for large quantities of materials. While writing about this decision making a third choice has occurred to me. All three choices and their respective trade-offs are described in the following table.

CHOICES AND TRADE-OFFS IN USE OF MATERIALS

ACTIVITY	CHOICES	TRADE-OFFS
Comparison of sets of objects and their representation (a) unequal sets, and (b) equal sets as illustrated in Chapter 7 on page 167.	Package sets to be compared in separate bags, using color as a distinguishing property. Order the bags in a sequence with unequal sets preceding sets of equal size. Each child has his own materials.	-Maximum involvement of individual children. -Maximum impact of the confrontation with the need to represent equal sets. The sudden appearance of these sets following an expectation for unequal sets is a discrepant event that might engage their thinking. -Lots of materials are required. -All children will not arrive at the discrepant event at the same time.
	Pair up children for a "handfuls" game in which the size of the sets to be compared is randomly generated.	-Impact of confrontation with equal sets is reduced by random exposure. Possibility exists for children to avoid the confrontation by adding another object to their handful. -Fewer materials are needed.
	Teacher demonstrates sets to be compared and children represent the comparison.	-No child involvement with materials. -Sequence is controlled for maximum confrontation with need to represent equal sets. -Fewest materials required. -Possibilities for interaction on the problem of representing equal sets are increased as all children are asked to deal with it simultaneously. -Children are forced to function at the same rate.

In selecting the first approach I magnified the intensity of an encounter to engage the children's reconstruction of existing notions of equality and its representation. As I asked the children to make displays of the sets together with their equality sentences, their work space quickly became cluttered. Needless to say, in a whole-class situation this choice would have been impractical and I would have examined the trade-offs involved in the other approaches.

The impact of blocks hitting the hard table and

floor surfaces at the independent learning centers was a major source of distraction from concentrated work with the children at hand. The noise factor of these materials was minimized by providing large cardboard trays that served as both containers for materials and as work space. The trays not only muffled the sound but also reduced the number of blocks hitting the floor. Although this management problem was minimized, the next section will reveal that it was not solved totally. In our search for the best way of doing things we often find that there is no best way—only alternative ways and trade-offs.

ON EXPECTATIONS AND TEACHING

Another teacher dilemma is how to deal with the apparently contradictory demands of planning lessons and responding flexibly to children's spontaneous learning attempts.

We need to plan activities based on what we know about children.

We need to anticipate what will happen in order to maximize learning and to prevent problems.

In order to operate we need a degree of uniformity/predictability.[2]

It is difficult to listen to children and to appreciate their responses when you're thinking ahead to what has been planned.

It's easy to overlook children's unusual responses or to overlook their needs when trying to achieve the objectives of the lesson plan.

In my work with second graders I usually arrived overplanned in terms of activities available and general expectations for our session. By overplanning I set myself up for disappointment. Although I had written on the evolution of mathematical ideas in children's minds, I often left the session frustrated at the slow rate of progress. Having only twelve teaching/learning sessions available added to both my feelings of frustration and to my overplanning tendencies. In reviewing the videotapes each night I usually was pleased with the children's progress and expressed continued surprise at the level of their interaction. This videotape feedback allowed me to listen more closely to the children's responses to the tasks. This was accompanied by an ability to detect changes in the level of the children's interpretations of the many faces of equality. From this vantage point I was able to ease up on my expectations to some degree and be more relaxed during the teaching/learning sessions. In reflecting on the pressures of my own expectations, I gained a feeling for what it must be like for teachers who usually have expectations placed on them by parents, administrators, standardized tests, and so on, and how this pressure affects their teaching and their enjoyment of children.

Planning lessons allows for reflective decision making that results in some intelligent choices. This is best accomplished in the comfort of an armchair. However, many classroom decisions must be made in a split second, with no time for reflection. I was reminded of the latter type of decision making faced constantly by teachers throughout the school day by the following incident. One rainy day, I met the children after they had been confined to their classroom during both recess and lunch-hour periods because of the weather. At this point in the day the sky was beginning to clear and the playground had some dry areas between the puddles. Faced with some challenging tasks at the opening of our session, the children were restless and generally unresponsive. At this point I decided we all should take a break outdoors. After five minutes of playful running, the children returned to the classroom breathless, warm, and in no condition to resume serious activity. Rather, they suggested a bathroom break. At this point, two of the children bolted for the door. Although I called them back and tried to get started again, the rest of the session was not very productive, a situation compounded by one of the children dropping a large number of blocks. This experience reminded me of the need for split-second decisions and their cumulative stress on the teacher. Furthermore, it made me conscious of how expectations can interfere with recognizing the needs of children. If I had paid closer attention to the children on this rainy day, we could have taken the bathroom break to cool down, and could have revised plans by switching to a less demanding activity.

---

2. Wheeler et al. *Notes on Mathematics for Children*. London: Cambridge University Press, 1977.

ON TEACHING IN THE FACE OF RESISTANCES

The second graders in the teaching/learning experiment were studying addition combinations to 20 and computation of two-digit numbers in their math textbook, with little exposure to materials. Thus, the introduction of equality situations with both a focus on small numbers and materials resulted in some initial resistance. Randy commented that it was "baby stuff," and Ted frequently ignored the number sentence as a representation of the physical model and thought up unrelated ones. On one occasion, Ted wrote 1000 + 2000 = 3000, perhaps in silent protest of our focus on small numbers. To counteract these resistances, I introduced activities as dealing with "big ideas about small numbers." I also tried to make them aware of the progress the group had made, such as being able to accept specific formats of the equality sentence that had previously been rejected. The emphasis on modeling with physical materials was rationalized for children unaccustomed to working with them as something needed by me to help me understand how they thought about numbers.

I usually met the children outside their classrooms and walked across the school yard with them to reach our room. This provided an opportunity for casual conversation between us or between the children. Although the children were not always talkative en route, they became talkative just at the instant that the activities were to get underway. The girls, in particular, initiated interactions among themselves as if to avoid getting started. Being aware of some of my own ways of avoiding difficult tasks, I sensed a parallel in their diverting behaviors. On three occasions I expressed my frustration at not being able to get started sooner. Since their participation in our sessions was voluntary, I even offered the option of terminating the activities if they really didn't want to do them. In each case, the children agreed that they wanted to continue and to return the following day despite the fact that the sessions took place after school for the boys and during the physical education activities for the girls. Although this discussion brought an immediate focus on the tasks at hand, the ritual continued as the children returned the next day. Once the children got started, however, the level of our interaction was both intense and productive.

Unexpected resistance came from the girls at being paired up with a boy to form subgroups for alternating concentrated teacher-led activities and independent activities. After introducing the pairing in terms of a "need to work with others who don't think like you do so you can think about each other's ideas," the girls reluctantly accepted the pairing. Prior to encouragement to get into the range of the TV camera, the boy and girl usually sat at a maximum distance of separation. Again, once the activities got underway the mixed pairs worked well together.

Randy showed the greatest resistance to modifying his ideas of equality when exposed to a variety of novel contexts. There were a number of occasions where he was the only child in the group who clung to an old notion of unidirectional equality. Children with less confidence in their own ideas might have distrusted their own thinking and gone along with the group. Rather, Randy frequently expressed dissent from or confusion about the other children's ideas. Despite occasional withdrawal from participation, he would bounce back and maintain the autonomy of his thinking, even when it was a minority view in this group. At one point he became aware of some compelling reasons for changing his ideas and discussed them with his father. Perhaps his father's authority may have influenced his thinking at this stage. However, this discussion did not result in merely assimilating some information, but rather in modifying basic ideas of equality that later served as a departure for further refinement. Additional insight into Randy's resistance can be gained from Sinclair's thoughts on the virtues of resisting new ideas, particularly as suggested by others (1977):

> It is very important to hold on to your theory until you've got good reason to let it go or to change it in some way. . . . It is much better to stubbornly stick to a theory that is partly wrong and to refuse to see the counterexample than to give up your faith in them altogether. (pp. 69–70)

While persisting in following his own ideas, Randy was demonstrating faith in his ability to construct reliable theories of the world around him.

REFLECTIONS AND DIRECTIONS

Considering the complexities of teaching a class of children, the level of explanation and advice offered is usually too simple—with classroom variables discussed in isolation.

1. Reflect on the following relationship of teacher and child variables in classroom interaction as proposed by Rowe (1973) following a study of many classrooms. Discuss it with at least one other person.

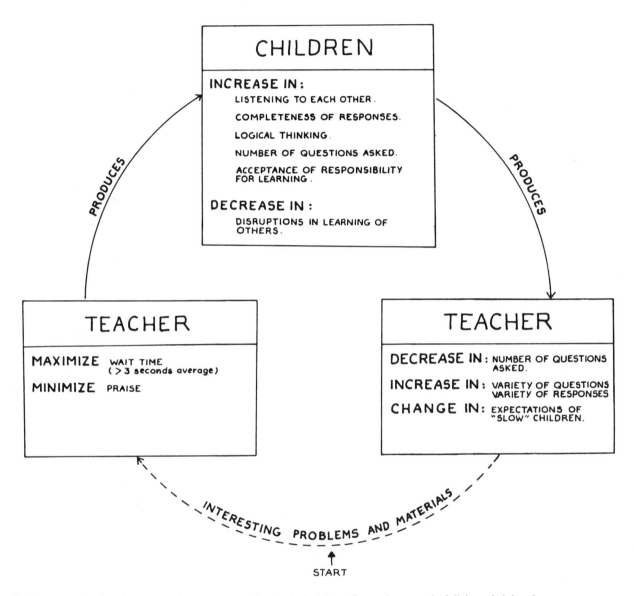

2. You are invited to test the proposed relationship of teacher and child variables in your own classroom. You might begin by tape recording a discussion with at least four children in which ideas are being developed and refined and then listening to the interaction on a playback of the tape.

PART THREE

PLACE VALUE FOUNDATIONS AND APPLICATIONS: Thinking, Learning, Teaching

10

COUNTING
BY ONES, TENS, AND HUNDREDS:
MIRROR OF
PLACE VALUE READINESS

PREVIEW

Children are actively engaged in organizing the world around them in an attempt to understand it better and to interact with it further. Their continued development of counting abilities is one example of such active organization. In dealing with the complexities of counting larger quantities, children adapt and reconstruct their counting procedures at higher levels of organization and efficiency—counting by tens and hundreds as well as by ones.

Earlier chapters have examined three phases in the development of young children's counting:

1. Acquisition of the conventional number-name sequence to thirteen.
2. Coordination of number names with objects being counted
3. Construction of number meanings to assign to the number names

Each phase has been observed to emerge gradually, with the construction of number meanings being the slowest to develop. This delayed acquisition of number meanings places constraints on children's counting as a problem-solving tool. Once the number meanings have been constructed for small quantities, children demonstrate the power of their counting abilities through reversible counting-on and counting-back methods in problem solving. Also accompanying the construction of number meaning is the invention of counting strategies requiring more mental involvement with less dependence on perceptual support from objects. In the continued development of counting abilities to counting by ones, tens, and hundreds, all three of the above phases can be observed once again at another level of complexity.

While children are constructing number meanings for small quantities and experiencing the power of their counting, they are simultaneously extending their conventional number-name sequences and coordinating these sequences with the counting of increasingly larger quantities of objects. Whereas previously the children acquired thirteen number names by memorizing the sequence, they now begin to extend the number-name sequence by abstracting sound patterns from spoken language combined with spontaneous practice in oral counting. As they count progressively larger collections of objects, they partition them into equivalent groups to facilitate both their counting and comparison. Now the question "How many?" can be answered by counting by ones, twos, threes, fours, fives, tens, and so on, depending on the size of the groups. However, just as young children experience difficulties in coordinating the assignment of successive number names to objects in a one-name-to-one-object correspondence, older children experience difficulties in assigning new lists of number names with intervals of one, three, five, ten, and so on to equivalent groups of objects in a *one-name-to-many-objects correspondence*. As the size of the collections continues to increase, eventually the process of organizing them is extended to forming groups of groups for easier counting and comparison. With the increased complexity of the counting process, children integrate different procedures for counting and construct relations for groups and groups of groups at a very gradual pace. Such construction and coordination of relational meanings represents another major intellectual achievement by children and forms the basis for their understanding of our place value numeration system.

In most classrooms, counting methods are discouraged in favor of the apparent speed of formal computational methods. Teachers tend to consider counting as a simple procedure and to view its use by older children as a sign of intellectual weakness. Despite being taught difficult formal methods of computing, children continue to use counting strategies throughout the grades. Many children have more confidence in methods of their own construction than in their ability to imitate the ready-made methods of adults. A better understanding of children's natural methods provides invaluable clues in our search for more effective means of teaching. It may be possible to use our knowledge of children's methods as a basis for teaching—not only encouraging their use, but also helping children to extend them to higher levels of competence and efficiency.

This chapter again will take counting out of the classroom closet, as it were, and explore how beginning third-grade children deal with the complexities of counting by ones, tens, and hundreds in a variety of different situations. The understanding of number relations that children's counting methods imply will be examined in the context of these tasks. Also, alternative teaching methods for extending children's counting of larger quantities to more efficient levels will be examined.

You are invited to place yourself in the children's shoes and to anticipate how they might deal with the following tasks.[1]

Generating Number-Name Sequences—Oral Counting

Count forward (by ones) from seventy-seven.

Count back (by ones) from fifty-four.

Count forward by tens.

Count forward by tens starting at thirty-four.

Count back from one hundred thirty by tens.

(Prompts were given if needed—initiating the sequence and asking the child to continue.)

Counting Objects

Count the number of cubes showing. There are more cubes under the box. As I uncover them, count to find the number of cubes altogether on the board.

1. Counting tasks used throughout this chapter have been adapted from the work of Leslie Steffe and Larry Hatfield conducted at the Project for the Mathematical Development of Children and the University of Georgia Mathematics Education Center.

There are forty-six cubes under this box. Count-on to find out how many cubes there are on the board altogether.

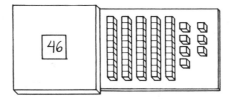

There are fifty-one tiles on this board. How many do you see? The rest are hidden here. Count to find out how many are hidden.

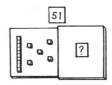

GENERATING NUMBER-NAME SEQUENCES FOR ORAL COUNTING

Children's ability to generate number-name sequences is based on their capacity to notice sound patterns in the spoken language of numbers. This same capacity accounts for much of children's early language development. A discussion of the child's active role in his language development will serve as a general introduction to generating number-name sequences for oral counting.

Children usually demonstrate a good grasp of spoken language by age five, independent of any formal instruction. This language development is reflected not only in the child's growing vocabulary but also in the application of many language rules. Although imitation plays a role in the young child's acquisition of language, his speech does not have a simple direct relationship with adult models he hears. Without formal teaching of language the child is exposed to isolated instances of spoken application. From such an exposure he gains an intuitive grasp of the rules for an invisible language system. In the gradual acquisition of grammatical structure in his speech, the child provides evidence of active constructions within the limits of language. His use of "he goed" and "I taked it" are suggestive of rule-governed behavior. The child's active search for rules is further suggested by his apparent dissatisfaction with the plural form of "foot," his active attempt to generate other viable possibilities, and, finally, his acceptance of "feet" as an exception to his rule. This cycle is reflected in the following possible usage pattern: "feet, feet, foots, feets, feetses, feet, feet, feet." The child experiments freely with words in an active search for patterns in adult speech (Sinclair, 1976).

Similarly, there is evidence of children's experimentation with number words and number-word sequences in an active search for patterns and a persistent effort to extend those sequences. Initially, since no sound patterns exist within the first twelve number names, these are learned by direct imitation of adults or older children. However, in hearing other number words in the course of everyday experience, children begin to notice repeating patterns and derive new number names to extend their sequence. Spontaneous practice in oral counting, active experimentation with the decade cycles noticed within the sounds of number names, together with a willingness to ask for information, result in a continuous extension of their number-name sequence (Ginsburg, 1977). Imitation gives way to active generation. Children's invention of strings of nonconventional number names provides evidence of their active construction of procedural rules for their generation. Thompson (1982b) writes that a child's extension of the number-name sequence beyond nineteen might result in unconventional words such as "tenteen, eleventeen, twelveteen." From this unconventional construction he would infer that the child had abstracted a rule for counting beyond twelve, such as the one below.[2]

> After twelve keep counting from "three", sticking on the word *teen* after it each time. (Watch out for the funny-sounding words, like *thir-* and *fif-*. They sound better if you say them fast.)

At both ends of the sequence of teen numbers generated by this rule, the child must incorporate information from external sources. This may require some modification of earlier attempts to make sense out of the sound patterns of number names.

> "Eleven" follows "ten" and "twelve" follows "eleven." After "nineteen" comes "twenty."

Now hearing "nineteen" in his oral counting sequence serves as a "bump in the road"[3] to remind him of this transition rule. Eventually, the child notices that one through nine sequences are repeated in the twenties, the thirties, and so on. In further extending his number-name sequence the child often skips decades before noticing another application of the one through nine cycle. Initially, he may count from one to nine in the forties and skip to the sixties for another cycle. Even-

Counting Forward by Ones from Seventy-nine

This task explored how children accomplished the century transition. Successful transition and continued counting not only requires a specific rule, but also an integration of earlier procedural rules within expanded ones.

> Remember, after "ninety-nine" comes "a hundred." Next, keep saying "a hundred" every time and start counting all over from "one."

tually he will abstract a rule for generating the sequence to ninety-nine such as

> Keep saying the word *twenty* and keep counting from one to nine. Each time you come to __ *ty-nine* change to the next counting word that goes in front of *ty*. Watch out for the funny sounds. For the next counting words keep saying the *(word)-ty* every time as you keep counting from one to nine.

In these extensions of conventional number-name sequences there is little evidence that young children have constructed number meanings. For them, "twenty-five" is the number coming after "twenty-four" rather than the number composed of 2 tens and 5 ones. Even though children are actively extending their sequence of counting words, these extensions appear to be based on intuitive procedural rules abstracted from sound patterns in spoken language, rather than on number meanings.

In the following tasks, beginning third-grade children are asked to generate forward and reverse sequences from different starting points, using different intervals. Here, the children interviewed are the same 29 third graders described in Chapter 1.[4]

Although most of the third-graders experienced no difficulty with this task, 6 of the 29 children made errors at transition points at or near "a hundred." One child omitted one hundred altogether. Another child repeated the tenth decade—"ninety, ninety-one . . . ninety-nine, ninety, and so on" before making the transition. Four other children made the century transition but produced an error following "one hundred nine." These errors are illustrated on the facing page.

---

2. The examples of procedural rules that help to explain children's counting behaviors have been adapted from computer program statements capable of generating the counting sequences (Thompson, 1982).

3. Robert Wirtz (1981).

4. This was the first of a series of three interviews focusing on counting by ones, tens, and hundreds, place value concepts, and standard addition and subtraction procedures conducted during October and November. The 29 children were taken from 2 neighboring schools in a white, upper middle-class community in the suburbs of Los Angeles. Each child received parental permission for participation.

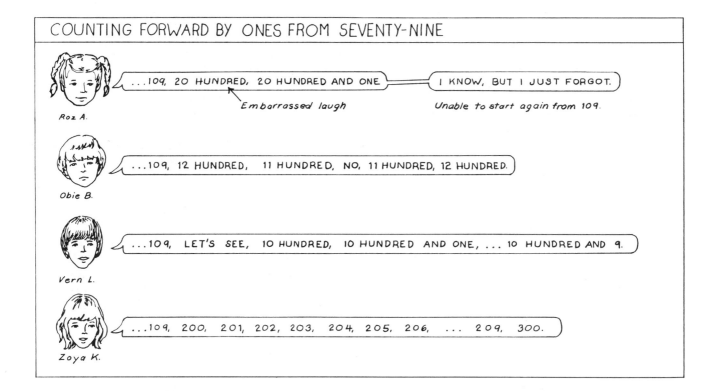

COUNTING FORWARD BY ONES FROM SEVENTY-NINE

Roz A.
...109, 20 HUNDRED, 20 HUNDRED AND ONE. I KNOW, BUT I JUST FORGOT.
Embarrassed laugh *Unable to start again from 109.*

Obie B.
...109, 12 HUNDRED, 11 HUNDRED, NO, 11 HUNDRED, 12 HUNDRED.

Vern L.
...109, LET'S SEE, 10 HUNDRED, 10 HUNDRED AND ONE, ... 10 HUNDRED AND 9.

Zoya K.
...109, 200, 201, 202, 203, 204, 205, 206, ... 209, 300.

Once counting beyond "one hundred" the hundred part of the name is held constant and the remaining parts are incremented. Vern produced the appropriate number-name components but reversed their order (hundred ten vs. ten hundred).[5] Following this transition he continued to increment by ones. Obie's error was of a similar nature, although he skipped a step in the sequence. For Roz and Zoya, reaching nine appeared to signal another major transition point. Rather than increment the ten portion of the name, they incremented the hundred portion and continued from there by ones. Whereas some beginning third graders experience difficulties in making the century transitions, others are capable of oral counting beyond one thousand. In a study of a comparable group of third graders, Bell and Burns (1981) report that half of them were able to count from nine hundred ninety-seven to one thousand three. Based on the observations made of children's oral counting just beyond the century transition, a truer test of their abilities would have been to extend the counting span beyond 1,010. It is interesting to note that classroom teaching pays little or no attention to counting once children are able to count orally to one hundred.

The procedural rules that children abstract from the patterns noticed in the sounds of number names are intuitive in nature. Yet it is interesting to speculate what kind of mental organization might account for the production of sequences of number names. Thompson (1982) posits a multilevel organization for the production of number names. Notice that since words (or their homonyms) appearing in the lowest level also appear in the highest level, the one through nine cycle reappears in twenty through ninety and one hundred through nine hundred.

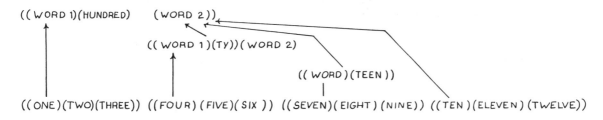

((WORD 1)(HUNDRED) (WORD 2))

((WORD 1)(TY))(WORD 2)

((WORD)(TEEN))

((ONE)(TWO)(THREE)) ((FOUR)(FIVE)(SIX)) ((SEVEN)(EIGHT)(NINE)) ((TEN)(ELEVEN)(TWELVE))

---

5. Use of *and* in compound number names as shown by Roz and Vern is considered bad form. Rather, the parts of the number names should be run together with the *and* used only to indicate the position of the decimal point. At this stage in their counting the additional convention is unimportant for them.

Thompson indicates that such an organization not only can generate a number-name sequence to 999, but can make comparisons as well. "Two hundred" precedes "five hundred" because "two" precedes "five." Similarly, "one hundred ten" precedes "one hundred seventy" because "ten" precedes "seventy." But to compare "five hundred" with "seventy," reference must be made to the hierarchy—*ty* precedes "hundred." It must be noted, however, that at this point in the child's development these comparisons are based solely on successor information—the number names that come next—rather than on actual quantities.

Counting Backward by Ones from Fifty-four

The next task explores the reversibility of children's counting procedures. Based on earlier observations of second graders (Chapter 4), a lag is expected in the development of backward counting procedures. In counting backward many third graders were observed to slow down considerably when approaching the decade and to speed up the pace between decades. The slowdown may have been in anticipation of the transition or for rehearsal of the forward sequence. After shifting down to the preceding decade the word *ty* is held constant while counting from nine to one.

Twelve of the third graders omitted the (word) *ty* name altogether or inserted the wrong word preceding *ty*. It may be inferred that this error was made in anticipation of the next reverse decade. The children were so preoccupied with what they were doing, they lost sight of the goal. The Bell and Burns (1981) study reports that 9 out of 30 beginning third graders were able to count back in all three of the following ranges: 101 to 91, 203 to 193, and 1,003 to 993. As well as reporting the surprising ease of reverse counting for capable counters, it reports a lag in the backward counting abilities of other children.

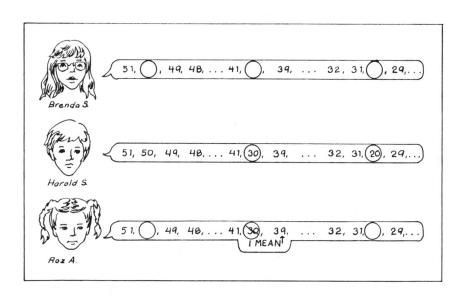

Counting Forward by Tens

Since the focus of the next task was on the children's century transitions while counting at intervals of ten, their counting was stopped at two hundred twenty.

To *skip-count* by tens the children had to abstract procedural rules focusing on specific parts of the earlier number name sequence. The following rules can produce such a skip-ten sequence.

After "ten" comes "twenty."	10, 20
Now keep counting and say *ty* after each word.	30, 40, 50, 60, 70, 80, 90
After "ninety" comes "one hundred."	100
Now keep saying "one hundred" and start counting all over from "ten."	110, 120, . . . 190
After "any hundred ninety" make it the "next hundred."	200
Now keep saying the "new hundred" and start counting all over from "ten."	210, 220

After reaching "one hundred ten" the hundred is kept constant while incrementing by tens according to earlier procedures. This repetition is more information to keep in mind.

Although the majority of children found this to be an easy task, some children hesitated at the century transitions or made errors that were self-corrected. Still, 8 of the 29 children showed errors at the century transitions that weren't corrected. In the following example, Teddy D. appears to be following a rule to increment the first part of the number name. This rule is productive till the century transition is reached. By continuing to increment the first-named portion by tens he is now incrementing the hundred portion. Teddy's rule needs modification.

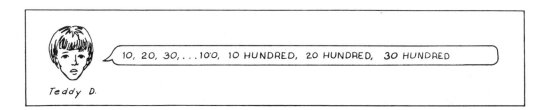

Most of the other children skip-counted successfully to "one hundred ninety" but had problems at the next century transition. Three of the children lacked a procedural rule to get from "any hundred ninety" to the "next hundred." Not knowing what came next they invented a number name. What followed from this point varied with each child.

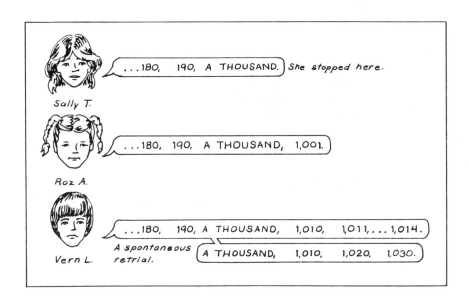

After making the correct transition to "two hundred," the next child, Bert, called on the wrong procedure but soon realized his error and recounted from that point.

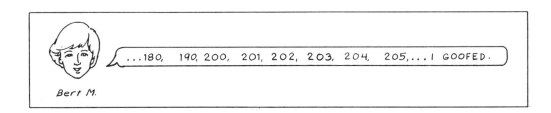

In the Bell and Burns (1981) study, 9 out of 30 beginning third-graders were unable to skip-count by tens from 180 to 210. At the same time, 7 of these children were able to do so across the following spans: 480 to 520, and 980 to 1,020. While some third-grade children have yet to abstract procedural rules for century transitions, others are able to make the next major transition across the thousand mark.

Counting Forward by Tens from Thirty-four

Starting a skip-count sequence of tens from a number other than some multiple of ten is difficult for some children, but they are able to continue following a prompt for the next number name in the sequence. Beyond getting started with the count, the next difficulty occurs immediately following the century transition.

The following procedural rules are capable of generating the required sequence.

From "four" next is "fourteen."	4, 14
From "fourteen" next is "twenty-four."	24
Now keep saying the next counting word followed by *ty-four*.	34, 44, . . . 94
After "ninety-four" next is "one hundred four."	104
Now keep saying "a hundred" and start counting over from "fourteen."	114, 124, . . . 194
After "any hundred ninety-four" go to the "next hundred four."	204

Following the century transition, the procedure continues by holding the outer portions of the number name constant while incrementing the middle *ty* portion. In addition to having three variables to keep track of, incrementing the middle portion of the number name may be a source of difficulty for children.

This task was difficult for most of the third graders interviewed. As many as 20 of the 29 children showed counting errors in the transition to or from "one hundred four." Some of the errors appeared bizarre initially, yet I assumed that they were serious attempts to extend and coordinate counting procedures.

You are invited to put yourself in the children's shoes. Imagine how each of the following responses makes sense from the child's perspective. Look for patterns in their sequences.

Counting Backward by Tens from One Hundred Thirty

Twelve of the third graders experienced problems in making the century transition from the reverse direction. Six of these children omitted "one hundred" entirely in apparent anticipation of the next series of tys, but continued back to "ninety," and so on. Some of the children appeared so preoccupied with what to do next that they lost sight of one of the aspects of their goal—counting in reverse direction with intervals of ten. Two of the children shifted counting intervals upon reaching "one hundred."

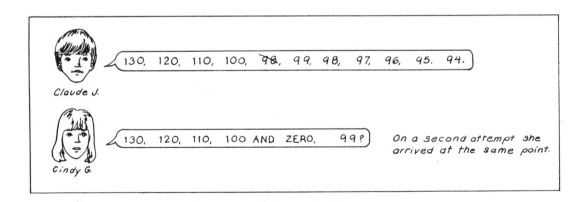

Another two children shifted direction upon reaching the century transition, with one also changing intervals.

Notice how running two number names together may have caused Roberta's reversal of direction. For her, "one hundred," "ninety" became "one hundred ninety."

Many children experience the decade and century transition points as bumps in their counting road, reminding them to shift procedures to continue the sequence of number names being generated. For unskilled counters these transition points resemble barriers to cross. As they approach the top of these barriers they hesitate, tend to be off-balance, and either stumble, lose the uniformity of their step, or slide back.

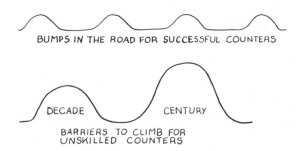

BUMPS IN THE ROAD FOR SUCCESSFUL COUNTERS

DECADE   CENTURY

BARRIERS TO CLIMB FOR
UNSKILLED COUNTERS

Whereas 20 of the Bell and Burns (1981) third graders were unable to count back by tens from 210 to 150, the remaining 10 children were able to do so for the following spans: 520 to 450, 1,020 to 950, and 3,010 to 2,950. While some children have considerable difficulties in counting backward by tens, others do so with ease.

The inclusion of comparative data on the same oral counting tasks for children from kindergarten through the third grade in the Bell and Burns study (1981) confirms the existence of a wide range in children's counting abilities, reported earlier by O'Brien and Casey (1982b). Some kindergarten children are better oral counters than some third graders. The ability of kindergarten children to generate number-name sequences far in advance of any formal teaching lends support to our contention that this ability is based on an active search for sound patterns in the spoken language of numbers. The errors children make show their struggles in coordinating the dimensions of the tasks.

The intuitive procedural rules invented by children for generating expanding number-name sequences exist initially in isolation. The following sections describe how children may integrate these procedures within a relational framework that provides a basis for understanding and operating on our base-ten number system.

COUNTING GROUPED OBJECTS—A CHALLENGE IN COORDINATION

Counting increasingly larger collections of objects becomes cumbersome and error-prone unless these objects are arranged in equal-sized groups (together with loose individual objects) for greater efficiency in counting. These groups may consist of twos, threes, fives, tens, and so on. The grouping process also permits easier comparison of large collections. This grouping organization can be extended by forming groups of groups, thus increasing the potential efficiency of the counting system and forming the very foundation of our place value system of representing numbers. However, this efficiency is not realized until the children are able to coordinate assigning new sequences of number names with intervals appropriate to the size of groups being counted in a *one-name-to-many-objects correspondence.*

Related Research

Since my interviews were conducted with third-grade children, samples from the work of other investigators will serve to illustrate some less sophisticated approaches to counting large collections of ungrouped and grouped objects.

The interview research of Stake (1979) describes primary graders counting a collection of objects over and over and not being bothered when their answers were different nearly every time. One third grader, Sally, counted an array of 25 poker chips and arrived at answers of "twenty-five," "twenty-seven," "twenty-nine," and "twenty-six." Initially the array was random

but later the interviewer made it a circular array. Each answer was considered correct until the next one came along. Sally nodded her head in rhythm while saying the number names rather than physically touching or separating the objects being counted. In the circular array she did nothing to physically separate or mark the first chip counted. Since she was depending on visual partitioning of the objects into those counted and those left to be counted it is not surprising that some chips were missed or recounted. Although this approach to counting objects may have worked for a small collection, it was error-prone and not reproducible. This ineffective approach to counting was typical of most primary graders at one school.

What puzzled Stake was the degree of confidence that these children had in their procedures. On closer observation of the videotapes it appeared that the children subitized groups of twos, threes, fours, or fives based on the spatial distribution of the objects. Once each new group to be counted was subitized, its number was confirmed by the rhythm of the oral count—for instance, three objects were confirmed by extending the counting sequence by three number names while nodding the head three times.

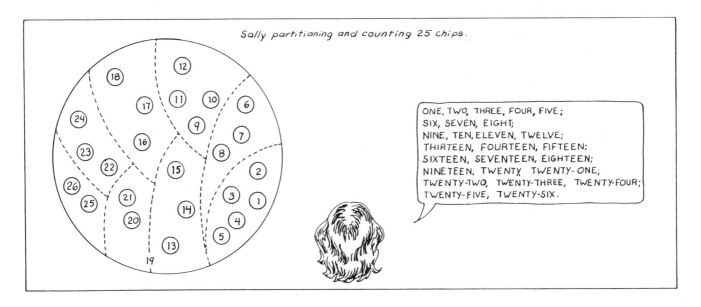

Sally partitioning and counting 25 chips.

ONE, TWO, THREE, FOUR, FIVE;
SIX, SEVEN, EIGHT;
NINE, TEN, ELEVEN, TWELVE;
THIRTEEN, FOURTEEN, FIFTEEN;
SIXTEEN, SEVENTEEN, EIGHTEEN;
NINETEEN, TWENTY TWENTY-ONE;
TWENTY-TWO, TWENTY-THREE, TWENTY-FOUR;
TWENTY-FIVE, TWENTY-SIX.

The children's focus was on accumulating small groupings and confirming the size of each. Hence their confidence in their answer. With repeated confirmation of subset size with the number of number names used, there appeared to be no need to compare the results of consecutive counts. In this approach to counting the children got so absorbed in the details of the process that they lost sight of the goal of finding out "How many?" But still, in this ineffective approach to counting large collections of objects we get a glimpse of what may be children's first attempts at grouping objects. Here, the size of their grouping was not uniform. Rather, it was determined by the spatial arrangement of the collection.

Another general observation made by Stake (1979) about primary-grade children's counting of large collections of objects was that they skip-counted by matching one object to one name in the sequence. This early level of attempting to coordinate the skip-counting of objects was clearly illustrated in the consecutive counts of Katia. First she counted the same set of 25 random objects individually as "twenty-five." When asked if she could count in a different way, she counted by fives to "one hundred eight." She then recounted the collection as "one hundred five." Asked if she could count them by twos, she obtained an answer of 'fifty-six." With the aid of some prompting by the interviewer, Katia recounted by fives to obtain "twenty-five." Then Katia attempted to count by fives again and started by matching the first three number names ("five," "ten," "fifteen") to the first three groups of five objects before reverting to assigning other names in the same number-name sequence to the remaining individual objects. Her answer this time was "eighty." In most instances she was counting in the same way, assigning one name per object. Only the list of number names had changed. After each task the answer was recorded. At the end of this sequence the interviewer attempted to have Katia focus on the discrepancies between the different answers.

> I: Now does the number of cookies stay the same or does it change?
>
> K: The cookies don't change, the number does.
>
> I: How does that happen?
>
> K: You count by different numbers

She had used lists of different number names and arrived at different answers. Again, the child appeared to be so engrossed in the process of generating the number names and assigning them to individual objects that she lost sight of the goal. For Katia there was no relationship between the final number name in the sequence and the quantity of objects in the collection. This early level of skip-counting objects is reminiscent of young children's first attempts to count objects. Fuson and Hall (1982) noted that much of young children's counting is done for practice in the counting act rather than for determining the number of objects included in a collection. At this later level of complexity in counting, much practice is needed, not only to generate new number-name sequences, but also to coordinate them with groups of objects in a one-to-many correspondence.

Excerpts from interviews by Thompson (1982b) with seven children provide us with a glimpse of children's early attempts to count objects by tens. The children were a mixed group of beginning first and second graders. All of the children could skip-count orally by tens but could not start at "two" and count by tens.

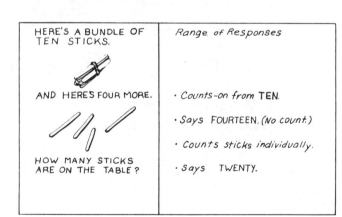

HERE'S A BUNDLE OF TEN STICKS.

AND HERE'S FOUR MORE.

HOW MANY STICKS ARE ON THE TABLE?

Range of Responses

- Counts-on from TEN.
- Says FOURTEEN. (No count.)
- Counts sticks individually.
- Says TWENTY.

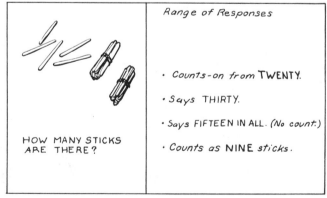

HOW MANY STICKS ARE THERE?

Range of Responses

- Counts-on from TWENTY.
- Says THIRTY.
- Says FIFTEEN IN ALL. (No count.)
- Counts as NINE sticks.

Place 33 sticks on the table.	Range of Responses
 CAN YOU USE TENS TO HELP YOU FIND HOW MANY?	• Makes three bundles of ten sticks, then proceeds to count the sticks individually. • Counts individual/separate sticks as TEN, TWENTY, THIRTY, FORTY, FIFTY. (Stopped by interviewer.)

	Range of Responses
Ten bundles of ten are placed on the table. CAN TENS HELP YOU TO COUNT ALL THESE STICKS IN THESE BUNDLES?	• Counts by tens to one hundred. • Counts 10, 20, 30,... 90,...20. Says TWENTY IN ALL. • Counts the sticks individually. • No attempt.

Regardless of grade level, some children are quite successful at counting objects by tens, while others have considerable difficulty in their attempts or make no attempt. In counting by tens, the children are subitizing the quantity. One child instantly recognized bundles of sticks as tens and counted-on from "ten" and "twenty." Also, the child who recognized a bundle and four sticks as "fourteen," appears to have been subitizing tens and ones. Subitizing is an important ability that becomes integrated into the act of counting. The less mature counting behaviors included the following:

> –counting ready-made bundles of sticks by individual sticks even when the value of the bundles had been given.
>
> –recounting self-grouped bundles of sticks by individual sticks
>
> –counting individual sticks as tens
>
> –counting both loose sticks and bundles as ones

Such early attempts at counting by tens are confirmed in the research of Stake (1979) and Behr (1976). The preceding excerpts should orient you to what might be anticipated for the counting of our third-grade children.

Before returning to the third graders, I want to mention some areas of needed research in children's counting of increasingly larger collections of objects. How do children realize the importance of uniform grouping? What sized groups do children tend to arrange initially? When does the notion of ten occur to them? When does the value of grouping the groups occur to them? Is the idea of grouping made easier by grouping into a smaller number of objects initially? Should grouping by tens be delayed until grouping in general is understood? By providing them with ready-made groups to count we fail to learn how children construct groupings naturally. Most available research describes children's counting of ready-made groups.

Counting Grouped Objects— A Preliminary Task for Third Graders

In my interviews with the same 29 third-grade children, the following two preliminary tasks gave the children the opportunity to first construct their own groups or to confirm the size of ready-made groups. When asked to show a specific quantity of objects they could select the grouped or ungrouped objects for counting. By not presenting materials that were all pregrouped, I created the possibility of observing a greater range of responses.

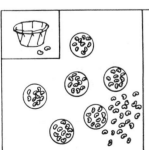	• HOW MANY BEANS ARE IN HERE? • I COUNTED TEN BEANS IN EACH OF THESE OTHER CUPS. • SHOW FORTY-FIVE BEANS IN THE EASIEST WAY YOU CAN. • HOW DID YOU DO IT SO QUICKLY?

Unifix cubes	
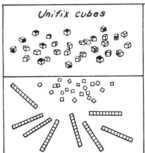	• THESE CUBES CAN BE CONNECTED TO BUILD STACKS OF TEN. WOULD YOU BUILD SOME STACKS OF TEN? I'LL DO THE SAME. • After the child and interviewer construct one or two stacks each, additional ten-stacks are introduced. • SHOW FIFTY-TWO CUBES IN THE EASIEST WAY YOU CAN. • HOW DID YOU DO IT SO QUICKLY?

Following introduction to the task, nearly all the children were able to subitize each group as ten and to coordinate their skip-counting with four cups of beans. Upon reaching "forty" they switched counting procedures from the one-to-many correspondence to a one-to-one correspondence of the individual beans until "forty-five" beans had been counted. However, three of the children counted out 45 individual beans despite the availability of the known groups and prior demonstration of oral skip-ten counting abilities. Tina had considerable difficulty with her counting. All four of her attempts are illustrated below. She seemed intent on counting by intervals of five despite available groupings of ten.

On both tasks Tina had difficulties in counting by tens beyond the initial ten-group that she had counted/constructed. Despite lining up the ten-groups in the second task, once counted, Tina was unable to subitize each of them as ten and to coordinate this with her oral skip-ten sequence. The perceptual clue of length had little significance for her. It seems as though ten was not a constant to her, so she went one by one. Some children need considerable experience before realizing that counting by tens is the same as counting one by one ten times. Another child, Rena, joined all of the interlocking cubes to form a long train while counting by tens and ones.[6] In doing so the groups of ten lost their identity. Rena's construction of the train suggests that she didn't think of "fifty-two" as 5 tens and 2 ones.

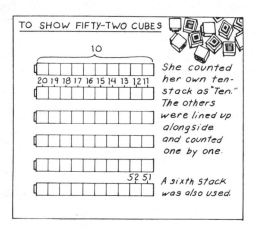

Thus far in this section we have seen ample evidence of how children structure their experience. Even though objects are grouped in a particular way, children do not necessarily recognize the quantity of this grouping. They see only what they understand. In order to recognize these groupings, some reorganization of ideas must first take place. Part of children's continual recounting of the individual objects in the groupings may be necessary for their mental construction of the group or unit by which the quantity of collections can be measured. Although most of these third graders have been able to construct units of tens and ones in their counting of objects in these tasks, they may not be able to do so consistently on the tasks that follow.

Counting Boards for Totaling Addends Composed of Tens and Ones

Just prior to the presentation of the following counting board tasks, the children were asked to set aside their self-constructed ten-group as a reference for future tasks. This served to facilitate children's subitizing of ten-groups and their integration into the skip-counting procedure.

6. Student interviewers report that some third graders take apart the interlocked groups of ten in counting one by one.

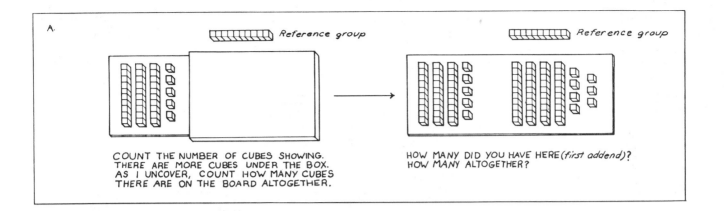

COUNT THE NUMBER OF CUBES SHOWING. THERE ARE MORE CUBES UNDER THE BOX. AS I UNCOVER, COUNT HOW MANY CUBES THERE ARE ON THE BOARD ALTOGETHER.

HOW MANY DID YOU HAVE HERE *(first addend)*? HOW MANY ALTOGETHER?

It is interesting to note that 9 children counted 35 cubes in the first addend by tens and ones but then reverted to counting-on by ones when the second addend was uncovered. Two of these children did count by tens following a prompt. Eventually, 23 of the 29 children were able to find the total through a variety of counting-by-tens-and-ones strategies.[7] Three such strategies are illustrated below in increasing order of sophistication. Although the last strategy is the most difficult, 15 children applied it successfully.

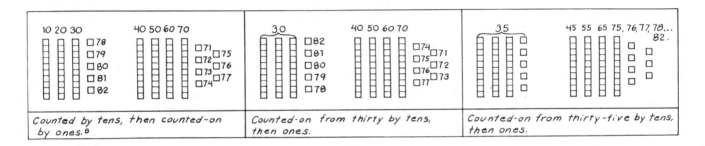

Counted by tens, then counted-on by ones.[8]

Counted-on from thirty by tens, then ones.

Counted-on from thirty-five by tens, then ones.

Five of the children exhibited some unusual counting methods. Zoya's counting appeared to be influenced by her initial counting unit. Lacking flexibility in her counting procedure, she continued to count by the initial interval even when matching groups were exhausted, rather than switching to a new list of number names. Within the same task she counted ones as tens and tens as ones.

Miscounts ones as tens.

Following a prompt with the reference stack she repeated her original count.

I WONDER HOW MANY CUBES YOU HAVE HERE? ...SO, *(pointing)* 30, 31... She continues from prompt.

ALTOGETHER? *(Uncovered)* DO I START FROM 35? WELL, THAT'S SOMETHING YOU CAN DECIDE... SO YOU HAVE 35 HERE, HOW MANY ALTOGETHER?

Starts to miscount tens as ones...stops. She starts again, counting the ten-group by component ones (43-52).

---

7. It is also possible to subitize the 35 cubes at a glance. The task's stress on counting may have discouraged some children from doing just that.

8. One child reversed the process—counting-on by ones and then by tens.

As the following excerpt shows, Rena continued to count by tens even though the tens groups were exhausted, so that she miscounted individual cubes as tens. It's as if she got caught up by the rhythm of the skip-count. However, once she reached "one hundred," the bump in her counting road caused her to shift to another procedure—counting by twos.

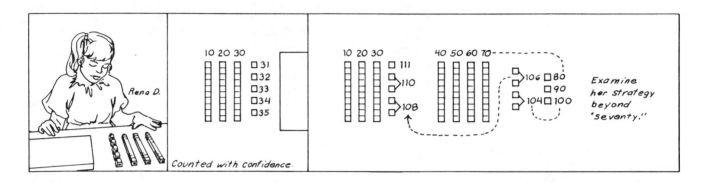

Although a similar shift was made at this transition point by other children, they usually switched to counting by ones. As a result of her inflexibility in counting mixed groups, Rena's last number name bore no relation to the quantity of objects.

As illustrated in the next panel, Obie showed considerable task persistence that eventually paid off with a successful count. After several false starts and numerous prompts, he decided to look at the problem from a different perspective. By visually partitioning the total collection into two groups consisting of tens and ones, he was able to arrive at a correct total.

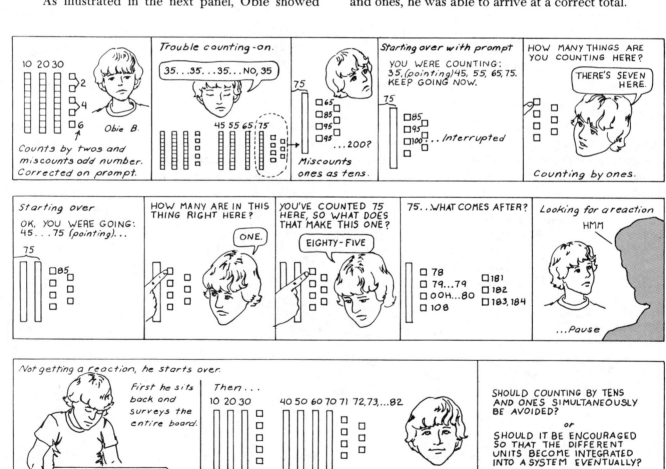

By neither reacting to his miscount nor rushing off to the next task, I provided Obie with an opportunity to judge his own procedure and to start again from a new perspective. His new focus did not require counting-on by tens and resulted in a successful count. Here, Obie reduced the problem to one that could be solved using available schemes.

A second counting board requiring the totaling of addends consisting of tens and ones presented the children with only one visible addend and oral information (together with a written reminder) on the other addend.

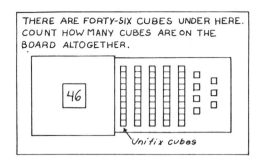

Rather than have the opportunity to count the first addend by tens and ones, the child must consider the given quantity—"forty-six" (46)—as already composed of tens and ones.

Only 13 of the third graders were able to count-on from the first addend by tens and ones using one of the two strategies illustrated below.

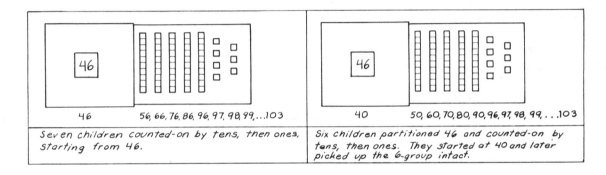

46 56, 66, 76, 86, 96, 97, 98, 99,...103	40 50, 60, 70, 80, 90, 96, 97, 98, 99,...103
Seven children counted-on by tens, then ones, starting from 46.	*Six children partitioned 46 and counted-on by tens, then ones. They started at 40 and later picked up the 6-group intact.*

The other 16 children could only count-on from 46 by ones, counting-on the components of the ten-groups individually or approaching the task in other unique

ways. Some of these unique methods are shown in the following panels.

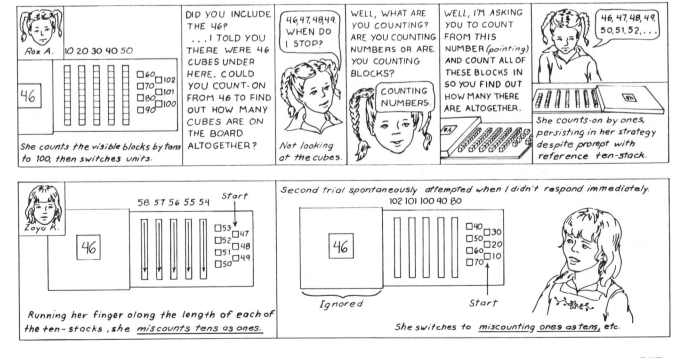

Both Roz and Zoya ignored the addend that was not perceptually available and counted only the visible objects. They then miscounted the ones as tens or the tens as ones when unable to shift to a second counting procedure at a transition point. It is interesting to note that on reaching the century transition both girls shifted from counting individual cubes as tens to counting them by ones.

A third counting board presented a mystery box situation. It displayed one of the addends as groups of tens and ones and provided oral information (with written reminder) about the total. The children were asked to find the missing addend by counting.

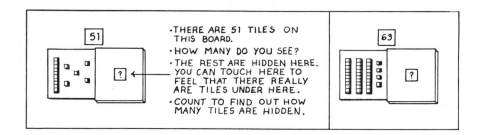

These tasks could be approached by counting forward or backward by tens and ones. Children who were successful in one direction were encouraged to try it in another way to check their answer. Eleven children spontaneously counted-on by tens and ones while two more did so following a prompt. Three children spon-taneously counted-back by tens and ones and five more did so with a prompt. Four of these children showed successful bidirectional strategies for counting-on by tens and ones. Some of these strategies are described in the following panels.

Although all of the efficient counters did not show obvious signs of tracking their count, their success suggests either an internal tracking procedure or use of visible tiles for tracking—as in counting back 34 from 63. The remaining children attempted to count-on by ones, twos, or fives.[9] Some did so unsuccessfully, not realizing the need to keep track of their count.

Once children are able to count-on and count-back grouped objects by tens and ones, their counting becomes an effective tool for dealing with large quantities. They have the flexibility to select the most efficient methods and the option of checking their solutions by an alternative procedure.

9. One boy was quite adept at counting by fives and converting to tens as needed. He used his fingers to keep track of the fives counted.

Related Research—continued

Thus far we've become familiar with how children approach counting objects by tens and ones. Examples from the research of others will provide a glimpse of how children count objects by hundreds. Resnick (1982b) reports that when shown displays of blocks representing hundreds, tens, and ones (Dienes blocks) most primary-grade children would count starting with the value of the largest block—for instance,

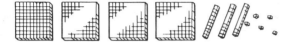

stance, "one hundred, two hundred, three hundred, four hundred, four hundred ten, four hundred twenty, four hundred thirty, four hundred thirty-one . . . four hundred thirty-five." A few children counted all the blocks by ones and then "multiplied" by the appropriate value as in "one, two, three, four—four hundred; one, two, three—four hundred thirty; one, two, three, four, five—four hundred thirty-five." However, Resnick also reports that in one specific study of third-grade inner-city children, more than half the group became confused when shown displays composed of two or more different blocks. Another interpretation of the block display is to count all the blocks by ones, giving "eleven" as the total. The latter response has been reported as well by Stake (1979) and can also be observed in the responses of the third graders reported in Chapter 1.

Stake's study (1979) also includes a detailed description of Marcia's and Katia's nine successive attempts to count 25 hundred-blocks while working

together. These third graders experienced considerable difficulty at the thousand transition points. Here are descriptions of two of their attempts, in which they arrived at "hundred twenty-five" and "two thousand forty."

1. Counting the hundred-blocks to "nineteen hundred, hundred twenty, hundred twenty-one . . . hundred twenty-five."
2. Counting the hundred-blocks to "ten hundred/ one thousand, thousand ten, thousand twenty, . . . thousand ninety, two thousand, two thousand ten, . . . two thousand forty."

Other examples of children's instability in counting beyond "a thousand" can be found in Chapter 1.

Counting Boards with Random Appearance of Ones, Tens, and Hundreds

In the preceding counting board tasks, children have tried to coordinate different number-name lists with various groupings of objects. The success of many children on these tasks suggests that mental organization of the different procedural rules for sequencing number names had taken place. This hypothesized coordination was put to a further test in the tasks described next. Here, rather than exposing all the objects in various groupings so the children could count all of one grouping before continuing to others, I showed them different groupings of objects in a random order, at a gradual pace.

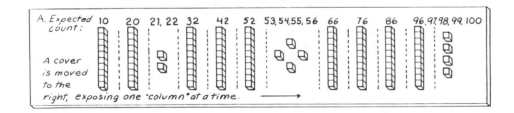

These counting boards required numerous shifts from one counting procedure to the next, in synchrony with changing groupings. The tasks simulated the counting of money in mixed denominations prior to sorting out the bills into piles of the same denomination. At the same time that they were expected to present further difficulties for the children, the tasks tested the limits of their counting abilities for grouped objects.

Twelve of the third graders were successful in counting the objects with different units as needed. A

small number from this group needed a prompt for the first group despite the availability of their reference ten-group. Other children were unsuccessful even after this prompt. In the next excerpt, Rena's disequilibrium in deciding what to call the first group of interlocked Unifix cubes is not resolved by my prompt. This clearly illustrates the dilemma children experience in dealing with multiple meanings of ten. Should she count the tens or should she count *by* tens? How would you explain her final count?

Other difficulties experienced by the children will be discussed in the context of the next counting board with random appearance of ones, tens, and hundreds. Although similar in many respects, one of the random exposures is a hundred-group composed of 10 ten-groups. Prior to the task the children were asked to count a reference group to facilitate subitizing 100 and integrating it into the counting process. All the children counted the hundred-group by tens, with the exception of Zoya. She counted by tens to "one

hundred" while running her finger along one dimension of the square. Rather than stopping here, she began moving her finger down the second row, as she continued her count. However, once reaching the bump in her counting road, she switched to counting individual cubes by ones. Following a prompt with her previously constructed reference ten-group, she successfully counted by tens. If I had been in a rush to get into the actual task, I could have cut off her original count at "one hundred", assuming that she was counting rows rather than individual cubes.

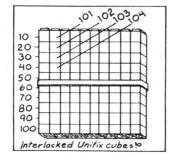

Interlocked Unifix cubes¹⁰

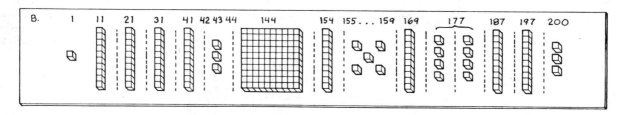

The second counting board began with a single cube followed by a ten-group. Nine of the children could not count-on by tens and persisted in counting-on by individual cubes composing the ten-groups. Zoya

was one child who experienced this problem. Furthermore, she had difficulty in deciding what to call the first cube.

10. Unifix cubes only interlock on two ends. To form a hundred-group, ten groups of ten were lined up side by side on a piece of cardboard backing and were fastened together by two rubber bands.

Cal picked up his reference ten-group and counted individual cubes along it as each ten-group on the board was uncovered.

Cal M.

Uncovering the hundred-group was met often by sighs and embarrassed laughter. Although the children had counted 100 cubes in the hundred-group during the introduction to this task and had it available as a reference, many needed to recount it or had difficulty in integrating it into their counting sequence.

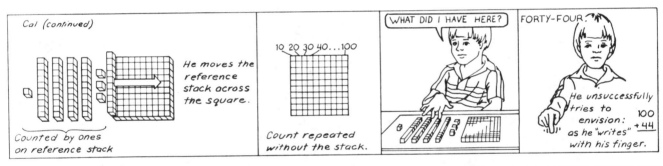

Some of the children were asked to work on construction and coordination of a hundred into their counting before they constructed and coordinated tens.

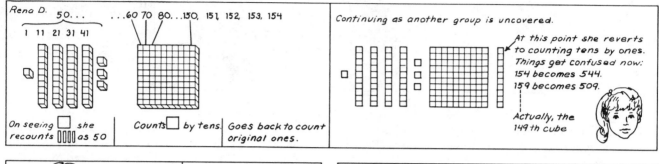

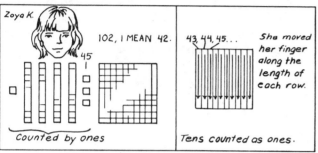

Five children completed the second task without any errors in counting and without reverting to counting by smaller units unnecessarily. The following is a summary of both problem areas and successes for the children:

Counted-on ten-groups by counting
   individual cubes
   (counting-on by ones) . . . . 9 children

Miscounted-on ten-groups as ones
   (counting-on by ones) 1

Miscounted-on ones as tens 4

Recounted the hundred-square 7

Subitized the hundred-square and
   counted-on by a hundred—
      first-try 12

      more tries 3

      counted-on by tens 2

      counted-on by ones 3

      unable to count-on 1

Miscounted immediately after count
   ing-on by one hundred 5

In the last group, three of the five children reverted to counting-on by ones when faced with a ten-group immediately following the hundred-square. The random exposure of different groupings of objects required numerous shifts from one counting procedure to the next, in synchrony with the changing groups. While some of the children were up to the challenge, others experienced disequilibrium when attempting to count-on by tens or hundreds or to shift down to counting-on by ones again. The novelty of counting-on one hundred presented the most difficulties.

The difficulty of counting objects by hundreds was demonstrated further on another counting board task requiring the children to count backward. The counting board contained two hundred cubes in groupings of ones, tens, and hundreds in random arrangement. The children were told how many cubes were covered, and then they observed as the groups were gradually uncovered. They were asked to count backward to determine how many cubes remained covered at each step. Four children completed the task with no errors. Of particular interest is how children dealt with counting back one hundred—going from "one hundred fifty-six" to "fifty-six." Although 20 children subitized the hundred-square, only 10 were able to count-

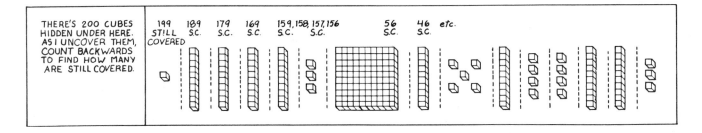

THERE'S 200 CUBES HIDDEN UNDER HERE. AS I UNCOVER THEM, COUNT BACKWARDS TO FIND HOW MANY ARE STILL COVERED.

199 STILL COVERED | 189 S.C. | 179 S.C. | 169 S.C. | 159,158, 157,156 S.C. | 56 S.C. | 46 S.C. | etc.

back by one hundred. Three of the children counted-back the square by tens or ones. The remaining 7 children were stymied at this point, being unable to count-back by any unit.

The latter group of counting board tasks proved to be a rigorous test of children's coordination of different counting procedures. Not only did these tasks confirm the difficulties experienced by children in counting-on objects by tens, but they also introduced children to further difficulties in counting-on and counting-back objects by one hundred. Although the children were able to subitize the hundred-group, all were not able to integrate it into their counting sequence. The latter children had not constructed a mental hundred-unit as yet. In isolation, children may have said "a hundred" as a rote association with the square, but in the counting process they saw and counted-on or counted-back what they understood.

COUNTING WITH MORE ABSTRACT UNITS OF TENS AND ONES

Thus far we've used counting boards to present tasks for the counting of grouped objects. Some third-grade children construct mental units of ones, tens, and hundreds with the perceptual support of the grouped objects. Will these children continue to do so on comparable tasks in the absence of this perceptual support? This section will describe tasks involving totaling addends or finding missing addends when the information is presented both orally and with numerals. The first pair of tasks involves totaling addends. The following illustrations describe both tasks together with the successful strategies of three children.

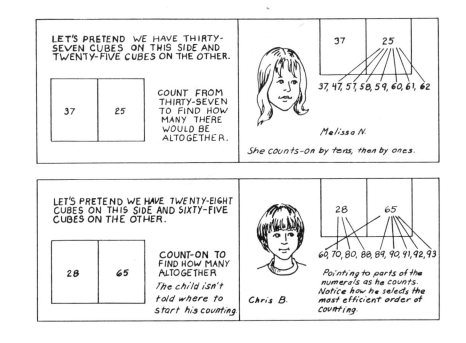

LET'S PRETEND WE HAVE THIRTY-SEVEN CUBES ON THIS SIDE AND TWENTY-FIVE CUBES ON THE OTHER.

| 37 | 25 |

COUNT FROM THIRTY-SEVEN TO FIND HOW MANY THERE WOULD BE ALTOGETHER.

37   25
37, 47, 57, 58, 59, 60, 61, 62

Melissa N.

She counts-on by tens, then by ones.

LET'S PRETEND WE HAVE TWENTY-EIGHT CUBES ON THIS SIDE AND SIXTY-FIVE CUBES ON THE OTHER.

| 28 | 65 |

COUNT-ON TO FIND HOW MANY ALTOGETHER

The child isn't told where to start his counting.

28   65
69, 70, 80, 88, 89, 90, 91, 92, 93

Chris B.

Pointing to parts of the numerals as he counts. Notice how he selects the most efficient order of counting.

These examples provide impressive evidence of children's organization of their counting procedures towards greater efficiency.

In the absence of perceptual support from grouped materials, some third graders were able to adapt and to invent strategies requiring more mental complexity. Twelve children counted-on by tens and ones using different strategies, including counting-on from the highest tens and the highest ones. The latter group provided evidence of organizing their counting procedures toward greater efficiency. Twelve other children counted-on by ones only. Some did so unsuccessfully without realizing the need to keep track of their count.

A second pair of tasks presented a missing addend situation.

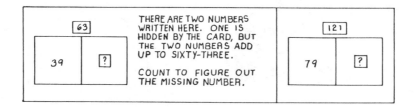

Six children were able to count-on by tens and ones to find the missing addend. One child found the missing addend using bidirectional strategies. A few additional children were able to count-on by tens and ones but made errors along the way. The successful strategies of two children are illustrated next. Notice how they manage to keep track of two different counting units.

In the absence of perceptual support from the grouped materials, fewer third graders were able to complete comparable tasks with counting-on or counting-back with tens and ones. However, some children were successful by inventing strategies requiring greater mental involvement than before. To distinguish the latter group from those capable of counting-on and counting-back by tens and ones only in the presence of grouped materials, we can say that the more sophisticated counters constructed more abstract units in the absence of perceptual support.

A THEORETICAL FRAMEWORK FOR COUNTING BY ONES, TENS, AND HUNDREDS

This section sets out to generate a plausible explanation for children's development of abilities to count by ones, tens, and hundreds—one that can serve to interpret their counting errors as part of the process of reconstructing their counting at a higher level of complexity.[11]

Children have a fascination with the names of large numbers. By abstracting intuitive procedural rules from sound patterns in number names through active experimentation and practice, they are able to generate ever-extending number-name sequences.

Difficulties experienced at transition points such as the century mark can be explained by faulty and incomplete procedural rules and their lack of coordination. Such coordination may involve recycling general procedures and integrating them within more extensive ones—for example, recycling procedures for generating one through nine at progressively higher levels. Coordination may also involve integrating varied procedures to permit counting from any starting point with skip-counts of different intervals. There is evidence that this process begins prior to any formal instruction. Despite children's impressive abilities to actively extend their sequence of number words, these extensions are based more on patterns of spoken language than on children's number meanings.

_____

11. The seeds for this model of children's construction of numeration concepts were planted in my reading the work of Behr (1976) and Stake (1979), and in conversations with Steffe and Easley. More recently, my model has been influenced by the work of Thompson (1982) and Resnick (1982).

Once children begin to group objects in dealing with large quantities, they begin to count objects at another level of complexity. In order to coordinate assigning different lists of number names to appropriate uniform groups of objects, children must restructure the counting process to incorporate a one-name-to-many-objects correspondence. This is accomplished following considerable practice in which they vacillate between one-to-many and one-to-one correspondences. During this period of disequilibrium, the last number name for a collection of objects often has no relation to quantity.

We have seen children who in dealing with counting ready-made groups have been able to subitize their value in isolation but not to integrate the value as an increment in their counting. Children who appropriately assign number names to a plurality of objects such as ten-groups in the course of their counting are mentally constructing units of ten. These units of ten are used to measure other numbers. For example, in

the missing addend was measured as 4 tens and 2 ones in the children's counting. The child's concept of ten now begins to incorporate multiple meanings. "Ten" is no longer just the number that comes after "nine," it simultaneously is a unit composed of other units (ones).

Children construct such number meanings at a gradual pace. During an extended period of disequilibrium they may produce answers that at first glance seem bizarre, careless, or flippant. Rather, they are usually serious efforts at coordinating procedures and constructing meanings. In attempting to extend number-name sequences, the children actively experiment before modifying their procedural rules to produce conventional number names. In coordinating varied number-name lists to appropriate group sizes, children vacillate between one-to-one and one-to-many correspondences before arriving at a stable, integrated system that consistently provides an appropriate match with quantities of objects. Furthermore, in dealing with the *multiple meanings of ten*, children are faced with the dilemma of whether to count by tens or to count the tens. Once these multiple meanings are integrated by the child, focus can be placed on either view of ten without losing sight of the other view. Hence the appropriate selection can be made for application in specific contexts. Until this time, the two views of ten may be seen as separate and competing for application in different contexts.

The above described sources of disequilibrium or uncertainty are useful in interpreting children's spe-cific counting errors. *Miscounting ten-groups as ones* may be the result of a child's internal conflict between the multiple meanings of ten and may indicate the ultimate decision to count the tens rather than count by tens. It has been observed most often when children count a collection of ten-groups and individual objects starting with the ungrouped objects. Rather than shifting procedures with a different number-name list, the children tend to continue to count the grouped objects as ones. The error may be due to a tendency to assimilate the "unity" view of the ten-group to the child's initial counting procedure. More simply, it may be attributed to a lack of flexibility in shifting procedures that are not well coordinated. *Miscounting ones as tens* occurs most frequently when children count a collection consisting of ten-groups and individual objects, beginning with the ten-groups. They persist in counting by tens even though the ten-groups have been exhausted. This error can also be attributed to lack of flexibility in shifting counting procedures. The children may become so preoccupied by the production of a skip-ten sequence that they fail to notice that the number names no longer bear any relation to the quantity of objects before them.

We have observed children subitize a hundred-group in isolation, yet not be able to increment their count by one hundred. Rather, they count-on by other component units, such as tens or ones. Once they can integrate the value of the hundred-group into the counting process by incrementing by one hundred, the children have constructed a mental unit of a hundred. At the same time that "hundred" is the number name that follows "ninety-nine," it is also a unit of units, being composed of 10 tens and 100 ones. The frequency of counting errors at the century transition is due to at least two factors. A specific rule is required to get from "ninety-nine" to "one hundred," requiring feedback from an external source. After "one hundred," old procedures are recycled and incorporated into more extensive new ones. Eventually, the number name will be increased to three components with some to be held constant while another is incremented. As children proceed further down the number-name sequence they reach points that are under experimentation or that have begun to be practiced. Since the children have had less experience with century transitions than with decade transitions, they are likely to make more errors at these points. A similar argument can be made for children's counting by hundreds. A major source of these errors is children's inflexibility in counting procedures for mixed groupings of ones, tens, and hundreds. Added to this is the children's grappling with the multiple meanings of hundred.

In the process of counting groups of objects by

ones, tens, and hundreds, children gradually construct meanings within a relational framework to assign to number names, and to number-name sequences. Furthermore, their counting procedures become increasingly more complex, efficient, and freed from dependence on perceptual support of grouped materials. With increased mental action in the counting process, the number ideas interact and become further integrated into a numeration system at a high level of abstraction.

Initially, children begin to have a feel for the magnitude of numbers experienced directly. For example, on hearing the number name "thirty-five," they can visualize groupings of objects to associate with it and respond to questions on the composition of that number. Eventually, children develop a feel for the magnitude of numbers they have never experienced, solely on the power of the abstract number relationships they have developed. Since the latter level of sophistication takes considerable time to develop, it may be unreasonable to expect many children to reach it in the elementary school. This kind of abstract reasoning is typical of Piaget's formal operational stage of intellectual development.

When number-name sequences are generated initially, they are based more on sound patterns in spoken language than on number meanings. The number names may be viewed as merely successors.

one — next → two — next → three — next → four — next → five — next → six → seven → eight → nine → ten → eleven (next back) → twelve (next back) →

ten — next → twenty — next → thirty — next → forty → fifty → sixty → seventy → eighty → ninety (next back) → one hundred (next back)

With the construction of number meanings the children convert them into "one more" and "ten more" relationships linking numbers. Since the relationships are reversible, the numbers are also linked by "one less" and "ten less."

one — one more / one less → two → three → four → five → six → seven → eight → nine → ten → eleven (one less) → twelve (one less)

ten — ten more / ten less → twenty → thirty → forty → fifty → sixty → seventy → eighty → ninety (ten less) → one hundred (ten less)

Furthermore, these relations become further integrated through mental interaction of ideas, capable of producing a matrix relating all numbers up to "ninety-nine". In this matrix, on the following page (after Resnick, 1982), numbers are composed of units of tens and ones.

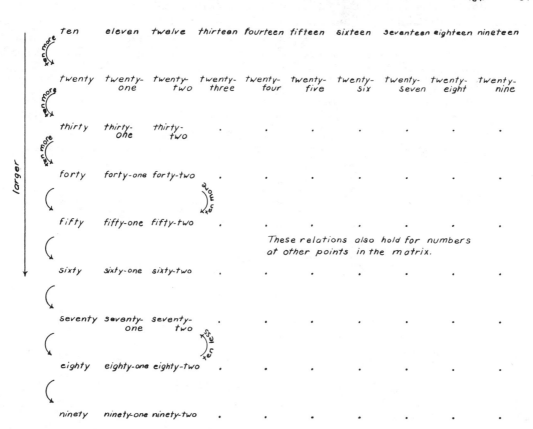

This matrix is a convenient two-dimensional illustration of how numbers can be simultaneously related by tens and ones.[12] Rather than this tidy organization existing intact in the child's mind once constructed, the matrix, or a part of it, may be generated as needed from a more intricate network of number ideas and procedures (Thompson, 1982b).

One of children's major constructions is relating each succeeding level of groups (groups of groups) as composites of ten of the preceding group. Children are able to respond to "How many tens in one hundred twenty?" based on the relation of 10 tens and 2 tens. Furthermore, they can infer that 1000 little cubes would match the large cube of the Dienes blocks without seeing the "inside" of the large cube (Chapter 1). The inference is made on the basis of ten-to-one relations of its dimensions, going beyond the visible.

By reflecting on the counting act and its numerical meanings, children eventually reconstruct relations at a higher plane detached from any physical contexts. Tens, hundreds, and thousands are no longer associated with any particular materials or shapes of groupings. Rather, they are abstract ideas that not only can generate endless number sequences, but also can operate on numbers in the sequence. The further interaction of existing number ideas with the place value notational system results in elaboration and extension of current understanding to larger numbers. Numbers such as a million are understood in terms of ten-to-one relations of the child's decimal numeration system.[13] Rather than existing "out there," this numeration system is constructed within children's minds. Although possible in theory, this advanced level of intellectual development may not be achieved by all children.

Third-grade children's range of methods on the counting board tasks provides us with some general information on their readiness for introduction to writ-

---

12. In support of such a matrix, Resnick (1982) reports that children are able to add or subtract 10 from any quantity more quickly than adding or subtracting other numbers except 0, 1, or 2.

13. This discussion is developed more fully in the context of place value notation in the next chapter.

ten place value notation and for introduction to standard addition and subtraction algorithms. Once the children are capable of constructing ten-units as demonstrated by counting-on and counting back methods by tens and ones, any place value notation for tens and ones becomes a natural representation of what they already understand. Children who only can count-on by ones do not have a conceptual basis for place value notation. They are likely to deal only with the face value of digits in different columns and to find answers by following a procedural rule for computation. Furthermore, children's responses of intricate counting strategies to counting board tasks involving informal addition and missing addend situations reveals the fact that they construct their own addition and subtraction methods. Place value notation, standard school methods of addition and subtraction, and nonstandard, child-constructed methods are the topics of future chapters.

ALTERNATIVES FOR TEACHING COUNTING IN THE CLASSROOM

Preceding chapters on counting (Chapters 3 and 4) have discussed approaches to teaching that are applicable here as well. They are based on the position that placing children in problem-solving situations in which old counting methods don't work will cause the children to adapt to these situations and invent new counting methods. The role of the teacher is to facilitate children's natural tendencies to count in ways that are at once more complex and efficient. Guidelines for these indirect teaching methods include:

1. Provision of problem-solving activities that encourage counting

2. Selective pairing of children to maximize interaction

3. Requests for children's clarification of their methods

4. Encouragement of invention and comparison of alternative methods

5. Modeling of an alternative strategy as a possibility

6. Encouragement of checking by alternative methods

You are invited to review this section in Chapter 4 (pp. 83–88) prior to reading further. Below, two classroom episodes illustrated from videotapes of lessons by two mathematics educators will serve as springboards to discussion of some of these guidelines for teaching.

The preceding guideline for modeling is aimed at those areas of counting where multiple methods are possible. Here, it is recommended that the teacher not model explicitly or repeatedly. Rather, on occasion, the teacher may model an alternative method to suggest a possibility—a germ of an idea for the child to consider. The following episode (Jordan, 1979) will help to qualify this guideline for oral counting sequences.

Although children can extend number sequences based on intuitive rules abstracted from sound patterns, they need direct information or modeling at specific transition points such as the century mark. Yet some of this information or modeling can come from other children. Choral counting, first by ones, from different starting points, and then by different intervals, provides not only practice for some children, but modeling for others who are unsure of the transitions. Although some direct input may be needed in specific transition points, the teacher need not be the only source of this information. The preceding episode illustrates this point.

The next episode (Steffe, 1979a) illustrates other guidelines from the preceding list.

Classroom counting activities can alternate between small group sessions with the teacher and paired activities of children. A teacher's observations during the paired activities can identify the focus of the next group session.

The following are some starting points for classroom activities that encourage a restructuring of children's counting abilities at higher levels of relations and efficiency. Many of the activities are suitable for work in pairs.

Counting Grouped Objects

Provide increasingly larger collections to count and compare to encourage grouping, practice in hearing number names in meaningful contexts, practice in coordinating number-name lists with grouped objects, and so on.

---

14. This episode is illustrated from hastily taken notes that became unintelligible following the last question. What follows is a hypothetical situation that illustrates what might have occurred and how I might have interacted as the teacher.

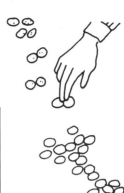

Arranging Beans [15]

Two people each take a spoonful of beans (or a handful of pennies, or any collection of small objects whose quantity is too great to be recognized "by eye" but not too large to be counted easily). Without writing, talking, drawing symbols in the air, mouthing words, or any other verbal or symbolic means, determine who has more beans. Then try to communicate how many beans each person has. You *are* allowed to move your beans around to make groups or patterns. (Don't arrange them in the shape of numerals!) Describe what you did.

1

Which has more? _____
How many whites? _____
How many blacks? _____

2

Which has more? _____
How many whites? _____
How many blacks? _____

3

Which has more? _____
How many whites? _____
How many blacks? _____

4

Which has more? _____
How many whites? _____
How many blacks? _____

---

15. Reproduced from Fuys and Tischler, *Teaching Mathematics in the Elementary School*, Little, Brown, 1979.

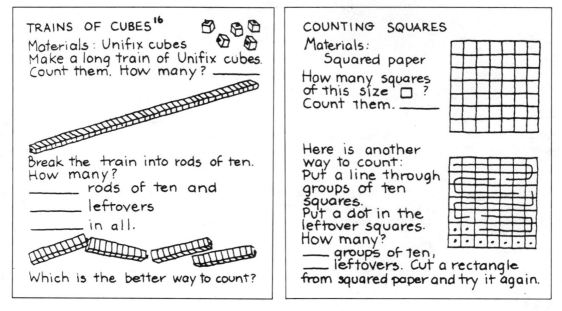

TRAINS OF CUBES[16]

Materials: Unifix cubes
Make a long train of Unifix cubes.
Count them. How many? _____

Break the train into rods of ten.
How many?
_____ rods of ten and
_____ leftovers
_____ in all.

Which is the better way to count?

COUNTING SQUARES

Materials:
    Squared paper
How many squares
of this size ▢ ?
Count them. _____

Here is another
way to count:
Put a line through
groups of ten
squares.
Put a dot in the
leftover squares.
How many?
___ groups of ten,
___ leftovers. Cut a rectangle
from squared paper and try it again.

PICKING PEGS

Materials: Pegboard, pegs

How many pegs can
you pick up in one
hand? Try it.
Guess how many?_____

Put your pegs in the pegboard
like this:

How many full rows?

How many left
over?

How many in all?

SPOONING BEANS

Materials: Jar of beans, 1 spoon
and 1 egg carton for each
person
Each person takes a big spoonful
of beans.
Guess how many?_____

Count the beans in a
spoonful like this:
    put ten beans in
    each cup of an
    egg carton.

How many cups?

How many
leftovers?

Who has the most beans? Who
has the fewest? Try it again.

16. Reproduced from Fuys and Tischler, *Teaching Mathematics in the Elementary School*, Little, Brown, 1979.

Use materials conducive to grouping at more than one level, as in grouping of groups:

MATERIALS	GROUPS	GROUPS OF GROUPS
beans	beans in souffle cups	cups in bowls
buttons	buttons on small safety pin	small safety pins on larger safety pins
beans	beans in small zip lock plastic bags	small bags in larger bags
sticks	bundles of sticks held by rubber bands	bundles enclosed in plastic bags
Unifix cubes	cubes interlocked in stacks	stacks arranged on square cardboard flats and held by rubber bands
buttons	buttons in egg cartons (cut down to ten places)	stacks of cartons

Notes on Counting Materials

Two prominent categories among the materials available for counting are the discontinuous, group-able objects just described and the continuous, preformed objects such as Dienes blocks. Differences between the two types of materials are set forth in the following table.

DIFFERENCES BETWEEN TWO TYPES OF MATERIALS

DISCONTINUOUS, GROUPABLE OBJECTS	CONTINUOUS, PREFORMED OBJECTS
Children can construct groups of tens and hundreds through *direct* action. Conversely, they can decompose these groups directly, as needed.	Children can "build up" each block through the *indirect* action of using ten smaller blocks as measuring units, then trading one form of representation for another. Larger blocks are decomposed indirectly by trading for ten blocks of the next smaller size.
The form and size of these groups is not precise. Yet the quantities in successively larger groups are precise and related by a ten-to-one system of grouping.	The form and size of progressively larger blocks are precise, with dimensions following ten-to-one relations.
Since larger groups are composed of discrete objects, 123 would be represented in a one-to-one correspondence—that is, with 123 objects. BEANS IN ZIP-LOCK PLASTIC BAGS	Since larger blocks are still composed of single pieces of continuous material such as wood or plastic, 123 would be represented by a combination of six blocks. Sometimes lines scored on the surface represent the individual components. Yet in the thousand block, the inner components cannot be scored. Scored lines DIENES BLOCKS
The direct activity of composing and decomposing groups appears to make relations more accessible to children.	The continuous nature of the materials, together with the indirect activity necessary to build relations, seduces children to count any block as one, suggesting that introduction of these materials be delayed.

A unique feature of Unifix cubes is that they combine characteristics of both types of materials. Although their interlocking ability encourages direct and systematic grouping of discrete objects, each suc-

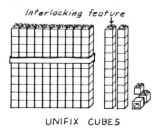

Interlocking feature

UNIFIX CUBES

cessive level of grouping is related by measurement of related dimensions. Therefore, Unifix cubes have the potential of serving as a *unifying model* (Davidson, 1982), or as a bridge from discontinuous to continuous objects. Although there is some basis for making distinctions between materials and identifying Unifix cubes as a unifying model, much research is still needed to study children's interaction with them.[17]

Counting Grouped Objects— Continued

Use a variety of materials to count. Sometimes the same activity with different materials presents another challenge. Generalization from different physical contexts is helpful in abstracting number relations.

Building Numbers.
1. *Making numbers one at a time.* As one child counts orally at a slow pace, the second child shows a matching set of objects. At transition points the objects are regrouped. Starting points can vary as well as intervals. In another variation, both children can build their own numbers simultaneously with contrasting materials.
2. *Make a hundred/thousand and prove it.* By asking children to "prove it" for someone who can walk into the room and check the quantity without counting the objects one by one, you are encouraging a display of grouping of groups.

17. Discrete objects permit observation of children's natural approaches to grouping—for example, the size of children's initial groupings, their awareness of the need for uniform groupings, the time of their first grouping by tens, and the time of their first systematized grouping. By contrast, unless a variety of multibase blocks are available, a particular arrangement tends to be imposed on children when using continuous materials.

3. *How many blocks do you see?* Children are given a collection of Dienes blocks and asked to work out tasks similar to those in Chapter 1, in collaboration. The focus can be an individual flat, the large cube, or any combination of blocks.

Comparing Quantities
Who has more? How many? Estimate/prove it. Compare handfuls, spoonfuls, bagfuls, and bottlefuls of the same or different type of material. Estimate, then group and count to check.

Counting Boards.
1. *Altogether.* Children take turns in setting up the problem and in counting. Two groups of objects to be counted are arranged and one of the groups is covered with a cloth or box. The second child

counts the first addend and counts-on the second addend when uncovered. He then recounts the objects, using another method. The first child can contribute a third method. Both the size and the number of addends may vary.
2. *Crazy, mixed-up counting.* Children set up problems for each other and cover with a cloth, then expose gradually (as on p. 259). The counter first counts the objects as they are shown. When the whole collection has been uncovered the counter recounts the collection in the easiest way to check his original total. Initially, the size of groupings and total quantity should be limited.
3. Mystery box (hide a handful). A collection of grouped and loose objects is displayed by one child for counting by the other. When the counter is looking elsewhere, some of the original objects are covered with a box, cloth, or hand. The counter is then asked to find the number of hidden objects. An alternative method of counting is requested prior to uncovering the hidden objects. (Similar to the missing addend task on p. 258).

Visualizing Quantities.
Picture it. Children are asked to envision grouped materials in response to oral number names. They are then asked the composition of these numbers—for example, "How many tens are there in fifty-three? seventy-one? seventeen? one hundred forty? two hundred? In case of uncertainty, they demonstrate the quantities with grouped objects to check their ideas.

Counting in More Abstract Situations

Number Cards.

1. *Altogether.* Children take turns in turning over two number cards face up and counting to find their total. Each of the cards has a two-digit numeral written on it. (Similar to task on p. 263). Alternative methods of counting are encouraged.

2. *Diffy.* When two number cards are turned face up, the child's task is to find the difference using alternative counting methods.

Play Money.

1. *Pay day.* Stacks of bills are counted when different denominations are mixed, and checked when they are sorted. Stacks and denominations are increased with skill. (Omit $5, $20, and $50 denominations from initial sets of bills to be counted.)

2. *Making change.* The "storekeeper" has a number of pictured items on display with price tags. The "customer" has bills in large denominations only. The "storekeeper" makes change by counting-on. The "customer" checks the change by using an alternative method, such as counting back.

According to Reid (1981), "activity consists not only of encounters with objects, people, and events, it consists also and primarily of encounters with oneself" (p. 69). This section has suggested starting points for counting activities that provide for both interaction between the child and the environment and within the child. First of all, the counting activities provide for interaction with a variety of materials. They also maximize interaction between children through selective pairings, simultaneous use of different materials, and emphasis on alternative methods. Becoming aware of viewpoints other than their own may provoke children to rethink counting procedures on the spot or may sow the seeds for later rethinking. Despite the input from materials and/or people, the encounter is ultimately with oneself—with one's own ideas. Experiencing the vacillation in counting mixed groupings of objects is an encounter with conflicting ideas, eventually leading to their integration at higher levels of organization and to the construction of counting methods at higher levels of efficiency.

REFLECTIONS AND DIRECTIONS

1. Locate some third graders and engage them in counting activities from this chapter. Compare your findings with those reported here. Reflect on the implications of these findings.

2. If you are unable to locate any children for interviewing, select one of the following children for closer study of their counting behaviors on different tasks.

   Rena: pp. 249, 256, 260, 263   Roz: pp. 245, 246, 247, 249, 257
   Zoya: pp. 245, 249, 255, 257, 260, 261

   Try to account for any consistencies or inconsistencies in a child's thinking.

3. Examine the following model of children's construction of simultaneous relations of ones and tens (C. Kamii, 1982b). Decide whether it is consistent with the theory discussed in this chapter and in Chapter 4.

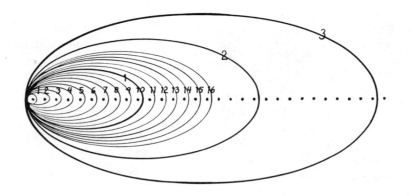

11

INTERPRETATION OF THE PLACE VALUE SYSTEM

PREVIEW

The preceding chapter examined children's representations of grouped and ungrouped collections of objects through oral number names. This chapter will focus on written representation of quantities of grouped objects and their corresponding number names from both the perspectives of mathematicians and children, prior to considering congruent methods of teaching. To set the stage for this study of place value notation, a brief history of numbers and their representation is in order.

A BRIEF HISTORY OF NUMBERS

Historically, many cultures, living in relative isolation, developed their own numeration systems. Since man's first calculator was his own fingers, most cultures tended to group collections of objects by tens. Some cultures, however, grouped them by fives (one hand) or twenties (hands and feet), while others had further variations in groupings. Each of these groupings formed the base for a different culturally-codified system for communicating number ideas. Just as the choice of base groupings was arbitrarily made by different cultures, so were systems for their representation.

Although systematic grouping by tens was the basis for most numeration systems, even in early civilization, different cultures invented progressively more efficient ways of representation. Major changes in these written notational systems came as a result both, of interaction between cultures and the need to represent increasingly larger numbers. Our place value system of representation evolved very slowly, only after simpler systems had been in use for centuries. Thus, our conventional place value notational system is not the only way to represent systematic grouping by tens.

Different levels of representation of our decimal numeration system are outlined below in the order of their development over several centuries in the history of mankind. The purpose of such a discussion is to encourage a greater appreciation of both the hidden sophistication of our conventional place value notational system and the potential difficulties faced by children when first dealing with the system, as well as to show that this system is one of many.

By grouping a large collection of interlocking Unifix cubes into hundreds, tens, and ones, as opposed to grouping by twenty-fives, and so on, we have represented the quantity according to a numeration system having a base of ten. Thus, although the objects are concrete, the grouping is representational. As seen in the preceding chapter, the grouped objects can be scrambled in a random array without altering the total quantity. Keeping like groups together merely facilitates counting.

CONCRETE REPRESENTATION:

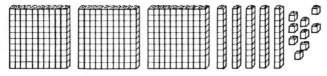

"THREE HUNDRED FIFTY-NINE"

The same grouped objects can be represented pictorially at different levels of detail. In the first example here, the actual blocks are shown. In the

PICTORIAL REPRESENTATIONS:

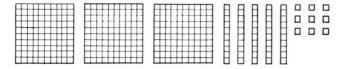

second, a drawing of the front of the blocks is given. In the third, the quantity of blocks is no longer depicted directly. Rather, the shapes and relative sizes of the

outlines serve as reminders of their value. The child must read values into the shapes based on prior experience with such materials and his current understanding of units of tens and hundreds. These pictorial representations also can be scrambled without affecting the total value depicted. Although this type of pictorial representation of quantity is cumbersome for illustrating large quantities, it provides a close mapping of the reality experienced by children.

At another level of representation, an arbitrary choice of symbol, such as a colored chip or letter of the alphabet, can stand for an entire group. Then there is no longer a resemblance to the shape of the grouped objects or a direct correspondence to their quantity. In the example shown, only seventeen symbols represent

SYMBOLIC REPRESENTATIONS

◎◎◎⦿⦿⦿ ⦿⦿⦿⦿⦿⦿ ⦿⦿⦿

h h h t t t t t o o o o o o o o o

a total of three hundred fifty-nine objects. For a child already thinking in terms of systematic groupings of tens, it is possible to infer the values encoded in the symbols for each group and to obtain the total quantity by addition. Although the one-to-one correspondence for the entire collection is lost, it is retained for the ones. Since the color of the chips or the shape of the letters encode their value, spacing or positioning

does not affect their total value. Spatial arrangement and positioning contribute only to ease of reading or counting. This level of representation is similar to the notational system invented and used for centuries by the early Egyptian civilization. Although this system adequately served their needs for recording small quantities, it was inefficient in terms of time and space required for recording large quantities. Although cumbersome in this regard, owing to repetitive use of symbols, the Egyptian notational system allows novices to adapt readily to it.

Just as a group of ten cubes can be represented by a single symbol, so can collections less than ten, in the familiar manner ·1, :2, ∴3 ::4, ·:·5, :::6, and so on. Once again, the one-to-one correspondence of symbol to object has been lost. Now, with this innovation, the repetition of the same symbol can be represented more economically by multiplication. A combination of five or six symbols represents three hundred fifty-nine objects with considerable savings in space and time required to record the numeral. The trade-off is that the symbols bear no direct resemblance to the quantity represented.

3h5t 9(0)⎯understood

The preceding level of representation resembles the notational system developed by the Chinese culture during the third century B.C.[1] Both this and the Egyptian notational systems were used primarily for recording results of computation completed on the abacus or on fingers. Neither system facilitated computations on "paper." Although more compact than the Egyptian notation, the Chinese system also became cumbersome with very large numbers and required the invention of new symbols at each new level of grouping, such as thousands, millions, and so on. Despite these limitations the Chinese notational system contains specific symbolic references to different groupings that assist a novice in reading meaning into the numerals.

Still further economy in representation is achieved by deleting any direct use of symbols for tens and hundreds, while encoding this information in the left-right positioning of a small number of basic symbols. Now any number can be represented by using only combinations of any ten digits (0–9). The same digits

---

1. Here, more familiar symbols are used to illustrate how the Chinese symbols are used in their system.

can be reused, representing a different number based on its relative position. In 333 each digit represents three of something different by virtue of its position. Since the value of each position progressively increases by multiples of ten from the rightmost position of ones, a value of a hundred is inferred for the third

position going left. The digit in that position now represents 300 or 3 hundreds. Using this place value system, three hundred fifty-nine objects can be represented by only three symbols—359. Although this system is quite economical in terms of symbols and space, the novice is now required to read more meaning into fewer symbols. A mathematically sophisticated adult can encode the following meanings into 359 for another mathematically sophisticated person to decode.

POSITION	THIRD	SECOND	FIRST
FACE VALUE	3	3	3
PLACE VALUE	3 HUNDREDS	3 TENS	3 ONES
TOTAL VALUE	300	30	3

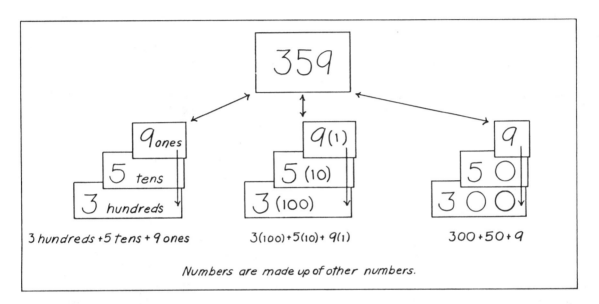

Numbers are made up of other numbers.

Notice that this system of place value notation incorporates both the additive and multiplicative features of the earlier system.

These early notational systems were used mainly for recording the results of computations completed on an abacus.[2] In these systems there was no need to depict the absence of any grouping in the written record. However, once specific symbols for tens and

hundreds were deleted, a placeholder was essential to distinguish between two hundred five, twenty-five and two hundred fifty in written notation. Since the written symbols corresponded to oral counting words, there was no symbol to represent the absence of number. A new concept was conceived and represented by "0".

Egyptian	ℓℓ||||	∩∩)||||	ℓℓ∩∩∩∩
Chinese	=百五	=+五	=百五+
Hindu-Arabic	205	25	250

2. The beads or chips on the abacus need not be color-coded for value. Rather, they can be identical in all properties since their relative positions determine their value.

In earlier Egyptian and Chinese systems of notation, it was possible to scramble the symbols without affecting the value depicted. Now strict adherence to both the specific order and values of positions and the use of zero as a place holder are essential to retain the integrity of the place value system of notation.

The conception of zero by Hindu scholars around 600 A.D. converted the Hindu-Arabic notation into a place value system, providing further economy in representation and facilitating paper-and-pencil computation. This event marked a major intellectual breakthrough that opened the way for considerable development of arithmetic by mathematicians.[3] Initially, this place value system found common usage only among mathmaticians and merchants. Despite its obvious advantages for representation and computation, it took centuries for this efficient system to be accepted widely.

PREVIEW—CONTINUED

We have noted that our conventional place value notation is one of a number of representational systems that were invented and adopted by mathematicians. Also, at each succeeding level of more efficient representation, the symbols are fewer and more detached from concrete reality. Although this conventional system of place value notation is quite economical in terms of symbols and space required, this economy may itself be a source of difficulty for children.

Current methods of teaching introduce the reading and writing of two digit numerals in the first grade. Since using any system of representation assumes a shared framework of ideas for both encoder and decoder, introduction of the place value notational system in the first grade assumes that these children are already thinking in terms of systematic groupings by tens and have concerns for most efficient representation of these ideas. Furthermore, the introduction of first graders to a ready-made, efficient system of representation assumes that the children are not only ready for such a formal system, but that they have no ideas of their own on the pictorial or written representation of large quantities. However, Piaget observed that the history of evolution of scientific and mathematical ideas is often paralleled by the levels of children's understandings as they construct their own ideas. This general trend suggests that, given the opportunity to develop their own methods of representation for the systematic grouping of objects, children will develop ways similar to the lower levels of representation described earlier.

By exposing children to place value notation in the first grade, we do not allow children time to construct their own place value relations based on systematic groupings of tens and to express their own ideas for representing these groups. In doing so we impose on children an already-constructed and sophisticated system that eluded scholars for centuries.

The purpose of this detailed introduction is to prepare you to view children's interpretation of our place value notational system with a critical eye. You are invited to place yourself in the children's shoes and to anticipate how they will deal with the following tasks.[4]

---

3. In the Chinese system there was a symbol used as a gap filler but its meaning signified "and" as in "two hundred *and* five."

4. The interview tasks are adapted from Behr (1976) and West (1978).

1. Write the following numbers: seventeen, seventy-one, ten and two,
   two hundred eighty-nine.

2. Read these numbers: 342   243   432

   What does the 3 stand for? How do you know?

   Write the number that is one more than 342.

   Write the number that is ten more than 243.

3. Read these numbers: 231   198.   Which of these two numbers is the larger one?

   How did you decide?

WRITING NUMERALS FROM ORAL NUMBER NAMES

Children's everyday experiences include exposure to many conventional number symbols. Although children's television experience may present such symbols in a numerical context, other experiences use these symbols in a nonnumerical context. Numerical symbols found in telephone numbers, addresses, room numbers, uniform numbers in different sports, television channel numbers, license plates, and so on have no quantitative meaning. They are merely identification labels. Bell and Burns (1981) report that all beginning first graders in their study, conducted at a suburban school, were able to read 100 and a large proportion of the group could write it prior to any formal teaching. Owing to this constant exposure to the conventional place value notation, it is difficult to study children's natural tendency to use simpler representations, particularly in school settings. However, observing children outside the context of standard school tasks, as in observing children while they keep a record of the frequency of randomly occurring events, keep a record of points accumulated during a game, or "writing a message to Snoopy" (Chapter 6), may provide glimpses of other representational methods. Further glimpses of these tendencies to use simpler representational methods are available from observing children's early attempts to deal with conventional place value notation.

Since there is no pattern in the number names from "one" through "twelve" and no direct mapping between the numerals and the quantity of objects, number names and their numerals must be associated by rote memory. Considerable help is needed from external sources to achieve this association. However, once children become exposed to numerals for larger quantities together with their number names, they begin to notice correspondences in their order of representation and search for further regularities. Soon they develop a procedural rule for abstracting digit names from position labels in the order of oral number names and for writing the corresponding numerals:

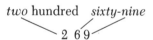

Thompson (1982) points out that in processing a number name, the child must keep the temporal order of the position labels in mind, so that if it is missing for the name, a "0" is written. The following processing may take place for "five hundred four."

Let's see you write this number:

two hundred eighty-nine

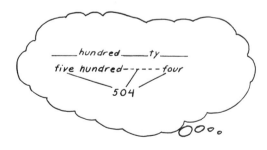

Although children begin to abstract digit names from the oral number names quite readily, their initial attempts reflect procedural rules in need of refinement.

A common "error" made by children in representing numbers such as "two hundred eighty-nine" is to represent them in a nonconventional manner. Their natural tendency is to write combinations of separate symbols for each part of the number name in a manner similar to the pre–place-value notational systems. These systems had separate symbols for hundreds and tens. Thus rather than 289, you may see 200809 or 20089.[5] In interacting with conventions that conflict with their natural tendencies, children arrive at a compromise solution, writing 2089. They may vacillate between these approaches for some time before accepting the compact conventional notation as having incorporated the same information. Just when they appear to have accepted the convention, they are likely to revert to their natural tendencies when asked to represent a large number requiring an extension of their existing procedural rule. Bell and Burns (1981) report that 14 of 35 beginning second graders made such errors—instead of writing 489, eleven of the children wrote 40098 or 4098. In my own interviews, only 3 of the 29 beginning third graders made such errors, writing 2089 rather than 289. The remaining children appear to have constructed a procedural rule for producing conventional numerals. It would have been interesting to have tested the limits of their rule.

Write the following numbers:

seventeen

seventy-one

ten and two.

Although oral number names offer useful patterns of sound clues for writing most numerals, the -teen numbers in the English language are misleading. Here, the ones are sounded out prior to the -teens. Furthermore, the sound of the -teens and the -tys are very similar. Owing to this inconsistency in order and similarity of sound, children may represent "seventeen" as 71 or 70. The inconsistency in oral number names prompts children to reverse digits in writing both -teen numerals and two-digit numerals ending with one as part of an active search for some overall consistency. The observable result of this active search is a rash of reversals followed by vacillations over a period of time prior to acceptance of -teen number names as an exception to their general procedural rule for writing numerals. Another example of this search for consistency is illustrated by the following anecdote of a British primary school child (Wheeler, et al., 1977).

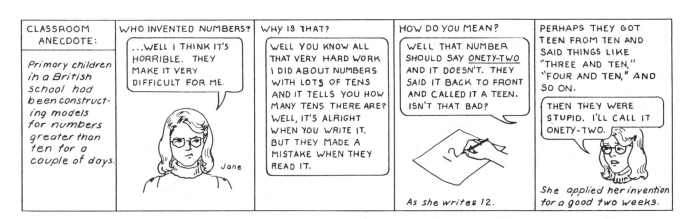

Jane not only invented an oral number name that brought uniformity to both the naming and the written notation, she insisted on using her invention for several days before accepting a more irregular convention. The anecdote suggests that she worked backwards to the teens from the consistent pattern extending beyond

it. To reduce this specific difficulty in writing -teen numbers, teachers in some British schools do, in fact, encourage the use of language such as "onety-two." Other teachers avoid writing -teen numbers until the pattern of writing -ty numbers has been firmly established in the presence of increasing collections of grouped objects.

Since all of the 29 third graders interviewed were able to write correct numerals for "seventeen" and "seventy-one", it appeared that they had accepted the

5. Bell and Burns (1981) report that 20089 occurs more frequently than 200809.

irregularities in oral number names of the -teens. However, a child's lack of spacing between his numerals—1771—suggests that he did not appreciate the positional features of the conventional system. The request to write *"ten" and "two"* also highlighted this lack of awareness as 7 of the 29 children wrote 102. Two other children responded to this atypical request by writing a more acceptable 12. The remaining children spaced out the numerals to retain the integrity of their representaton. M. Kamii (1981) points out that at a time when children are realizing that the order in which they count a collection of objects and their spatial arrangement don't affect its number, they must make sense of a place value notational system in which spatial arrangement and position are critical to the values assigned to them. In the preceding examples, asking children to read their run-on numerals may have provoked a rethinking on their part or provided further information on their views of the importance of spacing of digits.

The following data from interviews with first and second graders by Thompson (1982b) illustrate some of the children's tendencies just described in writing numerals. The data compare the numerals written by three children for selected number names.

CHILDREN'S REPRESENTATIONS FOR SELECTED NUMBER NAMES

ORAL NUMBER NAME	CHILDREN'S REPRESENTATIONS		
	KAPPA (SEVEN YEARS)	DELTA (SEVEN YEARS)	ALPHA (SIX YEARS)
thirteen	30°	13	13
forty-one	41	41	41
eighty-four	84	84	84
one hundred one	1001°	101	101
one hundred seven	17°	107	107
one hundred fourteen	114	114	114
one hundred twenty-one	120 121 (rewritten)	121	121
one hundred seventy-eight	178	178	178
two hundred nine	2009°	209	209
two hundred nineteen	291°	2019°	2019°
two hundred sixty-seven	267	2067°	267
nine hundred thirty-four	90034° "I don't know this one."	9034°	934

Thompson infers that when Kappa wrote 17 for "one hundred seven", he was in disequilibrium—feeling uneasy with the "funny look" of his preceding 1001, and overcompensated on the length of the numeral. Thompson's direct observations of the children's process of writing numerals provides further insight into some of their errors. He noticed that Delta said "hundred" as he wrote 0 in 101, 107, 209, and likewise for 2019, 2067, and 9034. The zero was part of his notation for hundred. Thus Kappa's rewriting of 120 as 121 can be viewed as a suppression of that "expanded" notation. Thompson's observation of Alpha's method of writing -teen numbers may also account for his only error. Following the oral sound pattern of "eighteen", Alpha wrote the 8 first and then the 1 to the left of it upon hearing "teen". Thus when given "two hundred nineteen," he first wrote 209 before inserting the 1 upon processing "teen." Without crossing out 0, his end result was 2019. Thompson's work clearly indicates the value of observing children during the process of writing numerals rather than attempting to follow their reasoning solely from their written products.

There is evidence that the development of procedural rules for writing numerals for oral number names is begun by some children prior to any formal teaching. The abstraction of digits from temporal sound patterns in the number names takes place at a relatively mechanical surface level without underlying number meanings. The reconstruction of the counting process for systematic groupings of tens and the application of oral counting names and written notation in such physical contexts is critical to the development of meaning. Here, the written notation is a record of the number of each type of grouping.

READING NUMERALS

Read these numbers

342

243

432

Children's initial attempts at reading numerals may involve reading successive individual digits. Although 28 of the 29 third graders read these numerals with ease, one child did read them as individual digits. She did so despite demonstrating the ability to write numerals for numbers such as "two hundred eighty-nine".

Although she focused on the numerals as unrelated digits for some time, her eventual successful reading suggests that she had abstracted a procedural rule for reading numerals. She was vacillating between a digit and a numeral focus. After all, her method of reading would be acceptable for addresses, telephone numbers, license plates, and other situations in which digits are used as identification labels.

The next two children, described in Thompson's study (1982b), appear to be in the midst of the process of abstracting a procedural rule for reading numerals.

Delta, a second grader, is able to recognize instantly that the three-digit numeral starts with "one hundred". He does have difficulty in associating the next digit name with a *ty* word. However, with some hesitation he is successful. Similarly, Sigma, another second grader, usually ended up saying the correct name, but only after some false starts. However, he appears to be experimenting with partitioning the numeral in different ways prior to deciding on a standard procedure of starting from the left.

Thompson's Kappa (1982b), another second grader, wrote reversals for -teen numbers and now reads them in a similar manner. His incomplete rule for writing numerals appears to influence his reading of numerals as well. Next, we see that Kappa has conflicting procedures for reading -teen numerals and two-digit numerals ending with "1." The interviewer's questions help Kappa become aware of this conflict.

Other reading errors demonstrated by children (Thompson, 1982b) include reading 143 as "fourteen three" and 201 as "twenty hundred and one." These errors reflect faulty partitioning of the numeral into individual digits and not knowing how to interpret "0." Bell and Burns (1981) report that 11 of the 30 beginning third graders read 5004 as "five hundred four," reflecting a procedure consistent with their early attempts to write such numerals. Thompson (1982b) notes that until such time as procedures are firmly established, children are likely to misapply them or to include incomplete or overgeneralized rules in the process of their formation.

According to Thompson (1982b), successful reading of numerals requires instant recognition of the leftmost starting position name and succeeding position names to its right, together with instant recognition of digit names. The following processing accounts for the reading of numerals according to our place value convention.

Recognition of starting name:	IF IT'S A THREE-DIGIT NUMBER IT STARTS WITH A HUNDRED NAME.
Numerals:	243 ... 304
Position names from pattern of oral number names:	___ hundred ___ ty ___ ___ hundred ___ ty ___
Digit name inserts:	two hundred forty-three three-hundred (say nothing) four

Once the leftmost starting position is named and the successive ones are known, the number name of the digits is recursively inserted into the positional sound pattern of the oral number name. Note that a specific rule for reading zeros must be incorporated into a general procedural rule for reading numerals.

Thompson (1982b) cautions that the procedures for reading and writing numerals are basically nonnumerical. In other words, they are based on the form of the notation rather than on numerical meanings. These meanings must still be read into the number names and numerals that result from these procedures. Thus, although our third graders appear adept at reading and writing numerals, we cannot accept these skills as reliable indicators of understanding our numeration system. Since systematized grouping forms the underlying rationale of place value notation, their abilities to count-on by tens and hundreds (Chapter 10) should indicate whether they are able to read meanings into numerals. Other tasks in this chapter will go beyond mechanical processes dealing with patterns in surface features and rote associations to explore children's interpretation of numerals.

INTERPRETING PLACE VALUE NOTATION

We have observed that our third-grade children are quite adept at reading and writing numerals up to three digits. Now we'll examine the extent to which they are able to identify the place value of digits in a numeral and to give an underlying rationale.

> 342 243 432
>
> What does this 3 stand for?
>
> How do you know?

Twelve of the 29 children were able not only to identify the place value of the "3" in each numeral, but also to verbalize a rule. Sometimes the child's response prompted a follow-up question. What other questions could be asked?

A spontaneous extension

The quick response prompted a follow-up question.

His use of "middle" prompted a follow-up question.

"First number" was tested by a follow-up question.

While some of the children spoke of hundreds, tens, and ones on the basis of a rote association with positional names, others made reference to the underlying rationale of place value notation. Further probing would be needed to distinguish clearly between these children.

Ten other children neither identified the place value of the 3s nor provided any underlying rationale. These children identified the total values as "three hundred" "thirty" and "three," but could not go further to interpret the values as hundreds, tens, and ones. The following excerpts illustrate how two of the children resisted the interviewer's efforts to have them view the numerals from another perspective.

Although these children have abstracted procedural rules for reading and writing numerals based on surface features, they have difficulty in attaching deeper meanings to the symbols. Their focus on thir*ty* allows them to identify the position based on the order of naming numbers. However, a focus on *tens* would identify both the position and the grouping represented at that position.

The remaining seven children either gave unusual rationales or invented place values. The next child's response sounds plausible initially, but when extended to another example it suggests a lack of place value understanding, including the significance of zero as a place holder.

The other six children had difficulty recalling the conventional place values, making reference to zeroes, twenties, or to a rearranged order of values.

How would you explain their reference to the zero column?

It is interesting to note that all of the latter group experienced difficulties in counting-on by tens and hundreds in the tasks described in Chapter 10. It would appear that they have not sufficiently restructured their counting to establish a basis for the symbolism of place value representation. All but one of these children eventually corrected the place values after prompting. It will be interesting to observe whether Roz persists in using her own place value scheme in other tasks. In searching for an explanation of the children's reference to the zero position, I examined their math textbook and located examples of computation in which columns were labeled as

h	t	o

The end result gives the impression of mindless manipulation of written symbols. Rather than indicate a learning disorder on the part of the children, this example is more suggestive of a *learned disability* resulting from faulty teaching. Hendrickson (1983b) makes reference to many children being curriculum disabled, textbook disabled or teacher disabled.

Similar observations of children's resistance to reading meanings into numerals have been reported by other researchers. Some of these are described below. Behr's extended teaching experiment (1976) with second graders contains repeated anecdotal observations of this difficulty. At the conclusion of his experiment, Behr identified five levels of number concept. Identifying the place value of the numeral 24 as 2 tens and 4 ones marked the highest conceptual level in his sequence. Referring to total values of the digits—20 and 4—was placed at the next lower level. Interpretations characteristic of still lower levels are found in the following description of M. Kamii's work (1980).

M. Kamii explored children's interpretation of place value notation through informal representation. Children were asked to draw a representation that would communicate how many cars could be outfitted with sixteen wheels. Some of the children just showed sixteen wheels in a row while others showed groups of four wheels. The children were then asked to interpret the numeral 16 in terms of their drawing. Different levels of interpretation are illustrated on the facing page.

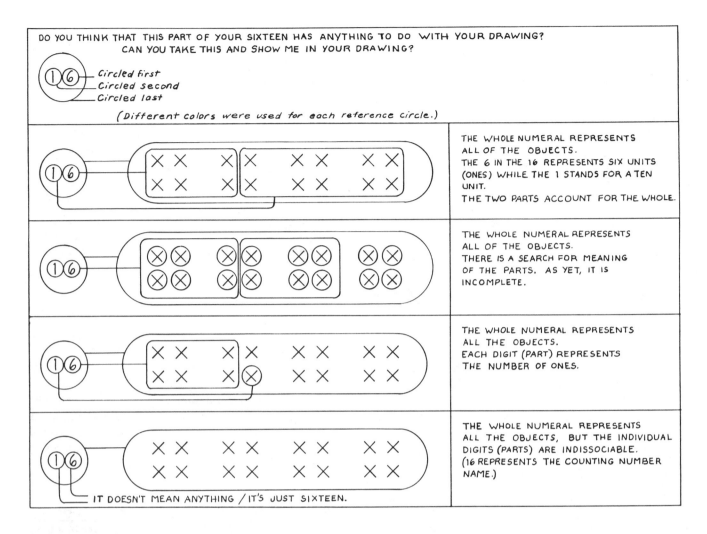

DO YOU THINK THAT THIS PART OF YOUR SIXTEEN HAS ANYTHING TO DO WITH YOUR DRAWING?
CAN YOU TAKE THIS AND SHOW ME IN YOUR DRAWING?

Circled first
Circled second
Circled last

(Different colors were used for each reference circle.)

THE WHOLE NUMERAL REPRESENTS ALL OF THE OBJECTS. THE 6 IN THE 16 REPRESENTS SIX UNITS (ONES) WHILE THE 1 STANDS FOR A TEN UNIT. THE TWO PARTS ACCOUNT FOR THE WHOLE.

THE WHOLE NUMERAL REPRESENTS ALL OF THE OBJECTS. THERE IS A SEARCH FOR MEANING OF THE PARTS. AS YET, IT IS INCOMPLETE.

THE WHOLE NUMERAL REPRESENTS ALL THE OBJECTS. EACH DIGIT (PART) REPRESENTS THE NUMBER OF ONES.

THE WHOLE NUMERAL REPRESENTS ALL THE OBJECTS, BUT THE INDIVIDUAL DIGITS (PARTS) ARE INDISSOCIABLE. (16 REPRESENTS THE COUNTING NUMBER NAME.)

IT DOESN'T MEAN ANYTHING / IT'S JUST SIXTEEN.

Having interviewed children from preschool through age nine, Kamii found that from age seven on all the children indicated that two-digit numerals represented all of the objects drawn. From age seven, some of the children spoke of ones, tens, and hundreds, but not all could apply this information/knowledge in their representations. It was not until the nine-year-old group that half of the children consistently gave place value responses.

As part of a broader task, Bednarz and Janvier (1982) asked children to construct the numeral for *four hundred forty-five* with the following tags (presented to them in random arrangement):

Only twenty out of seventy-five third graders gave meanings to ones, tens, and hundreds in terms of groupings by putting together the following tags:

| 4 HUNDREDS | 4 TENS | 5 ONES |

The largest proportion of children worked exclusively with digit information on the tags while ignoring and giving no meaning to the words *hundreds*, *tens*, and *ones*. For example, one child put together

| 4 ONES | 40 TENS | 5 HUNDREDS |

and wrote the numeral as 4405. Although an improvement was noted in the group of fourth graders studied, still, less than half focused on hundreds and tens and ones groupings. Despite the prominent position of place value notation in the primary school curriculum, the findings of Bednarz and Janvier indicate that hundreds, tens, and ones convey only an order for reading and writing numerals for many children rather than systematized groupings. For these researchers, this limited interpretation following two to three years of exposure to place value notation casts doubt on existing teaching methods.

COMPARISON OF NUMERALS

As children accumulate experience with place value notation they develop procedural rules for their comparison. These rules may take the following forms.

997 vs. 2467   The more digits the bigger the number.

231 vs. 198   When you have the same amount of digits, comparing the size of the first digits (2 vs. 1) shows the bigger number.

Note that successful procedural rules for comparison of numbers may be based solely on the surface features of numerals or number names.[6] However, it is possible to read meanings into these rules.

The following task was given the third-grade children to examine their ability to compare the relative magnitudes of numbers represented by a pair of numerals and to provide some underlying rationale.

The range of responses from third-grade children illustrates both faulty and reliable procedural rules, as well as indications of deeper meanings.

Faulty procedural rules are developed when children view numerals as merely collections of unrelated digits. Two of the third graders made comparisons either by considering all the digits in each numeral or by arbitrarily partitioning the numeral.

Given an opportunity to interact with Dienes blocks in a follow-up, both children altered their perspectives of the numerals to a focus on the hundreds place.

Most of the third graders focused on the leftmost digits, thus appearing to have abstracted a procedural rule. Yet these children often were unable or unwilling to give an underlying rationale. This level of successful comparison is illustrated by Zoya's response.

6. Johnson (1975) finds this comparison routine to be parallel to finding which of two words comes later in the dictionary.

Some of the children still viewed the numerals as collections of independent digits at one time, while at another time they gave more significance to the wholeness of the numeral. Thus, they were seduced by the size of randomly positioned digits one time and the focus on the leftmost digits at another time. When these shifts of focus occur with long periods intervening they may go unnoticed. Yet when occurring rapidly in a short span of time, these vacillations are recognized readily as a state of *disequilibrium*. In the following panel, Claude vacillates several times in making his decision. Once he finally makes his decision, he is unable to justify it in terms of place value.

A handful of children were able to provide some rationale for their decision. While some were quick to justify their responses, others needed time to ponder the question. Given time, one girl was able to rethink her approach and provide some rationale for her decision.

While some of the children's rationales were quite abbreviated, others were rather extensive. The abbreviated rationales focused on the relative values of the digits in the leftmost position while the more extensive rationales took a broader view.

The children who selected 231 as being greater than 198 were given a follow-up task to test their place value conviction.

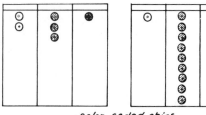

These chips stand for a certain number.

What number do you think they stand for?
The chips stand for ones, tens, and ____. (if needed)

Which board stands for the bigger number?

How do you know?

-color coded chips
-boards presented individually

Countersuggestion 1: The other day a boy/girl said, "Look at all of these chips here. There's eight green chips here and only one there. Nine yellow chips here and only three there. So this board stands for the larger number." What do you think about his/her answer?

Countersuggestion 2: (Original numerals are brought out.) You told me earlier that 231 was the larger number and now you think 198 is . . . How come?

Although the children had no prior experience with the color-coded chips, 13 children read 231 and 198 from the displays, based on contextual clues and interaction with existing place value ideas. An equal number of children said the chips stood for the number "six." These children were given the following prompt: "The chips stand for ones, tens, and ____" (simultaneously pointing and moving from right to left). Seven children responded with "fives" or "twos" for the colored chips in the leftmost column and needed considerable help in reading the boards. The task was terminated at this point for three of the children.

Some of the children were misled by the total number of chips on the board (or in a particular column), thus magnifying the problem already existing in formal place value notation. Five children wavered in their decision making but eventually justified

231 being the larger number. Four children either reverted to 198 in the face of countersuggestion or were not able to justify their choice of 231.

This is another task that not only highlights the instability of children's place value understanding, but also the limitations of the chips as a place value model. Chips and abaci are almost as representational as formal symbols. It would have been interesting to observe how the children would have responded if all the chips had been the same color.

Thompson (1982a) used an extension of the comparison task involving ordering several numerals. In addition to providing information on children's abilities to relate numbers in a series, the task also provided support for his position that children mentally process the "sound images" of number names rather than the numerals themselves.

PUT THE CARDS ON THE BOARD FROM THE LARGEST TO THE SMALLEST.	*The results of Alpha's ordering.*	*Alpha read.*
143 134 124 117 113 107 103 97 *Alpha* *The cards were shuffled prior to presentation.*	107 143 134 103 124 117 113 97 *Alpha read the numerals both while ordering and upon completion of the task. His consistent reading provides an important clue to his mental processing.*	107 HUNDRED SEVENTY 103 HUNDRED THIRTY *Thus, if you had listened to an audio-tape recording of the interview, you would have concluded that he ordered the cards correctly – HUNDRED SEVENTY, HUNDRED FORTY-THREE, HUNDRED THIRTY-FOUR, HUNDRED THIRTY, HUNDRED TWENTY-FOUR, etc.*

WRITING NUMERALS THAT ARE ONE MORE THAN/TEN MORE THAN

When place value notation is understood as a representation of systematized grouping by tens, it is possible to alter a numeral to another showing one more/ten more/a hundred more, and so on, by incrementing the appropriate digit by one. The following questions explore third graders' abilities to represent such increments. They are presented below in the context of the interview sequence.

Read these numbers: 342   243   432

What does the 3 stand for? How do you know?

Write the number that is one more than 342 / ten more than 243.

(The interviewer's finger encircles the numeral as the question is asked.)

Fourteen of 29 third graders wrote 343 as one more than 342. The remaining children gave a variety of responses. The most common of these was 453, given by 5 children.

One more than 342

1,342

453

442

452

352

5,342

You are invited to explain the results. You might—
- examine these responses
- find the difference between 342 and each response
- determine if the numerals vary from 342 in any other way
- explain these responses in terms of children's perspectives of "one more"

Based on the children's interpretation of one more than 342, predict the range of children's responses to ten more than 243.

Ten of the third-grade children wrote the expected 253 as the number that was ten more than 243. However, not all of these children gave an immediate response. Eye, lip, and finger movements suggested that some children were counting-on ten more by ones. The remaining children gave a wide range of responses again. The most common response, written by four children, was 1243.

Again, you are invited to explain the results.

Now that you have some hunches about how the children interpreted the initial "more than" questions, what follow-up questions would you ask to gain further information and to test your hunches?

Two such follow-up questions were planned in advance and asked of each child.

Now write the number that's one ten more than 432.[7]

Write the number that comes after ___ .

Write the number that comes ten after ___ .

The first question has some obvious ambiguity built in, in a conscious attempt to get another look at the child's thinking. It often had the effect of prompting children to rethink their original "incorrect" views that had been consistent for the initial "more than" questions. By rephrasing the "more than" questions in terms of "comes after," nearly all of the children gave the expected answer. This answer was often preceded by further evidence of children counting-on by ones. Asking these additional questions served to enrich the data base for explaining children's interpretations of this task. Such explanations follow, together with a closer look at the responses of individual children.

The third graders' performance on the place value tasks was not always consistent. Often the children appeared to understand place value concepts at one minute but not the next. Although some children appeared quite competent on the "reading 3s" task (p. 286), they gave surprising responses on the "more than" task. On the other hand, other children had difficulty "reading 3s," but gave the expected answers to the "more than" task. Zoya's responses, characteristic of the latter group, are illustrated on the facing page.

7. The "one ten more" question was used in Behr's teaching experiment (1976).

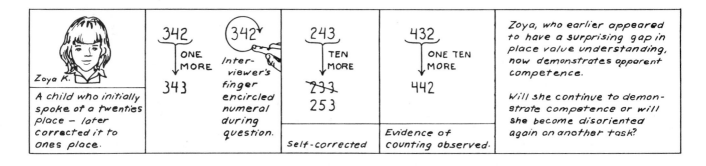

Zoya K.	342 ↓ ONE MORE 343	(342) *Interviewer's finger encircled numeral during question.*	243 ↓ TEN MORE ~~233~~ 253	432 ↓ ONE TEN MORE 442	Zoya, who earlier appeared to have a surprising gap in place value understanding, now demonstrates apparent competence. *Will she continue to demonstrate competence or will she become disoriented again on another task?*
A child who initially spoke of a twenties place — later corrected it to ones place.			*Self-corrected*	*Evidence of counting observed.*	

Also, Brenda shifted from one low-level perspective to another and back, yet appeared to take a place-value perspective on the next task.

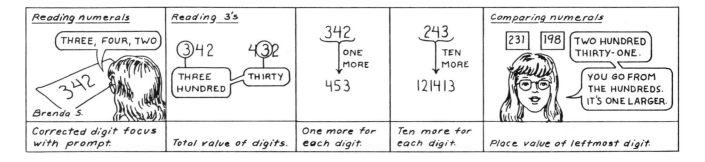

Reading numerals	*Reading 3's*	342 ↓ ONE MORE 453	243 ↓ TEN MORE 121413	*Comparing numerals* 231 198 TWO HUNDRED THIRTY-ONE. YOU GO FROM THE HUNDREDS. IT'S ONE LARGER.
THREE, FOUR, TWO 342 *Brenda S.*	③42 THREE HUNDRED 4③② THIRTY			
Corrected digit focus with prompt.	*Total value of digits.*	*One more for each digit.*	*Ten more for each digit.*	*Place value of leftmost digit.*

In writing about the extraordinary inconsistency in the arithmetic behavior of individual children, cognitive psychologist Ginsburg (1977) takes a Piagetian viewpoint. He writes that children's errors are seldom capricious or random. Rather, he views children's behavior as meaningful, with their errors having sensible origins. From a child's perspective, each of Brenda's responses makes sense. At this point in her development of concepts of number and place value notation, her major problem is deciding which perspective to choose for a given task.

Despite the general inconsistency of many third graders' responses across place value tasks, brief periods of consistency were also observed. Within the "more than" task, the children's perspectives were fairly consistent for the two initial questions. Considering the range of responses to these questions within the group of children, the amount of consistency may surprise you. Such periods of consistency within a longer span of general disequilibrium argue against dismissing such responses as random errors.[8] Not only are most of the children's errors sensible, they also serve to remind us of the complexities of the place value system of notation and the underlying concept of systematized grouping. The next sample responses illustrate both the consistency in perspective within the "more than" task and inconsistencies across tasks. Furthermore, it illustrates the effect of follow-up questions in prompting a shift in this perspective.

			Follow-up questions	
Paula L.	342 ↓ ONE MORE 5,342	243 ↓ TEN MORE 2,530,687,413	WRITE THE NUMBER THAT'S 432 ↓ ONE TEN MORE 2,734	243 ↓ WHAT NUMBER COMES TEN AFTER? 253
A girl who spoke of ones, tens, hundreds and thousands places.	*One more place*	*Ten places (as opposed to ten more places.)*	*One more place*	*Ten more*

---

8. Yet the children's range of responses does include some that are interpreted less readily and may indeed be random. The complete list also includes the following: 666 for 1 more than 342; 1212, 1264, 284, 268, and 8713 for 10 more than 243.

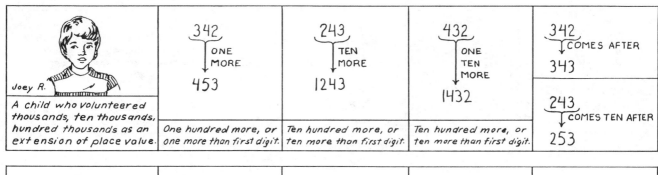

	342 ↓ ONE MORE → 453	243 ↓ TEN MORE → 1243	432 ↓ ONE TEN MORE → 1432	342 ↓ COMES AFTER → 343 / 243 ↓ COMES TEN AFTER → 253
Joey R. A child who volunteered thousands, ten thousands, hundred thousands as an extension of place value.	One hundred more, or one more than first digit.	Ten hundred more, or ten more than first digit.	Ten hundred more, or ten more than first digit.	

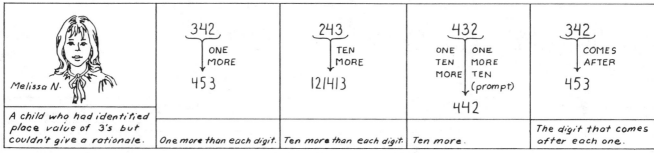

	342 ↓ ONE MORE → 453	243 ↓ TEN MORE → 121413	432 ↓ ONE TEN MORE TEN (prompt) → 442	342 ↓ COMES AFTER → 453
Melissa N. A child who had identified place value of 3's but couldn't give a rationale.	One more than each digit.	Ten more than each digit.	Ten more.	The digit that comes after each one.

Sometimes the errors that appear the most bizarre have the most sophisticated rationale.

	342 ↓ ONE MORE → 453	243 ↓ TEN MORE → 1227 Changes to → 121413	432 ↓ ONE TEN MORE → 425 Explained as → 41312	432 ↓ COMES AFTER → 442 ↓ Counting prompt → 343
Peter W. A child who read 3's in terms of place value. After the "more than" task he compared numerals with a place value rationale.	One more than each digit.	Ten more than each digit, arbitrarily compacted.	Ten more than arbitrary digits / compacted arbitrarily.	Comes after first digit ↓ Comes after last digit.

One child, demonstrating an unexpected place value order in the "reading 3s task," systematically applied her incorrect order rule in the "more than" task.

	342 ↓ ONE MORE → 352	243 ↓ TEN MORE → 253 (looks for approval)	432 ↓ ONE TEN MORE → 433	342 ↓ COMES AFTER → 352
Roz A. A child who had rearranged the place value system: hundreds, ones, tens.	One more (her system)	Ten more (conventional system)	Ten more (her system)	Next ones (her system)

The unusual and generally inconsistent responses across different place value tasks reflect an instability in both children's structuring of number ideas and their interpretation of a sophisticated system of place value representation of these ideas. The relative consistency of perspectives observed at different times within a period of general disequilibrium and their reoccurring characteristic indicate that children have structured ideas for different perspectives of number and numerals. What these children have yet to do is to integrate these ideas into a coherent organization. A similar viewpoint on children's disparate behaviors on numeration has been expressed by Thompson (1982a). He writes:

> The fact that some children do behave inconsistently on seemingly related problems suggests that their knowledge of numeration is more or less compartmentalized, and that inconsistencies might vanish when those compartments become meaningfully related. (p.2)

He goes further to posit that children develop whole number numeration through the interaction of these compartments or domains of knowledge.

Initially, children's multiple perspectives of number ideas and their representation in a formal system exist in relative isolation and compete for application on different tasks. The range of children's responses to the "more than" question can be explained in terms of these multiple perspectives.

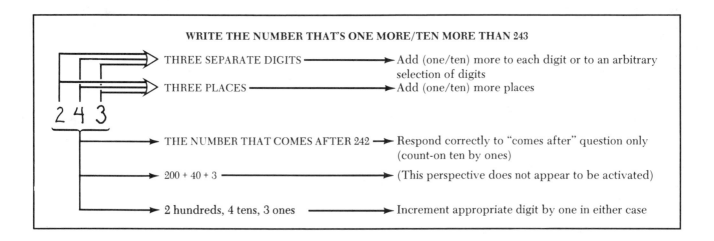

WRITE THE NUMBER THAT'S ONE MORE/TEN MORE THAN 243

- THREE SEPARATE DIGITS → Add (one/ten) more to each digit or to an arbitrary selection of digits
- THREE PLACES → Add (one/ten) more places
- THE NUMBER THAT COMES AFTER 242 → Respond correctly to "comes after" question only (count-on ten by ones)
- 200 + 40 + 3 → (This perspective does not appear to be activated)
- 2 hundreds, 4 tens, 3 ones → Increment appropriate digit by one in either case

Until such time as children have integrated the multiple interpretations of numerals with underlying concepts of systematized grouping of ten, different place value tasks are likely to trigger any of the competing perspectives. The ambiguity of the follow-up question dealing with "one ten more" may serve to bring isolated perspectives into interaction. It is only through such interaction that integration eventually occurs. Not only must interaction take place between a child and collections of objects, between his own viewpoint and those of other children, but also between competing ideas within himself.

Initially, children relate number words in terms of "comes after" relations derived from sound patterns. This nonquantitative connection between adjacent number names must be restructured in terms of quantitative "one more than," "one less than," "ten more than," and "ten less than" relations (Chapter 10, p. 268). The construction of such a coherent structure not only provides the underpinnings for understanding place value notation, but also supports children's performance on the task in question. Now, writing a numeral showing "one more" or "ten more" requires only that an appropriately positioned digit be incre-

mented by one. The third graders' performance on this task suggests that less than half had constructed the necessary framework of ideas. The remaining children had to cope with the task with a choice from a number of unrelated ideas.

Piaget's systematic observations of children on conservation tasks of number, area, weight, and volume also provide some insight into the difficulty of the "more than" task. He observed that most seven-year-old children understood "more than" and "less than" in the context of a number conservation task involving seven to ten objects. He pointed out that a comparable understanding for larger collections of objects was still under progressive construction. He also noted that children's understandings of these comparative terms varied with age for specific physical contexts. Although most eight-year-olds can understand "more than" in the context of a conservation of area task, they are not likely to do so in a conservation of weight (age ten) or a conservation of volume (age eleven) task.[9] Over half

---

9. Yet based on the children's performance on the task in Chapter 1, it is likely that some of these third graders are capable of conserving more than number and area.

of our third graders (eight-year-olds) appeared to still be progressively constructing their "one more than" and "ten more than" relations for numbers up to 432. Despite the availability of the appropriate vocabulary, "more than" and "less than" are limited to specific applications by children's existing organization of ideas.

Internal logic gives meaning to these words and the construction of this logic appears to be under developmental constraints. Thus, we note that children structure "one more than" and "ten more than" relations for large numbers in their own time.

ONE MORE THAN / TEN MORE THAN: SECOND LOOK FROM THE PERSPECTIVE OF CONCRETE MATERIALS

Behr's teaching experiment with second graders (1976) documents children's difficulties with "more than" and "less than" when working in a variety of representational modes: Dienes blocks, abaci, pictures of concrete materials, and place value notation. Although children sometimes experienced difficulties when the tasks were presented in the context of Dienes blocks, these materials provided the most useful mode of representation for children to make sense out of the tasks. Making the blocks available to the children when they were unable to respond to the task in other modes often helped them out of the difficulty. The following excerpts from Behr's anecdotal records illustrate the range of effectiveness of the Dienes blocks in "more than" and "less than" situations.

> Some children seem confused about whether to increase or decrease the (block) display and also about whether to use a representative of 10 or 1. (p. 66)

> All exercises which called for me showing a picture and having the children find the picture to show 100 more, 10 more, 1 less, 10 less, 1 more, 100 less were extremely difficult. Strangely enough, the comparable exercise at the enactive (concrete) level went very well. Apparently children were not ready for these kinds of activities at the iconic (pictorial) level. (p. 69)

> Finding 10 less than 309 from a picture was hard. When blocks were brought to illustrate, Calvin knew what to do. (p. 70)

The observation of children's difficulties with such quantitative comparatives supports the earlier position that the origin of these difficulties goes beyond place value notation to its underlying conceptual foundation.

The third-grade children experiencing difficulty with the "more than" task were given a comparable follow-up task with Dienes blocks. The latter task was not planned originally, but the need for it soon became apparent. It was given several minutes later, after the children had gained some familiarity with the blocks (through the task described in Chapter 1). The spontaneous follow-up was both an attempt to gain a better understanding of children's thinking and to test the widely accepted power of concrete materials.

Will these children see the tasks differently now? Will they know what to do?

Using the blocks prompted most of the children to revise their original views of the "more than" task. Yet all of these views were not the ones expected. Out of 14 children, 10 were now able to demonstrate "one more than" and 8 could do so for "ten more than." Although some of these children knew what to do immediately, the task was not instantly obvious to others. The initial question was often greeted with puzzled expressions. How would you clarify the task for these children?

How did the other children see the problem? The range of perspectives is illustrated next.

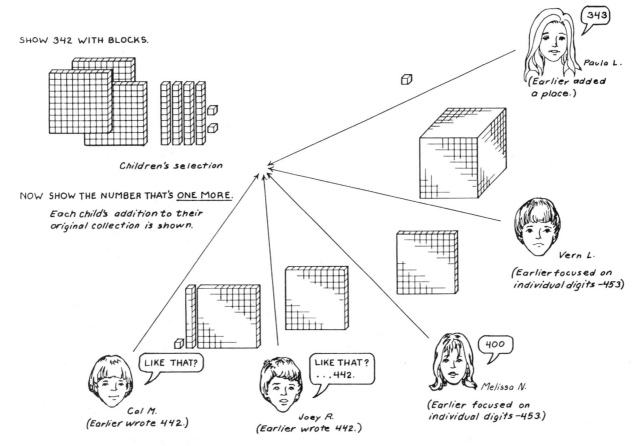

SHOW 342 WITH BLOCKS.

Children's selection

NOW SHOW THE NUMBER THAT'S <u>ONE MORE</u>.

Each child's addition to their original collection is shown.

Paula L.
(Earlier added a place.)

343

Vern L.
(Earlier focused on individual digits – 453)

Melissa N.
(Earlier focused on individual digits – 453)

400

Cal M.
(Earlier wrote 442.)

LIKE THAT?

Joey R.
(Earlier wrote 442.)

LIKE THAT? . . .442.

All but Joey changed their perspectives from the initial question. These children interpreted "one more" as just that, or as "one hundred more," "one thousand more," "one block more," and "one of each block more."

The range of children's perspectives on the "ten more than" task with blocks follows.

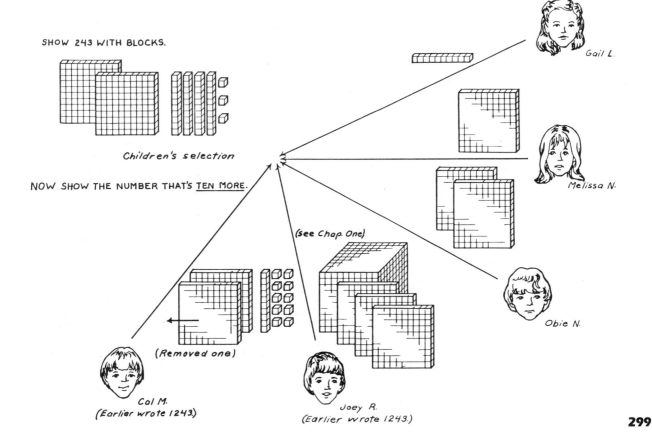

SHOW 243 WITH BLOCKS.

Children's selection

NOW SHOW THE NUMBER THAT'S <u>TEN MORE</u>.

Gail L.

Melissa N.

Obie N.

(see Chap. One)

Cal M.
(Earlier wrote 1243.)
(Removed one)

Joey R.
(Earlier wrote 1243.)

299

All of the 14 children were able to construct a block display of the reference quantity, yet not all could extend their display to show "one more" or "ten more." Just one more block was needed in each case, but deciding on the appropriate block was not easy for all the children. Melissa added the same block in response to both "one more" and "ten more."

Working on these comparable tasks with materials, all the children, with the exception of Joey, changed their original perspectives. Although Joey and Cal earlier gave the same written answers, they now showed different block displays.

	JOEY		CAL	
one more than 342	442	one hundred blocks more	442	one of each block more
more than 243	1243	ten hundred blocks more	1243	(one/ten) of each block more
pattern focus	consistent focus on (one/ten) hundred more in written notation and materials		Inconsistent focus as perspective changed with introduction of blocks.	

The boys' identical written responses had dual interpretations—they either focused on incrementing the first digit or on showing (one/ten) hundred more.[10] Joey's consistent focus regardless of the context may rule out the alternative interpretation of his thinking. The use of materials not only helped some children to make sense out of the task, it also provides us with a basis for further inferences on their thinking.

Although "one more" and "ten more" can be shown by adding only one more block, two of the children were very indecisive about which block to add. Attempts to clarify their understanding resulted only in continuing disequilibrium.

---

10. A few minutes earlier, Joey viewed the large cube as 600 (Chapter 1). Thus showing the large cube and three flats can be interpreted as an attempt to show ten hundred more for the "10 more" question.

Often children who appear the most confused initially make the most progress over an extended period. By becoming aware of the conflict between alternative perspectives, children are motivated by the discomfort of the disequilibrium to integrate isolated ideas at a higher level of understanding. Now, rather than vacillations in which opposing perspectives appear to exist side by side, an advanced, integrated perspective appears consistently. Children who are not aware of any conflict continue to select different isolated perspectives on each new task. For this reason, a valuable teaching strategy is to ask questions that have the potential of intensifying the interaction between a child's existing ideas, promoting awareness of contradictions and provoking a reorganization of these ideas.

As observed earlier, some tasks raise more questions about children's numerical thinking than they answer. This task highlights a previously unrecognized pattern of behavior in need of explanation. It is the tendency of four different children to show *two* blocks of the same kind in response to "one more" or "ten more." Although at a loss to explain this behavior, I would be curious to see how the same children would respond to "a hundred more."[11] Another question prompted by the preceding task involves the effectiveness of concrete materials in helping children construct an understanding of number and a connection with the conventional system of place value notation. Since these questions extend beyond this specific task, they will be discussed separately in the next section.

CONCRETE MATERIALS, NUMBER RELATIONSHIPS, AND PLACE VALUE NOTATION

Mathematics educators consider concrete materials such as Dienes blocks, Unifix cubes, and abaci as models for large numbers and for the place value numeration system, referring to them as *embodiments* of such relationships. Numbers and their relations are abstractions that no one has ever seen. However, the terms *model* or *embodiment* are interpreted by many as meaning that these materials have the capacity to make these abstractions concrete or directly perceptible. Yet Chapters 1, 10, and 11 provide ample evidence that many third-grade children do not automatically recognize these relationships:

Thousand–blocks are seen as showing six hundred or counted as ones..

Hundred–blocks are read as thousands, tens, or ones.

Groups of ten objects or ten–blocks are read as ones or hundreds.

Individual objects are counted as tens.

Our internal organization of number ideas determines what we are able to see or read into the materials. We see what we understand. Although abstract number relations become more accessible through interaction with the materials, they are not externally available as in the simple abstraction of color. Rather, these abstract number relations are constructed internally through extended interaction of the child's existing relational framework of ideas with action on these materials. Eventually, the child is so comfortable with these newly constructed number relations that he is likely to exter-

nalize this understanding to the materials, as if it had always been in existence "out there." Only then do the materials serve as models or embodiments of his understanding.

Piaget (1971) cautioned teachers about the misuse of concrete materials for brief demonstrations before moving quickly to written notation. Rather, children must act on them extensively in situations that engage their minds. Piaget (1964) wrote,

> Knowledge is not a copy of reality. To know an object, to know an event, is not simply to look at it and make a mental copy, or image of it. To know is to modify, to transform the object and to understand how the object is constructed. An operation is thus the essence of knowledge; it is an interiorised action which modifies the object of knowledge. (p. 8)

Thus, Piaget clarified the common misconception of the power of both embodiments and physical activity. He emphasized the interaction between mind and materials—a coordination between physical and mental activity—as essential for reconstructing reality at a level of abstract relations.

To construct a progression of ten-to-one relations through counting grouped objects and to represent

---

11. Thompson (1982b) observed a related behavior in his interviews with Sigma. During a counting board task involving counting different Dienes blocks in random arrangement as they were gradually uncovered, he counted the following blocks □ | | □ as 100, 110, 120, 140. In response to probing about the source of the last count, Sigma repeated his count and also explained it. He said, "One hundred and twenty plus one hundred is forty. Cause one hundred counts as *two* more (tens) instead of one more."

these relations through conventional place value notation, children require considerably more time than currently allowed in schools. However, once having done so, children can use the materials that now serve as models of their number relations as tools for extending their understanding. For example, these materials serve as vehicles for interaction in constructing understanding of operations, such as addition and subtraction, with large numbers. Furthermore, the interaction of the child's relational framework of ideas (including understanding of place value notation) with these models serves as a springboard to the construction of extended relations for much larger numbers. For example, direct experience with quantities like a million are difficult to obtain and too cumbersome. Yet through the interaction of existing number relations with the sequence of number names and both pictorial and notational representations of progressive levels of groupings by tens,[12] it is possible for an older child in Piaget's formal operational stage to notice recurring patterns and to gain a sense of magnitude for a million in relation to known numbers.

You are invited to extend your own understanding of numbers such as a million, a billion, and a trillion through interaction of ideas, models, and place value notation.

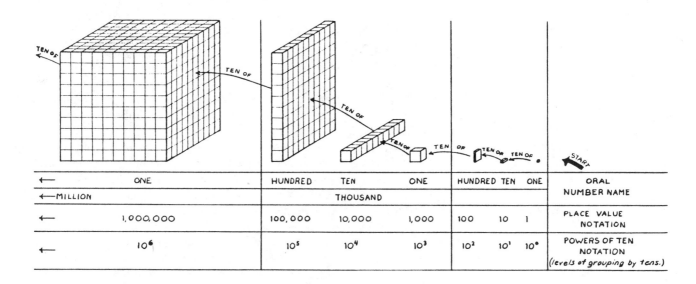

ONE	HUNDRED	TEN	ONE	HUNDRED	TEN	ONE	ORAL NUMBER NAME
←—— MILLION		THOUSAND					
1,000,000	100,000	10,000	1,000	100	10	1	PLACE VALUE NOTATION
10^6	10^5	10^4	10^3	10^2	10^1	10^0	POWERS OF TEN NOTATION *(levels of grouping by tens.)*

Large numbers are impractical to represent with concrete objects.
How many thousand-blocks would be needed to show a million?
How many levels of systematized grouping by tens would you go through in counting out a million objects?

If the next large grouping after a million is a billion,
How many millions make up a billion?
What shape would the block model of a billion be?
How many levels of systematized grouping by tens would you go through in counting out a billion objects?

Suppose you were given a million dollars to spend at a rate of a thousand dollars per day.
Approximately how long would it take so spend the money?
At the same rate of spending, how long would it take to exhaust a billion dollars?

What questions can you ask yourself about a trillion?

---

12. The second notational system has been developed since place value notation to represent very large numbers with greater efficiency. This powers of 10 notation indicates the level of systematized grouping by tens. One hundred is represented as 10^2 indicating the second level of systematized grouping.

TEACHING METHODS AND ACTIVITIES

The preceding chapter has highlighted the extended time needed by children in developing ideas of systematized grouping by tens. This chapter has brought into focus both the sophistication of place value notation and its roots in grouping ideas. Available research on children's numerical thinking points to the delay of any place value ideas until grouping by tens is understood, and the introduction of these ideas in the presence of grouped materials. Yet the classroom experience of many teachers and children suggests an alternative. Although research has not established whether children's groupings by threes, fours, fives, and so on facilitates their understanding of grouping by tens, both Hendrickson (1983a) and many teachers of *Mathematics Their Way* (Baratta-Lorton, 1976) report success in teaching beginning place value ideas with small groups of objects as early as the first grade. Both approaches discussed below can be used separately or in combination, as you think best.

A Gradual Introduction to Grouping and Place Value Notation

The place value activities in *Mathematics Their Way* involve split boards, colored differently on each side, to familiarize children with counting objects to a specified quantity, constructing groups, and keeping groups and individual objects separated. As the children add one object to the board at a time, they orally describe the number of groups and ones on the board at each step (see activity A in the next illustration.) The activities are repeated with different materials and later with group sizes being gradually increased according to the capabilities of the children. The activities can also be experienced in the reverse direction. Once the children repeatedly anticipate the next move in the activity, it is time to increase the size of the group or to replay the activities at a beginning level of representation.

At the next level, a running record of the groups and loose objects is kept with number flip cards located at the head of each half of the split board. This notation is coordinated with oral description of the boards at each step (see activity B). Here the grouping and positioning of objects is associated with the corresponding position of digits.

At the third level of these activities, they are replayed and eventually extended to the children's written notation and search for repeating patterns (see activities C, D, E). The teacher organizes children's boards in a stop-action visual display by removing one child's board after each step in the activity. From this overview of the action and running record on the flip cards, children begin to notice repeating patterns. This search for patterns is extended as the teacher records the children's results on similar charts, and loops identified patterns. With repetition of this level of activity with small groupings, eventually the children are prepared to undertake their own grouping, written recording, and search for patterns in working with tens.[13] The patterns noticed in their hundred-chart are critical to understanding the whole number numeration system and applying it in addition and subtraction of large numbers (See Chapter 14, p. 384). Rather than place value ideas being dealt with in isolation, they are related both to grouped materials and other ideas, with the potential of provoking children to construct an integrated system.

Although Baratta-Lorton's place value activities can be initiated in the first grade, it's unlikely that children's level of number knowledge is sufficiently developed to achieve grouping by tens.[14] Although most children can conserve quantities as large as eight or ten by age seven, they may not be able to conserve 25 objects. Children regrouping 25 Unifix cubes (tens and ones) prior to taking away six may believe that there are more cubes after regrouping. Conservation of quantities to infinity develops gradually. For this reason, it is advisable to pace the activities gradually, according to the children's capabilities. This deliberate and systematic *concrete-to-abstract* approach avoids the problem of children moving too quickly into numbers that are beyond their understanding. At the same time it helps them to develop a foundation for understanding place value notation for our decimal system.

The children's introduction to these games is highly structured for large group situations. Each child adds another object and names the collection on signal, having no opportunity for spontaneous exploration. Yet the children appear to enjoy the activities. Furthermore, at advanced levels the children are prepared for independent or paired grouping of objects by tens, recording their own hundred-chart and searching for patterns within it. According to Piaget (1964), "Teaching means creating situations where structures can be discovered." Sometimes this is achieved with a blend of direct and indirect teaching methods.

---

13. Variations at this level include adding by twos, threes, and so on or random adding based on a spinner.

14. Although the majority of first graders aren't ready for grouping by tens, the range of counting abilities indicated in previous chapters suggests that some first graders are ready for the challenge.

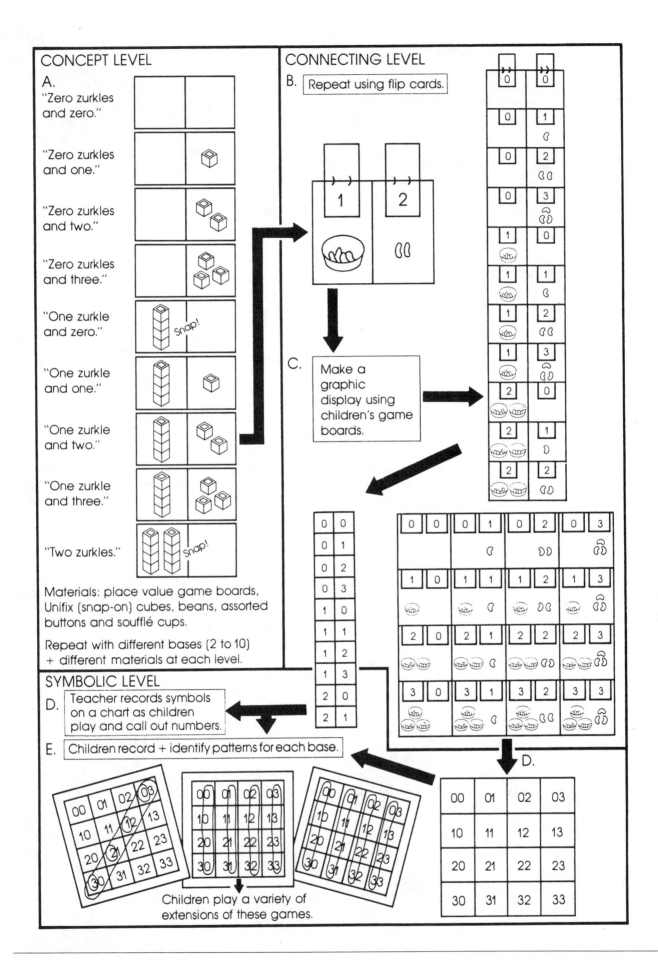

CONCEPT LEVEL

A.

"Zero zurkles and zero."

"Zero zurkles and one."

"Zero zurkles and two."

"Zero zurkles and three."

"One zurkle and zero."   Snap!

"One zurkle and one."

"One zurkle and two."

"One zurkle and three."

"Two zurkles."   Snap!

Materials: place value game boards, Unifix (snap-on) cubes, beans, assorted buttons and soufflé cups.

Repeat with different bases (2 to 10) + different materials at each level.

CONNECTING LEVEL

B.   Repeat using flip cards.

C.   Make a graphic display using children's game boards.

SYMBOLIC LEVEL

D.   Teacher records symbols on a chart as children play and call out numbers.

E.   Children record + identify patterns for each base.

D.

00	01	02	03
10	11	12	13
20	21	22	23
30	31	32	33

Children play a variety of extensions of these games.

Maximizing Interaction Between Different Representations of Number

Representations of number take many forms—concrete objects, oral number names, written number names, pictorial representation, expanded written notation, and place value notation. Classroom experiences must provide exposure to many different representations for building relations and integrating perspectives.

CONCRETE ↕ ORAL NUMBER NAME	*Unifix cubes*	HOW MANY CUBES DO WE HAVE? (SHOW/TELL)	THIRTY-TWO *Regrouped*
CONCRETE ↕ ORAL NUMBER NAME		HOW MANY? CAN YOU SAY HOW MANY IN ANOTHER WAY?	TWENTY-THREE TWENTY AND THREE ONES TWO TENS AND THREE ONES
ORAL NUMBER NAME ↕ CONCRETE		SHOW FORTY-TWO CUBES	
PICTORIAL REPRESENTATION ↕ ORAL NUMBER NAME		SHOW HOW MANY DOTS THERE ARE. TELL HOW MANY THAT SHOWS.	THIRTY-THREE
ORAL NUMBER NAME ↕ ORAL NUMBER NAME		WHAT'S ANOTHER NAME FOR FORTY-THREE?	FOUR TENS, THREE ONES FORTY AND THREE ONES.
ORAL NUMBER NAME ↕ ORAL NUMBER NAME		WHAT'S ANOTHER NAME FOR FIVE TENS AND SEVEN ONES?	FIFTY-SEVEN
CONCRETE ↕ WRITTEN NOTATION		WRITE DOWN THE NUMBER OF TILES SHOWING. CIRCLE THE DIGIT THAT SHOWS THE TENS.	③6
PICTORIAL ↕ WRITTEN NOTATION	: ___tens, ___ones ___ + ___	WRITE THE NUMBER OF CUBES SHOWING.	5 tens, 2 ones 50 + 2 52
WRITTEN NOTATION ↕ ORAL NUMBER NAME	④3	WHAT NUMBER DOES THIS SHOW? (Read this number.) WHAT DOES THE FOUR STAND FOR? HOW DO YOU KNOW?	FORTY / FOUR TENS FOUR IS IN THE SECOND PLACE FROM THE RIGHT...THE TENS PLACE.
WRITTEN NOTATION ↕ WRITTEN NOTATION	67 7 6 5 4 / 70 60 3 / 50 40 20 / 3 4	USE THE CARDS TO SHOW WHAT THIS NUMBER MEANS.	60 7 / 6 7
WRITTEN NOTATION ↕ WRITTEN NOTATION	73 ___tens, ___ones ___ + ___	WRITE THE DIFFERENT NAMES FOR THIS NUMBER.	7 tens, 3 ones 70 + 3 73

The preceding table has focused on naming and composition of numbers from different perspectives. Other place value activities to include are comparison and sequencing of numbers, and "ten more/less than" relations. Many of these activities can be adapted to small groups in which the situations are randomly generated and interaction is maximized. Following are two game activities played with place value number strips (Rathmell, 1979).

(Each type of card could be colored differently or have its numbers written in ink of different color.)

Making numbers (for three or four players). Each player has a complete set of place value number strips. They take turns picking a task card off the deck and reading the number to be built. The remaining children combine the appropriate cards and compare their results. The card reader keeps his numeral information secret unless the entire group has made a wrong decision.

War (for two, three or four players). The players mix up their sets of strips and then sort them into three face-down piles. They take a number strip from each pile, build a number, and read the result. The players then decide on the largest number. The builder of the largest number captures the other player's number strips. In case of a tie for the largest, those involved play another round. Variations include: lowest number wins, in-between number wins (in case of three players), closest to 500 wins.

To maximize the interaction between different representations of number and between children, the games can involve children using different representations—concrete materials, place value number strips, and paper and pencil—for the same task.

BUILDING NUMBERS

SHOW:
FOUR HUNDRED FIFTY-THREE

SHOW TEN MORE

SHOW THE NUMBER THAT IS TEN MORE THAN 258.

LARGER

SHOW THESE NUMBERS AND DECIDE WHICH IS LARGER:
SEVENTEEN
SEVENTY-ONE

IN-BETWEEN

SHOW THESE THREE NUMBERS AND DECIDE WHICH IS
LARGEST
SMALLEST
IN-BETWEEN.
TWO HUNDRED THIRTY-ONE
ONE HUNDRED NINETY-EIGHT
TWO HUNDRED EIGHTEEN

TASK CARDS

Initially, children can work from concrete to abstract; that is, the child working with the materials answers the question by reference to the blocks, followed by answers from the children using less abstract methods. At another level, all the children can respond in terms of written notation and justify by reference to at least one other representation. Interaction of different representations and children's perspectives of the same tasks stimulates rethinking of a child's own viewpoint at a

higher level of understanding, and an integration of representations.

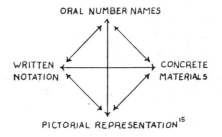

ORAL NUMBER NAMES

WRITTEN NOTATION

CONCRETE MATERIALS

PICTORIAL REPRESENTATION[15]

REFLECTIONS AND DIRECTIONS

1. Again, you are invited to find your own children to interview and to reflect on the implications of the outcomes for the classroom.

2. If you are unable to locate your own children, you might follow the thinking of two children on both counting tasks and place value interpretation to see if and how it is related.

	Chapter 10	Chapter 11
Roz:	pp. 245, 246, 247, 249, 257	pp. 288, 290, 296
Zoya:	pp. 245, 249, 255, 257, 260, 261	pp. 288, 290, 295

15. Aviv (1980).

Now that we have examined place value foundations, the next chapters will focus on their applications in methods of addition and subtraction of large numbers. Chapters 12 and 13 will explore third graders' computation with standard school methods of addition and subtraction, and their understanding of these methods. Chapter 14 will explore children's self-constructed nonstandard methods of computation.

12

UNDERSTANDING OF THE STANDARD ADDITION ALGORITHM

PREVIEW

Paper-and-pencil methods of computation dominate the elementary school mathematics curriculum. These methods of computation, or algorithms, have a long history of development and codification (Ginsburg, 1977). Over the centuries, the work of mathematicians has boiled down the early cumbersome methods to a minimum of well-defined, precise steps in powerful algorithms that are guaranteed to produce correct answers. Properly applied, the addition algorithm will work for all examples, regardless of the size of the numbers or the quantity of numbers to be added. These powerful codified methods are taught to children as ready-made algorithms, stripped of their messy human origins. In deciding to pass on these finished products of mathematical endeavor, educators assume that children are capable of assimilating them as presented. Furthermore, they assume that children are incapable of becoming involved in the process of inventing their own algorithms.

Although a number of different algorithms are possible, textbooks provide a standard approach. In nearly every classroom in the country, children are taught to add down, starting from the right, and to carry to the left when the sum of the digits in the column exceeds 9. Also typically, second graders are

expected to add two-digit numbers and third graders to add three-digit numbers, like the following:

$$
\begin{array}{r} 37 \\ +28 \\ \hline \end{array}
\qquad
\begin{array}{r} 345 \\ +278 \\ \hline \end{array}
$$

The focus of this chapter is to explore not only whether third graders are capable of performing such computations using the standard algorithm, but also the extent to which they understand what they are doing. It will examine the children's place value concepts in relation to vertical alignment of the numerals, and both the place value and conservation concepts underlying the process popularly referred to as *carrying*. The chapter will also describe an alternative semi-direct method for teaching the standard addition algorithm in the context of meaning.[1]

---

1. A separate chapter will be devoted to children's invention of computational algorithms and to indirect teaching methods that support this process. See Chapter 14.

Before beginning, you are invited to put yourself in the children's shoes and to anticipate how they will respond to the following tasks.

COMPUTATION	EXPLANATION	DEMONSTRATION	CONNECTION
READ THE PROBLEM. LET'S SEE YOU DO IT.	TELL ME HOW YOU GOT THE ANSWER.	SHOW ME HOW YOU WOULD DO THIS PROBLEM WITH CUBES.	POINT TO SOMETHING ON THIS BOARD THAT THIS "ONE" STANDS FOR.

ADDITION OF TWO-DIGIT NUMBERS

The interview task was composed of four subtasks in order to reveal different levels of children's understanding of the algorithm. Once a written record of the child's computation was obtained the child was asked to explain his procedure. An adequate place value explanation at this point might require no further exploration. However, an inadequate explanation was explored further by asking the child to physically demonstrate the task with grouped materials and to relate this experience back to the paper-and-pencil algorithm used in the original task. The results obtained from the interview subtasks will be discussed separately.

Computation

The computation exercise was presented in vertical format.[2] Each child was asked to read it before finding the sum. Of the 29 third-grade children, 23 of them wrote down the correct answer. All but one of these children appeared to follow the standard algorithm by working from right to left and recording evidence of carrying. The single exception proceeded from left to right, giving no indication of carrying. She counted to get her answer.

Unusual behaviors may prompt us to ask the children questions to explain their procedures and reveal their understanding. However, for those children who produce correct sums by following the standard procedure, it is easy to assume that they all "see" the exercise in the same way. The next three phases of the task will reveal how differently these children see it.

Explanation

In response to "Tell me how you got your answer," or "Tell me how you worked it out," most of the children gave descriptions of their procedures, speaking of the digits as individual entities. However, none of these children had difficulty reading the numerals or the addends or sum.

Although Betty was able to explain regrouping/renaming on request, she preferred to use her own counting method.[3]

---

2. The videotaped interviews were conducted with the same third graders described in Chapters 1, 10, and 11. The computational tasks were given in the third of a series of three interviews. The final interview was conducted in January and February.

3. It's interesting to note that although Betty could explain the regrouping process, she still preferred to use her own counting method.

Typical responses

I ADDED THESE TWO (8+7) CARRIED 1...BECOMES 6.
Charlie D.

7 AND 8 IS 15. 5 DOWN, CARRY 1. 3+2+1.
Steve R.

I COUNTED THESE TWO PROBLEMS UP AND PUT 5 DOWN HERE AND 1 UP THERE. I COUNTED THEM ALTOGETHER.
Zita K.

7 AND 8 IS 15, 5 CARRY 1, THAT'S 6.
Sally T.

$$\begin{array}{r} 1 \\ 37 \\ +28 \\ \hline 65 \end{array}$$

Only 2 of the 29 children explained their procedure in terms of place value ideas.

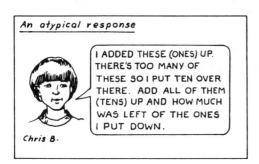

An atypical response

I ADDED THESE (ONES) UP. THERE'S TOO MANY OF THESE SO I PUT TEN OVER THERE. ADD ALL OF THEM (TENS) UP AND HOW MUCH WAS LEFT OF THE ONES I PUT DOWN.
Chris B.

Looking back at the question, I realized that it does not specifically request an explanation. The following question would have been more appropriate: "How would you explain this to a second grader who doesn't know how to do it?"

Perhaps the children responded in terms of an already-automated procedure, and probing questions will bring forth a place value rationale for their procedure. Even as adults, we are likely to omit reference to place value and refer only to the digits in doing the procedure. Yet we usually can provide a place value rationale upon request.[4] Will the children be able to do so when asked what the *carried 1* represents?

Decide which of the following responses reflect place value understanding.

How would you probe some of the responses further for clarification of the children's understanding?

---

4. Perhaps this is expecting more place value awareness than some adults have. Over the past few years, several student teachers have disclosed to me having no understanding of the carrying process prior to the math methods course.

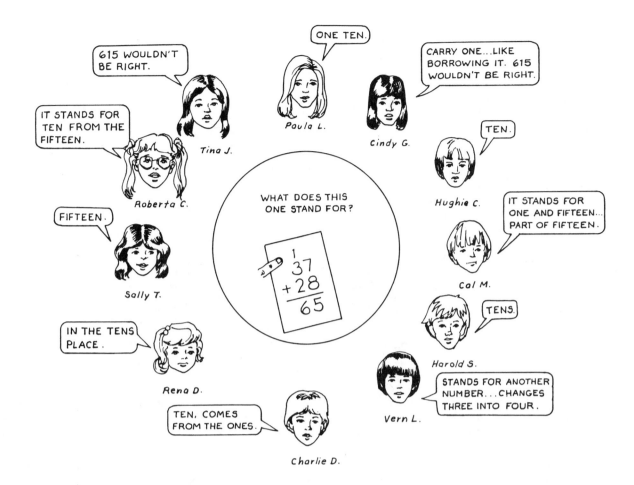

The probing question focuses on an essential ingredient for explaining the *regrouping/renaming of ones as a ten as the basis for carrying*. Although all of these children could read the numeral 10, identification of "1" in the left-hand column as representing the renamed ten requires a place value perspective. Perhaps a rephrasing of the initial follow-up question or further probing of a child's second response would coax out a reference to ten. Indeed, some of the children were asked further questions (after Ginsburg, 1977). These questions were directed both to children who had identified the "1" as representing a ten and those who hadn't. Some of the responses are illustrated next.

For children not yet identifying the carried "1" as a ten, attempts to clarify the question further proved to be unproductive.[5] Although children are introduced to the addition algorithm in the context of place value, many of them assimilate only the list of steps and operate on the digits. Nearly half of the 23 accurate computers appeared to fall into this category. Meanwhile, attempts to probe the understanding of children's reference to ten further uncovered some instability in their understanding and an inability to elaborate and provide a detailed rationale for the procedure. The brevity of their responses suggests that they are unaccustomed to thinking in terms of a rationale for the addition algorithm.

Computation and Explanation Revisited

Now let us examine the computational errors made by six of the children together with their procedures and explanations. Doing so will raise some interesting questions about answers and procedures.

Number Fact Errors. Three children made errors in combining 8 + 7 as 14 or 16. Two of these children worked from left to right in direct opposition to the standard right-left procedure. The same two children also gave explanations with spontaneous place value references.

Is it possible for a child who gave a wrong answer to have more understanding of addition with regrouping/renaming than a child who gave the correct answer?

---

5. Additional ways to coax out other views of the carried 1 include: "What is the 1 worth?" "What does it really mean?" (Russell, 1981).

Regrouping/Renaming Errors. Children's errors often reveal an aspect of the procedure or an interpretation of the underlying concept that is at least partially correct. Although three children made errors in regrouping/renaming, they all combined digits accurately. Two of these children worked from left to right.

```
 37
+28
───
515
```
Zoya added each column of digits separately. Although she has no "number fact" errors, there is also no awareness of place value or reasonableness of the answer. Actually, she is unable to recall the place value order, thinking in terms of tens, twenties, and hundreds to explain her procedure.

```
 37
+28
───
515
```
Vern obtained the same answer working from the right. In response to "Can you write it another way?" he revised his procedure to obtain the correct answer. He is unable to elaborate on the carried "one."

```
  1
 37
+28
───
 56
```
In working from left to right, Joey carries to the wrong column. He is unable to explain his procedure.

I DON'T KNOW WHAT HAPPENED WITH THE ONE.

Whereas errors made in combining numbers may be considered clerical, errors in regrouping/renaming are conceptual. How do you regard working from left to right?

Is working in a left-right direction actually wrong, inefficient, or merely different from the standard procedure?

Demonstration

To avoid hasty conclusions about a child's thinking based on limited samples of written work, more evidence can be obtained by observing and interacting with the child in the presence of concrete materials. After the written computation was explained, the answer was covered and the child was asked to do the following task.

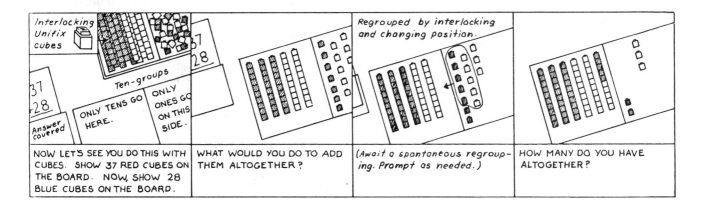

NOW LET'S SEE YOU DO THIS WITH CUBES. SHOW 37 RED CUBES ON THE BOARD. NOW, SHOW 28 BLUE CUBES ON THE BOARD.

WHAT WOULD YOU DO TO ADD THEM ALTOGETHER?

(Await a spontaneous regrouping. Prompt as needed.)

HOW MANY DO YOU HAVE ALTOGETHER?

Just as in the written computation, some of the children worked from right to left, while others worked in the opposite direction. In combining the addends, the interlocking feature of the Unifix cubes permits the joining of ten-groups without the introduction of new materials. Less than half (13) of the 29 children spontaneously joined 10 cubes when combined addends resulted in 15 loose cubes. Two of the children spontaneously traded 10 loose cubes for a ready-made ten-group from the materials box. They either may not have noticed the interlocking feature of the cubes or they may have chosen to avoid the slow construction process in favor of a quick exchange. Either way, the end result was the same. The following prompt was given to the children who did not regroup spontaneously.

Do something with your cubes so that if someone comes into the room he/she can just look and see how many cubes there are without doing a lot of counting.

This prompt was sufficient for 9 more children to regroup (7) or trade (2). The remaining children required two or three prompts for completion of the demonstration. Examples of prompts or reactions to a particular child's response are given below.

CHILD'S RESPONSE	MY RESPONSE/PROMPT
37 was represented by three cubes and seven cubes.	That could stand for 37. Put 37 cubes on the board.
Ten loose cubes are lined up without interlocking.	Can you show ten cubes in another way?
The sum is counted without regrouping.	Is there something else you can do with the ones? Can you show 65 in another way?
Cubes are interlocked by color only seven red eight blue	On this board tens go over here and ones go on this side. Do you have tens or ones?
Cubes are miscounted	Would you check that again?

Connection

The next phase of the task was used to find out whether the children could associate their combining and regrouping of materials with the written format of the standard algorithm. The connection phase is illustrated next with three possible responses highlighted.

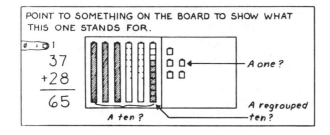

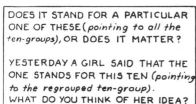

Fifteen of the accurate computers pointed to a ten-group. Seven accurate computers pointed to an individual cube, while one was unable to respond to the question. Some of the children identifying the "1" as a representation for a ten-group also demonstrated an awareness that it was a specific ten-group[6]—the one just regrouped (or traded). This awareness was demonstrated either spontaneously or following a prompt or countersuggestion. Other children resisted the countersuggestion, as illustrated next.

---

6. Initially, at the completion of regrouping, the addends and the regrouped ten were color-coded for easier recognition. Unfortunately, this feature was lost over a series of interviews as the children gradually mixed up the colors and I didn't take the time to reorganize them before each interview.

Some of the children pointing to a single cube were also given a countersuggestion. This was met with a range of reactions.

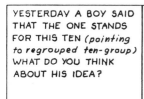

Vern's rapid shifts in focus from ones to tens in response to the interviewer's questions are characteristic of disequilibrium or uncertainty. Also, the frequent checks of the interviewer's expression suggest a lack of confidence in his own thinking. Vern appears to be switching in search of the response he thinks the interviewer wants.

Although 23 of 29 children computed accurately with the standard algorithm and demonstrated it with materials, only 15 could make a connection between the written notation and the action of combining and regrouping objects. For the remaining children capa-ble of computing accurately with the standard algorithm, a critical gap in understanding exists.

For the inaccurate computers, the connecting phase of the task served as a springboard for discussion of the written notation as a record of actions on objects. All six of these children arrived at 65 as the correct answer, but only one was able to make the connection between written notation and regrouped objects. An excerpt of the interview with Zoya illustrates her attempts at making such a connection. The following discussion took place immediately after her regrouping of the cubes.

Using the materials reminded Zoya of the place value groupings (previously thought to be tens, twenties, and hundreds). Her recording suggests a focus that extends beyond independent digits and now vacillates between place value and total value. Although she shows considerable progress in understanding and representing the regrouping process, she is unable to coordinate all of the relations involved.

ADDITION OF MULTIDIGIT NUMBERS

Following an intervening subtraction task, a more complex addition task was presented—involving addition of multidigit numerals requiring regrouping/renaming of tens as a hundred. The materials used in this task consisted of tiles—either individual ones or arrays of tiles (a row of ten or ten rows of ten)—glued to sticks or cards.

COMPUTATION	EXPLANATION	DEMONSTRATION	CONNECTION
READ THE PROBLEM. LET'S SEE HOW YOU WORK IT OUT.	TELL ME HOW YOU GOT YOUR ANSWER.	Fixed Tiles Fixed Loose 100 squares 10-sticks tiles	
152 +71	1 152 +71 223 WHAT DOES THIS "ONE" STAND FOR?	152 +71 SHOW ME HOW YOU WOULD WORK THIS OUT WITH TILES.	152 +71 223 POINT TO SOMETHING ON THE BOARD THAT THIS "ONE" STANDS FOR.

Based on what you've learned from the third-grade children, predict any problems they may experience in regrouping/renaming tens as a hundred.

Anticipate any problems they may experience in regrouping the new materials.

Computation

Of the 24 children presented the task, 21 computed the correct sum, with 19 using the standard algorithm.[7] Two other children worked from left to right, with one using a definite counting strategy. All three errors in computation were the result of difficulties with regrouping/renaming tens as a hundred.

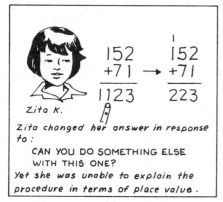

Zita K.

152 +71 = 1123 → 152 +71 = 223

Zita changed her answer in response to:

CAN YOU DO SOMETHING ELSE WITH THIS ONE?

Yet she was unable to explain the procedure in terms of place value.

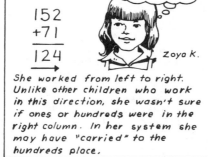

152 +71 124 →

7 AND 5 IS 12 CARRY ONE.

Zoya K.

She worked from left to right. Unlike other children who work in this direction, she wasn't sure if ones or hundreds were in the right column. In her system she may have "carried" to the hundreds place.

2 4 152 +71 205

7 AND 5. THAT MAKES A HUNDRED AND 2 (tens) LEFT OVER.

Teddy D.

He obtained a subtotal of 100 and 2 tens for the middle column, only to share it between two columns—leaving nothing.

7. On the same task, student interviewers report that one-quarter of one class of beginning third graders computed answers such as 88, 160, 143, 673, and 16. Since there was nothing to add to the 1 in the left column, the children arbitrarily combined it with other digits in creative ways by adding across the columns in some way.

Explanation

Once again, most children gave abbreviated descriptions of the procedure and spoke of the digits as individual entities with no underlying value. These descriptions resembled arbitrary rules for "getting the answer." Only 3 of the 24 children spontaneously explained their procedures in terms of place value ideas. It is interesting to note that two of these children had also worked from left to right and used their own methods of computation. Examples of both types of explanations are illustrated next.

To a mathematician the carried "1" in this task represents a hundred from the renaming of 10 tens (obtained from the partial sum of 12 tens). What does it represent to third graders? The initial probing question focused on the meaning of the carried "1". The range of responses is given below.

Decide which of these responses reflect place value understanding.

How would you probe some of the children's responses further to learn more about their thinking?

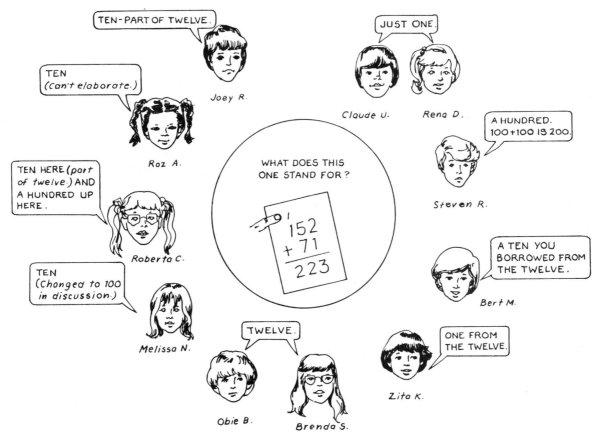

Many of the preceding responses seem to lack an awareness that their 10 and 12 are already 10 tens and 12 tens by virtue of their position. Perhaps a further probe will coax out an indication of place value awareness. Most of the children were asked at least one of the following questions that focused on the place value of the digits in the middle column.

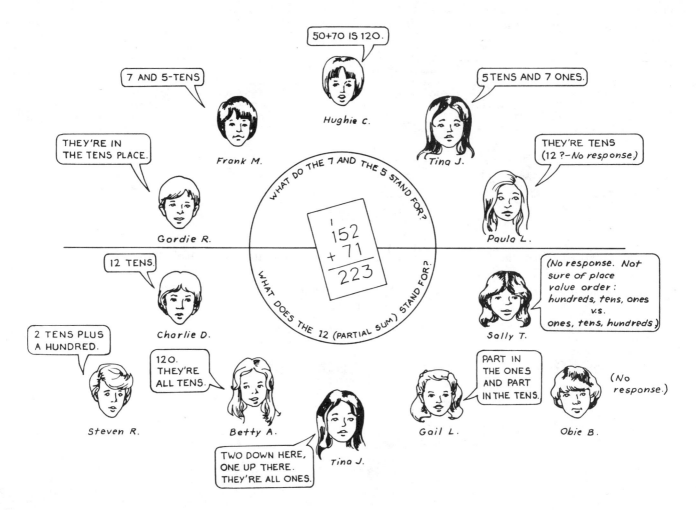

Although some children demonstrated insight into the place value concepts underlying the process of regrouping/renaming, half of the children were unable to give any indication of such awareness. In some instances the extent of children's place value understanding may have been masked by the lack of clarity of the initial question, problems of communication, or inability to ask the right question at the right time.

By not clearly requesting an explanation of the procedure at the start, I did not encourage children with place value understanding to construct an integrated response. In following up on the lack of clarity, resorting to a large number of focused questions may have created an impression that these children were unwilling to talk about an underlying rationale.

However, the focused questions did prove useful in clarifying a child's communication.

Also, the queries helped some children to rethink the algorithm in terms of place value.

Yet these focused questions were ineffective for half the children. Examining some of the interactions in retrospect suggests the possibility of other follow-up questions.

Asking the right question at the right time is the most challenging aspect of interviewing.

DEMONSTRATION

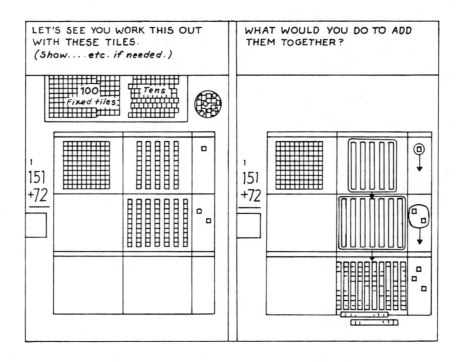

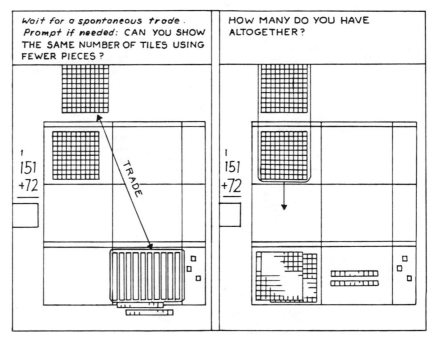

In combining the addends, eight children spontaneously regrouped by trading 10 ten-sticks for a hundred-square. Fifteen children made the trade following a prompt to show the same number of tiles using fewer pieces. Other children need further prompts to complete the demonstration. Their difficulties included:

- miscounting tens and ones
- miscounting the number of ten-sticks in the trade
- arbitrarily placing additional tiles on the board

One child completely circumvented the need for a "fair" trade, thus raising some interesting questions.

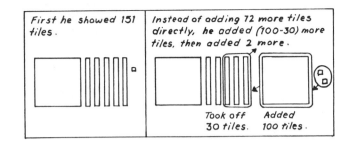

First he showed 151 tiles.	Instead of adding 72 more tiles directly, he added (100-30) more tiles, then added 2 more.
	Took off 30 tiles. / Added 100 tiles.

Do you find his method acceptable?

Is it significant that his method does not correspond to the standard algorithm?

Are there other ways to solve this task with materials?

Connection

Again, to determine whether the accurate computers were able to relate the action of joining and regrouping materials to the written notation of the standard algorithm, the following connection subtask was given. The task illustration highlights the possible responses.

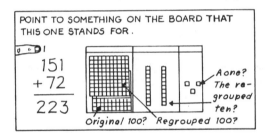

POINT TO SOMETHING ON THE BOARD THAT THIS ONE STANDS FOR.

151
+72
───
223

A one? The regrouped ten? / Original 100? / Regrouped 100?

Eleven of the 24 children pointed to a hundred-square, with 3 additional children indicating the specific hundred-square that was the object of a trade in the regrouping process. The remaining 10 children pointed to either a ten-stick or to an individual tile.

Two children making computational errors were able to clarify the regrouping process with tiles and, given prompts referring back to the materials, they were able to represent the process in written notation. The ultimate connection, however, was not made. The difficulty of making the connection between regrouping objects and renaming in written representation is demonstrated in the following sequences with accurate computers.

After a spontaneous trade and incomplete response to: CAN YOU TELL ME HOW YOU GOT THE HUNDRED? (Referring to symbolic form.)	WHEN ADDING 5 AND 7 ARE YOU ADDING THE ONES? YEAH. Sally T. The question was a follow-up on her "Twelve carry one" explanation.	WHEN YOU ADDED THE TILES ON THE BOARD, WHAT 12 THINGS WERE ON THE BOARD? TENS. I COUNTED TEN OF THEM AND TRADED THEM IN.	WHAT DOES THIS ONE STAND FOR? A HUNDRED...NO, WAIT A MINUTE...TENS.

Following no response to the connection question.	SUPPOSE I TELL YOU THAT IT HAS SOMETHING TO DO WITH TRADING, DOES THAT HELP? Brenda S. Silence.	WHEN YOU ADDED THESE TWO NUMBERS (5 AND 7) WHAT DID YOU GET? TWELVE.	WHAT TWELVE THINGS DID YOU HAVE ON THE BOARD? TENS.	WHAT DO YOU TRADE FOR? A HUNDRED.	CAN YOU TELL ME WHAT THIS ONE STANDS FOR? Silence.

Do they realize that the written symbols are supposed to correspond to systematically grouped objects, or do they see the two representations as unrelated? For many children there appears to be a critical gap between the two.

Related Research and Interpretations

Most of the third graders could compute sums of two- and three-digit numerals using the standard algorithm. Yet only about half could do so with understanding. Although a concern was expressed regarding an inability to coax out the place value understanding underlying the carried "1" for the latter group, most of the same children were also unable to connect it with an appropriate group of objects. Thus, for some children there is no stable place value understanding at this point in their development. In light of the inconsistent performances of the latter group of children on the counting and place value tasks in the preceding chapters, their lack of any place value rationale for the addition algorithm should come as no surprise.

Other researchers have also observed the critical gap between successful performance on algorithms and the underlying concepts of systematic grouping and their place value notation. Ginsburg (1977), Irons and Jones (1979), and Resnick (1982a) have identified this gap as a major source of difficulty in learning arithmetic. Irons and Jones report that children may demonstrate place value understanding independent of the computational procedure but still may not relate the two. Ginsburg (1977) refers to the rarity of finding children who understand the standard algorithm. Resnick (1982b) reports that virtually every third grader interviewed not only identified the carried "1" as a digit rather than by value, but also chose an individual one-block rather than a ten- or hundred-block to represent it. Thus, the third graders described in this book may be more advanced than most children in the third grade.

Chapter 11 has indicated how easy it is to be misled by children's apparent abilities in reading and writing numerals and in paper-and-pencil computation. Thompson's dissertation (1982b), "A Theoretical Framework for Understanding Young Children's Concepts of Whole Number Numeration," advocates that reading and writing of numerals can be done successfully at a superficial level of language and sound patterns, with no underlying place value understanding. The interviews described in this chapter have shown that successful performance with computational algorithms also can take place at a similar superficial level of manipulating digits according to procedural rules. The result may be a gross overestimate of children's numerical abilities and understandings and, ultimately, of the success of school mathematics programs. If one accepts the surface evidence without probing deeper, then disparities in children's performances on different tasks can be observed. For example, a child may show ability to compute three-digit addition with the standard algorithm, yet be unable to count-on grouped objects by tens and hundreds. Many teachers are misled by children's apparent abilities with written symbols and, in the absence of materials in the classroom, are unaware of disparities in understanding. Prior to the work of Thompson (1982b), researchers also may have misinterpreted abilities to read and write numerals as evidence of place value understanding, thus identifying a problem in children's ability to relate prior understanding to the computational algorithm. For some children there probably was no place value understanding on which to build.

Although textbooks devote time to developing place value concepts beginning in the latter half of the first grade, and refer to these concepts in introducing the standard algorithm, their objectives emphasize only correct answers and standard procedures. Since school district objectives are keyed to textbooks and their scope-and-sequence charts, as well as standardized tests, few teachers are encouraged to help children construct a conceptual base for understanding computational procedures. Rather, the district objectives pressure teachers to move ahead to more advanced topics at a level of superficiality. In Piaget's writings on education (1973), there appears the following caution against the dangers of moving ahead without listening to children:

> Without a doubt it is necessary to reach abstraction, and this is even natural in all areas during mental development of adolescence, but abstraction is only a kind of trickery and deflection of the mind if it doesn't constitute the crowning stage of a series of previously uninterrupted concrete actions. (p. 103)

ALIGNMENT OF NUMERALS

In search of further insights into the difficulties children experience in understanding the addition algorithm, two other areas will be explored—the alignment of numerals[8] and the conservation of number during regrouping/renaming.

At this point you are invited to anticipate the children's responses to the alignment of numerals task

> $$\begin{array}{r} 17 \\ 362 \\ +5 \\ \hline \end{array}$$
>
> Read these numbers.
>
> I wonder if you could add these numbers up.
>
> Read your answer.
>
> Does it sound right?

The task was presented to children with columns misaligned, thus violating the place value principle that allows the algorithm to work so well. Alignment arranges digits of like value in columns to be readily combined. Presentation of misaligned numerals gave the children opportunities to reject the task or to reject their answer as being unreasonable. Yet of the 21 children given the task,[9] only 5 were able either to identify the place value violation and rewrite it according to the algorithm or circumvent the violation by using their own procedures.

DOES THIS LOOK LIKE A GOOD ANSWER?

$$\begin{array}{r} 17 \\ 362 \\ 5 \\ \hline 1032 \end{array}$$

Steven R.

NO, BUT YOU PLACED THEM IN THE WRONG SPOTS.

I'D HAVE TO MAKE IT A 500...FILL IN ALL THE SPOTS.

$$\begin{array}{r} 170 \\ 362 \\ 500 \\ \hline 1032 \end{array}$$

Countersuggestion
WHEN YOU READ THESE NUMBERS, YOU READ THEM AS 5, 362, AND 17. I WONDER IF YOU COULD ADD UP THOSE NUMBERS.

YOU PUT THEM IN THE WRONG COLUMNS.

HOW WOULD YOU WRITE THEM?

$$\begin{array}{r} 17 \\ 362 \\ 5 \\ \hline 384 \end{array}$$

He quickly rewrote and recalculated the exercise.

HOW'D YOU GET THAT?

I USED MY BRAIN.

FIRST I ADDED THESE 2 NUMBERS (17 and 5). THEN I ADDED THAT (22) TO THIS (362).

$$\begin{array}{r} 17 \\ 362 \\ 5 \\ \hline 384 \end{array}$$

Hughie C.

$(17 \xrightarrow{+5} 22 \xrightarrow{+362} 384)$

17, 18, 19, 20, 21, 22, 382, 384

Chris B.

$(17 \xrightarrow[\text{counted-on}]{+5} 22 \xrightarrow{+360} 382 \xrightarrow{+2} 384)$

$$\begin{array}{r} 17 \\ 362 \\ 5 \\ \hline 384 \end{array}$$

Betty A.

After a long pause.

She had previously been observed counting by tens and ones from left to right rather than use the algorithm. Although her silent strategy wasn't checked on this task she was capable of using the same strategy.

8. The alignment task is taken from the work of H. Ginsbuirg (1977).

9. Since this task was given only on a time-available basis, the size of the sample was reduced.

It is interesting to note that most of the children able to solve this task were not dependent on the standard algorithm. They showed their own methods of adding while maintaining the integrity of the place value of the numerals regardless of alignment. The remaining 16 children either (1) seemed unaware of a problem for the standard algorithm by unhesitatingly adding the columns of digits and accepting the total, or (2) were aware of a problem but were unable to identify it.

Without place value understanding these children were unaware of the need to align the columns for the standard algorithm and had no basis for judging the reasonableness of their answers.

Since the alignment of numerals for the standard algorithm is based on place value principles, it is possible to conclude either that the number of children with place value understanding was overestimated earlier in this chapter, or that some children understand at one moment but not the next. A third interpretation is that the task presentation is biased toward the power of adult suggestion and that other avenues of gathering information on children's understanding of numeral alignment need to be explored. For example,

how would the children align the numerals if presented horizontally (17 + 362 + 5 = _____), dictated orally, or presented in the context of a number story problem?

Although none of these avenues was pursued with these third-grade children, information on other children's reactions to variations of the task are available. The next two interviews (Ginsburg, 1977) provide an opportunity to observe children's alignment from dictated number names prior to exposure to countersuggestions by the interviewer. From such interviews Ginsburg observes that arithmetic is an arbitrary game for some children.

Touger (1981) reports that, on being asked to write down numerals for addition exercises, less than half of a third-grade class was able to align the numerals correctly. This surprised the teacher since the children consistently turned in satisfactory workbook answers throughout the school year. This example clearly illustrates the importance of observing the children's computation outside the structure of worksheets with ready-made exercises. Touger points out the danger of fostering children who are capable of filling in correct answers, yet functionally illiterate in mathematics; that is, unable to set up the algorithm in response to dictated examples or, more importantly, in the context of real-life problems.

CONSERVATION OF NUMBER DURING REGROUPING/RENAMING IN ADDITION

Consider how the children might respond to the following conservation of number tasks.

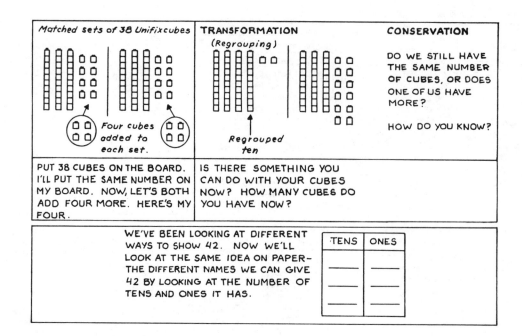

Matched sets of 38 Unifix cubes

Four cubes added to each set.

TRANSFORMATION
(Regrouping)

Regrouped ten

CONSERVATION

DO WE STILL HAVE THE SAME NUMBER OF CUBES, OR DOES ONE OF US HAVE MORE?

HOW DO YOU KNOW?

PUT 38 CUBES ON THE BOARD. I'LL PUT THE SAME NUMBER ON MY BOARD. NOW, LET'S BOTH ADD FOUR MORE. HERE'S MY FOUR.

IS THERE SOMETHING YOU CAN DO WITH YOUR CUBES NOW? HOW MANY CUBES DO YOU HAVE NOW?

WE'VE BEEN LOOKING AT DIFFERENT WAYS TO SHOW 42. NOW WE'LL LOOK AT THE SAME IDEA ON PAPER—THE DIFFERENT NAMES WE CAN GIVE 42 BY LOOKING AT THE NUMBER OF TENS AND ONES IT HAS.

TENS	ONES

Addition of Two-Digit Numerals

In this section we examine in further detail regrouping of objects and corresponding renaming of addends, together with the implied conservation of number.[10] Inclusion of yet another task was prompted by observations of third graders by a teacher—my wife, Shirley. In an example of *teacher as researcher*, she not only observed behaviors previously unreported by researchers, but also incorporated conservation questions in her teaching of both the addition and subtraction algorithms (Labinowicz, 1977).

Regrouping. Nearly all of the children grouped ten individual cubes by interlocking them. While 5 children spontaneously constructed a ten-group, the remaining 23 children required prompts. Three children required more than one prompt prior to doing so.

Conservation. On Piaget's classical number conservation task, most seven-year-old children are able to retain the equivalence of a matched set of 8 objects when one of the sets is rearranged. In a parallel task above, the quantity has been increased to 42 and the rearrangement is in terms of grouping objects. In the process of interlocking 10 individual cubes as one ten-group and carrying it to another area of the board, the number of cubes does not change despite the rearrangement of the parts. Are these eight-year-olds capable of conserving number in this context?

Nineteen of the 29 children recognized that the number of cubes remained unchanged during the regrouping and gave a justification for their response. The three children illustrated below have a rational basis for recognizing the equivalence of the collections without the need to count the interviewer's cubes.

He focused on equivalent actions preceding re-grouping, assuming nothing changed since.

SAME NUMBER BECAUSE YOU ADDED AS MANY AS I DID.

Harold S.

THEY'RE GOING TO BE THE SAME NUMBER UNLESS I TAKE SOME OUT OF HERE. WE HAVE THEM ALL HERE.

The identity of the blocks hadn't changed for this boy.

Vern L.

YOU HAVE 3 TENS AND I HAVE 4 TENS. YOU HAVE 12 AND I ONLY HAVE 3. SO IF YOU CARRIED THE 10 (from 12) THERE'S GOING TO BE 4 TENS AND ONLY 2 HERE.

He first considered the compensation of tens and ones on both boards.

Gordie R.

---

10. The regrouping, conserving, and renaming tasks were given in the second interview, several weeks prior to presentation of the computational algorithm in the final interview.

The remaining 10 children were transitional in their responses. Initially, they gave nonconserving responses but during their attempt to justify these responses, they changed their minds. Eventually, they were able to offer some justification for the conservation of number. It was as if the idea of preserving the number of cubes during regrouping had not occurred to them before and they needed time to mull it over.

Initially they focused on either the tens or the ones in both collections rather than consider them simultaneously. Coincidentally, some of these children had problems in counting the cubes. Following an initial disequilibrium in "reading" the cubes and focusing on the ones only, Tina eventually integrated both responses in terms of tens and ones.

The next child appeared to assimilate the conservation question to his existing routine for comparing numbers on the basis of the largest grouping. Unfortunately, this rule is restricted to standard representations of numbers in both collections.

The last child does not see *number* as the same as *amount*. Therefore, he conserves the amount of cubes during regrouping, yet he is unsure about number during this process. He may have a problem in accepting the possibility that the same number can be represented by different arrangements of objects or digits.

If Bert was thinking strictly numerically, he would have responded that he didn't need to do anything to make them the same number—they are both 42. Rather, he reacts to form instead of meaning. Although these children eventually conserved number (or amount), they did so only after counting both collections. They may continue counting after regrouping prior to recognizing the conservation of numbers spontaneously for rational reasons.

Renaming. To realize a rationale for the standard paper-and-pencil algorithm, children's understanding of conservation during regrouping of objects must be applied in a more abstract context. The following task explored whether third-grade children are capable of making such a transfer.

Ten children wrote nonstandard expressions of 42—that is, $\frac{\text{tens} \mid \text{ones}}{3 \mid 12}$—in which a number greater than 9 in a column is initially contrary to an intuitive or learned rule. Yet the possibility of having more than 9 in any column, even if temporarily, does exist during regrouping, and requires an expansion of the rule to include such situations. Unless the varied perspectives on digits, total value, and place value become integrated, any one of them may be applied to the task. The presence of a two-digit numeral, such as 12, in one of the columns may trigger a total value perspective resulting in the writing of 30 in the other column. Although 30 + 12 = 42, as Hughie said, this relationship needs to be rewritten to be consistent with the place

value headings. While Hughie realized the redundancy of 30 tens, other children saw no inconsistency. The third perspective—on digits—resulted in a reversal of digits, having no relation of 42. Children's difficulty in understanding that the same number could be expressed in different ways is also documented in Behr's teaching experiment with second graders (1976).

Addition of Three-Digit Numerals

A final task explored the children's interpretation of nonstandard notation through reading and representation with grouped tiles. It also took a second look at their understanding of regrouping and renaming 10 tens as a hundred.

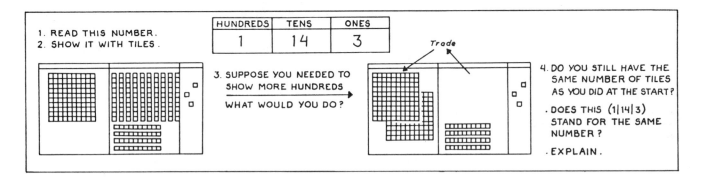

Although presented in isolation from the standard addition algorithm, the task still had the potential of providing yet another view of children's interpretations of situations that simulate a critical aspect of the algorithm—renaming a partial sum.

Reading Nonstandard Notation.

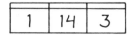

Representing Nonstandard Notation with Tiles. Although some children improved their interpretation of the nonstandard notation when given the pregrouped tiles, others were persistent in their misinterpretations

despite the availability of materials. Children who arbitrarily partitioned digits to read 117 also represented the same quantity with the materials or arbitrarily positioned the tiles while ignoring the column headings.

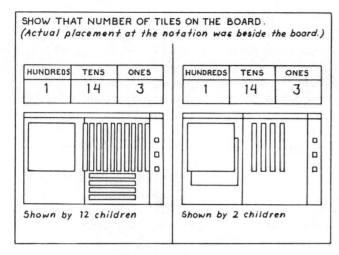

SHOW THAT NUMBER OF TILES ON THE BOARD.
(Actual placement at the notation was beside the board.)

HUNDREDS	TENS	ONES
1	14	3

Shown by 12 children

HUNDREDS	TENS	ONES
1	14	3

Shown by 2 children

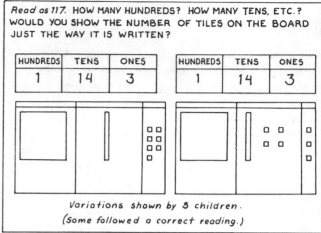

Read as 117. HOW MANY HUNDREDS? HOW MANY TENS, ETC.?
WOULD YOU SHOW THE NUMBER OF TILES ON THE BOARD
JUST THE WAY IT IS WRITTEN?

HUNDREDS	TENS	ONES
1	14	3

HUNDREDS	TENS	ONES
1	14	3

Variations shown by 5 children.
(Some followed a correct reading.)

The task was terminated for children unable to display the quantity of tiles represented by the nonstandard notation.

Regrouping and Renaming. Of the remaining children, several needed prompts to complete a fair trade. One child believed that the number had changed following the trade. When asked to explain the renam-ing during the regrouping of objects, again, some children spoke of place value groupings while others spoke of digits. Once more, the use of prompts in the presence of materials was only partially effective in coaxing out clarifying responses. The responses of four children are illustrated to show both detailed explanations and the problematic area of renaming 10 tens as a hundred.

HOW WOULD YOU EXPLAIN THIS TO A SECOND GRADER?

YOU CAN ONLY HAVE 9 TENS SO IT GOES TO ANOTHER HUNDRED.

Steven R.

HOW WOULD YOU EXPLAIN IT...?

IF YOU TAKE AWAY 10 TENS YOU COULD ADD A HUNDRED.

Frank M.

HOW WOULD YOU EXPLAIN IT...?

TAKE 14 OF THESE AND IT COMES TO 140 AND YOU ADD 100 TO THAT. THAT'S 240. THEN YOU ADD 3 ONES.

Chris B.

Betty needed a prompt to complete a "fair" trade.

She agreed that 243 represented the same number as

1	14	3

HOW WOULD YOU EXPLAIN IT TO A SECOND GRADER?

YOU TAKE THE ONE AND PUT IT OVER HERE.

Betty A.

WHAT DOES THE ONE STAND FOR WHEN IT'S OVER HERE?

A HUNDRED.

AND WHAT DOES THE ONE STAND FOR HERE?

ONE TEN.

Although Betty identified "1" as being carried to the hundreds column, she was unable to connect this to regrouping and renaming 10 tens as a hundred. She was able to identify the hundred in terms of position but not in terms of the "ten of" relations of systematic grouping represented between adjacent positions. Although on earlier tasks she had given clear evidence of having constructed this relation at the first level—ten is ten of ones—her performance on this task suggests that she has yet to extend this relation to the second level of systematized grouping expressed in symbolic notation.[11] Based on the performance of the third graders on this task and the related addition task described earlier, the renaming of tens as a hundred is a problematic area for more than half the group.

At the same time that the interviews show the value of materials in helping children to develop a conceptual foundation for the standard paper-and-pencil algorithm, they also show that children make connections in their own time. Yet they are asked to integrate many aspects of numeration at the same time. No sooner are children able to expand their string of oral number names to larger numbers, to read and write numerals that represent them, to group, count, and count-on objects by ones, tens, and hundreds, than they are expected to deal with numbers being represented in different ways and having multiple meanings. An analysis of the preceding task will illustrate not only the gap between concrete and written representations, but also the multiple meanings of "1". Perhaps it will provide insight into both the size of the gap and the time required to integrate the multiple meanings to bridge it.

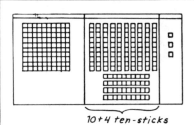

10 + 4 ten-sticks

They represent 10 ten-sticks and 4 ten-sticks regardless of position on the board. Ten groups of ten discrete objects are readily visible.

The positions on the board are important only for helping children to associate particular values with written notation.

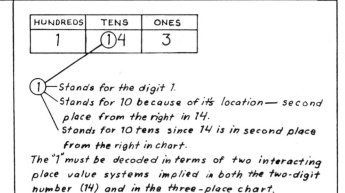

1 — Stands for the digit 1.
Stands for 10 because of its location — second place from the right in 14.
Stands for 10 tens since 14 is in second place from the right in chart.
The "1" must be decoded in terms of two interacting place value systems implied in both the two-digit number (14) and in the three-place chart.

The high level of abstraction demanded in dealing with the multiple meanings of "1" may explain why researchers have identified regrouping and renaming as a difficult area not only in primary grades, but also as late as the seventh grade. Dealing with "a system within a system" and its resultant multiple meanings may require the type of abstract thinking capacity characterized by Piaget's formal operational stage.

An overlooked variable in children's classroom learning is *time*. Children's mathematical ideas evolve slowly. In this regard, Piaget (1977) writes:

> . . . At all levels, including adolescence and in a systematic manner at the more elementary levels, the pupil will be far more capable of "doing" and "understanding in actions" than of expressing himself verbally. . . . "Awareness" occurs long after the action. (p. 731)

An even longer period is needed for expression in place value notation. If the move from concrete to abstract was treated as a long-range goal rather than a short-term, virtually instant objective, sufficient time would intervene for mental maturity to support the child's reconstruction of ideas at the next level of abstraction. Today's schools are not willing to provide this time.

---

11. Two months later, Betty demonstrated an understanding of renaming 10 tens as a hundred in the addition of multidigit numerals.

AN ALTERNATIVE APPROACH TO TEACHING THE STANDARD ADDITION ALGORITHM

The preceding interviews with third graders on addition with regrouping/renaming have shown that about half the children were unable to make the transition to the paper-and-pencil algorithm with understanding. This information suggests strongly that if we choose to teach the standard algorithm, we need to ensure that it is supported by a base of understandings gained through activities with grouped materials. In this approach we need to provide for a gradual transition from concrete activities to the written notation of the paper-and-pencil algorithm—treating the written symbolism as a long-term goal. The preceding interviews also provided evidence that at least some of the children are capable of developing their own computational procedures and do so despite being taught a standard procedure. This information suggests that we also need to explore children's abilities to construct their own algorithms as an alternative to presenting a standard algorithm to all children. Whereas the latter topic is the focus of Chapter 14, this chapter will close with a discussion of a developmental approach to teaching the standard addition algorithm.

Once the decision to teach the same algorithm to all children is made, it simultaneously reduces the amount of indirect/reflective teaching possible. Direct modeling of the standard procedure and imitation by the children become necessary to some extent. Since most teachers are already involved with such methods, how is the proposed alternative different? First, it pays attention to the gradual development of children's number ideas from a base of action on objects. The concepts underlying addition with regrouping are first taught in the context of activities with materials and related oral discussion prior to making connections with written representation of the standard algorithm. Second, it maximizes possibilities for indirect/reflective teaching within the limited context of a single procedure by creating opportunities for children to interact with the underlying rationale. Third, rather than being taught in isolation and practiced on struc-

tured worksheets, the algorithm is placed in real-life contexts.

The following sequence of teaching/learning activities is adapted from the work of a number of researchers, math educators, and teachers (Behr, 1976; Irons and Jones, 1979; Labinowicz, 1977; Resnick, 1982; Souviney et al., 1978; Swart, 1980; Thompson and Van de Walle, 1977; Wirtz, 1974). It is designed to assist children in making a smooth and effective transition to an understanding of the paper-and-pencil algorithm. By analyzing the standard written algorithm and working back to the concrete materials, it is possible to identify a routine of manipulations that directly parallels it. Each step in the manipulation of materials is structured to coincide with a step in the written algorithm, such as working from right to left. To ensure the organization of materials during the addition activities, all manipulation is done on a transition board with cells for the addends, the sum, and regrouped objects, in direct correspondence to the written algorithm.

Initially, the children find sums for number story problems using objects only. Later, they are asked to keep a record of their actions on objects, first pictorially, then in written notation. In this transition they alternate between actions on objects and recording the actions. Once the actions have been encoded the children are asked to decode their own records and those of other children. If each level is accompanied by opportunities for discussion, eventually the written algorithm can be completed with understanding in the absence of materials. The children can operate on visualized materials or on abstract numbers at the final level. While some children are capable of completing the sequence in several days, others may take several months to do so. Determining factors in the time required by different children include their conceptual foundation as reflected in their performance on tasks described in Chapters 10 and 11, and the grade placement of this developmental sequence.

Level One: Action on Objects and Discussion

Small-group lessons in which each child and the teacher have their own materials and transition board initiate the sequence. The teacher presents an addition task in the context of a number story problem to be worked out as a group. Although children are invited to show the addends and find the sum with regrouping as needed, the process is stopped at the end of each

step to identify specific constraints of the standard procedure. The teacher identifies particular cells on the board for the location of the addends, the sum, and the result of regrouping, as well as the particular right-left direction. To avoid blind imitation, the teacher invites the children to think about the meaning of the materials as well as their actions on these objects. For example, during the discussion, the children's vocabulary of objects based on size, shape, or color—for example, "longs", "squares", "white cubes", "big block"

—is accepted. At the same time, the children are encouraged to focus on number and to look at the quantities represented in different ways. While some children see a certain arrangement of materials as 23, others see that arrangement as 20 and 3 or 2 tens and 3 ones. By continually hearing different viewpoints, the children begin to integrate them as different names for the same number of an addend or sum. This group activity is described in greater detail below. Notice that a range of questions is suggested for teacher selection.

Introductory Lesson. Here is an example of an introductory type of activity provoked by a number story problem.

> *We have twenty-eight children in our class and Mrs. Clark's class has thirty-three children. If we all go on the picnic on Thursday, how many picnic lunches will we need?*

ORAL LANGUAGE (DIALOGUE)	CONCRETE MODELING
Exhibiting addends Show the two numbers on the board (in the middle rows). How do you see the 28 cubes on your board? (20 and 8, 2 tens and 8 ones.) **Addition of ones** Let's all combine our ones (by pushing them down to the answer space). (Later) Which pieces do we join first? How many ones are there altogether?	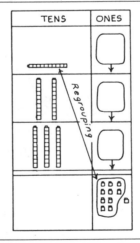 1. Teacher modeling with materials on her own board is important during introductory lessons. 2. Pause and check the children's arrangements at each step. 3. Counting errors can be reduced by clearing the area around the board of materials, counting by twos, or using a 2×5 grid in the ones column. The grid also allows children to anticipate regrouping.
Regrouping/Renaming (select) Can we show the same number of cubes in another way that's easier to count? Do we have enough ones to make another ten-group? Where will you put your ten-group? How many ones do you have now? Do we still have the same number of cubes on the board that we started with? (Or compare two boards at different stages of regrouping.)	4. Both the location and horizontal position of this ten-group emphasize its uniqueness. The location also corresponds with the standard written algorithm.
Addition of Tens Let's all combine our tens. How many tens do we have altogether? **Regrouping/renaming** Do we have enough tens to regroup (to make a hundred)? (Continue with selected questions as above if needed.) **Restating the problem** What problem did we jut solve? How can we say the problem as a number sentence? (28 + 33 = 61)	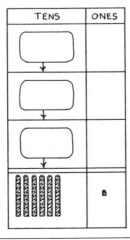 5. Since the addends are already joined, they need to be remembered in restating the problem. Here the need for a recording system is being established.

Following the introductory lessons, pairs or small groups of children engage in similar activities in a variety of contexts.

Collect and Group (Trade) Game. The goal of this game is to collect the most objects on the board in a prespecified time or to reach a specified number, like 10 tens or 10 hundreds. Each child rolls a pair of dice and places the same number of cubes on the board in appropriate cells, regrouping whenever possible. After two turns, each child has two addends to combine and regroup as needed. Prior to the third turn, the sum of the first two addends is moved up to become one of the addends.[12] At the end of each turn the child reads off his board. Play continues until the prescribed goal is reached. Variations of the game include using pregrouped objects such as Dienes blocks or tiles for trading. One of the children can play the role of a banker through whom all trades are made. As the target number is increased, a chance generator of larger addends is also needed. One of the possibilities includes shaking and tossing increasing numbers of lima beans marked with pictorial representations of tens (|) and ones (.) on either side (Fuys and Tischler, 1979).

Number Story Problem Cards. One child in the pair takes the top card from the deck and reads the number story aloud. The child's partner displays the addends with materials, performs the addition—joining the ones, then joining the tens, doing the necessary regroupings—and restates the problem in terms of an oral number sentence. Variations include using (1) audiotape cards and card readers for children with reading difficulties; (2) cards with blanks in the number story problems to be filled in by the reader; (3) blank cards for which the child constructs his own number story problem for his partner. (The card may specify limits to the size of numbers used.)

The Robot Game. Given a number story problem read aloud by one of the children, the child playing the robot director verbalizes the steps for addition to be followed literally by the robot with the materials. However, the robot is only programmed to respond to references to the materials as ones, tens, and hundreds—when ten-groups are referred to as stacks or by pointing, the robot can reply that he doesn't

know what those are. The idea of this activity is to encourage the director to become more conscious of both the procedure and the values of the materials (Swart, 1980a).

During these activities, the teacher's role is to circulate, observe the children, and initiate interactions that encourage children to think about what they are doing. For example, during the Collect and Group game, the teacher might interject a conservation question during regrouping—either asking the child to think about the quantity of objects before and after grouping or introducing another board to highlight the comparison (as on p. 326). Other children's views also should be encouraged to initiate interaction of contrasting ideas.

Level Two: Pictorial Representation of Action on Objects

As the teacher demonstrates the addition of two addends with materials on the transition board, the children are given outlines of the board and asked to represent each step of the procedure without use of numerals or words. They are asked to keep a record that could be sent to a foreign country and still be understood by the receiver. The children's pictorial representations are compared in terms of ease of communication. Their informal representations may not only provide you with insight into their level of understanding, but also allow you to build on their ideas in making the transition to written notation.

One of the follow-up activities for pairs of children can include an extension of the Robot Game. Here, the robot director is given the additional task of keeping a pictorial record of the robot's activities with materials.

Level Three: Written Notation of Action on Objects—Connection

Only after the children are comfortable with both the procedure of addition with materials involving regrouping/renaming and the underlying ideas, and have explored ways of informally representing the procedure, is the transition to written notation undertaken. The paper-and-pencil algorithm can now be approached naturally as a record for what is already understood at a concrete level.

During the introductory lessons illustrated next, the teacher presents a number story problem to the group and keeps a written record of the children's action on objects. At each step they pause as she involves them in considering what to record, as well as where and how to record different aspects of their

---

12. In a variation of the game, each player can begin with the same quantity of cubes representing the first addend prior to round 1.

procedure with materials. Following this encoding, the children are involved in decoding—interpreting the written symbols in terms of groups of objects and actions on these objects. The teacher's questions help the children to focus on the correspondence between the number of objects, their position and regrouping, and the numerals that represent them. Through this dialogue, connections begin to be made between existing ideas and new ways to represent them in a coded system.

Eventually, the children pair up, with one child doing the addition with materials while the second child does a simultaneous written recording on a sheet with outlines of the transition board. Later, individual children attempt concurrent manipulation and recording. Some teachers use large laminated transition boards allowing each child to record an erasable written record with crayon beside the materials (Thompson and Van de Walle, 1980). The materials can be removed from the board and the addition task can be reconstructed (decoded) with the materials again, based on the symbolic record. These activities can take place in game-like contexts.

Introductory Lesson. The teacher says, "This time I need you to slow down and work each step together so I can keep a record of everything that you're doing."

> In getting ready for our class party, Cheryl baked forty-seven cookies and Gloria baked fifty-six cookies. If they don't eat any or give any away, how many cookies will we have at the party?

CONCRETE MODELING	FORMAL SYMBOLIC RECORDING	ORAL LANGUAGE (DIALOGUE)
HUNDREDS / TENS / ONES (board showing 47 and 56 modeled with tens and ones)	4 7 / 5 6	(Select appropriate questions) What shall I write down here? What does this 4 stand for? How can you tell from my record?
HUNDREDS / TENS / ONES (board showing regrouping of ones into a ten)	1 / 4 7 / 5 6 / 3	What do I write down here? (Point to ones answer space.) How can I show the regrouping? Where will I write the "1"? What does this "1" stand for?

(continued)

(continued)

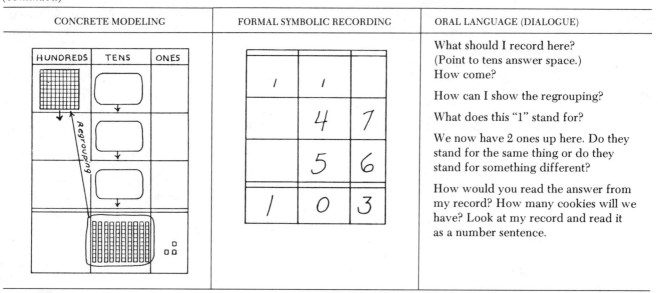

CONCRETE MODELING	FORMAL SYMBOLIC RECORDING	ORAL LANGUAGE (DIALOGUE)

What should I record here?
(Point to tens answer space.)
How come?

How can I show the regrouping?

What does this "1" stand for?

We now have 2 ones up here. Do they stand for the same thing or do they stand for something different?

How would you read the answer from my record? How many cookies will we have? Look at my record and read it as a number sentence.

Collect and Group (Trade) Game. While the children are playing the game with materials, they can also keep a record of their scores on prepared score sheets. (Fuys and Tischler, 1979)

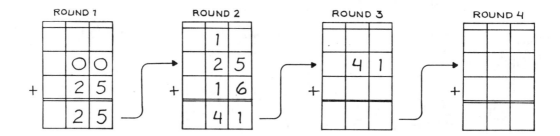

Robot Game. In this extension of the game, the robot director not only gives explicit directions for the robot to follow with materials, but keeps a written record of the activity.

Teacher (Swart, 1980a). Given completed addition records of other children (and teacher), including some with errors, the children use the materials to check the answers and mark them right or wrong, and determine the correct answer as needed.

Level Four: Written Notation of the Standard Algorithm in the Absence of Materials

Extended experiences with materials help children to develop mental imagery of the addition with regrouping process and its sequence. Some children initiate the visualization themselves and stop using the materials spontaneously (S. Labinowicz, 1977).

Other children may be encouraged to do so as a transition away from the concrete materials to working solely with the written symbolic representation. They can pretend to manipulate imaginary objects to decide on the sequence of numerals to record and to remind them of what the numerals represent. The use of an outline of the transition board for the written algorithm provides support for this process of visualization. Eventually, however, the child does not require this support and may be able to think in terms of abstract ideas rather than physical referents for them.

Follow-up activities include the following game-like situations.

Task Cards. Addition situations are presented on cards in a variety of ways. These include number story problems[13] and incomplete number sentences in both horizontal and vertical format. In the former situations, the child must set up the written algorithm by aligning the addends in the absence of transition board outlines.

Teacher Cards. Two types of tasks can be presented, one requiring children's error detection and the other an explanation of procedure to a partner (an imaginary first grader) in the absence of materials. These cards can be mixed in with the preceding task cards.

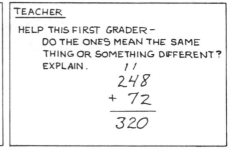

Detective Cards. Missing addend problems are presented in the context of the standard addition algorithm. The tasks are introduced as "somebody's record that had ink spill over it," and the children are invited to find the missing numbers.

During follow-up activities throughout the sequence the teacher's role is many-faceted. It includes the following tasks:

1. Observing the children during the activities
2. Observing a particular child more closely by playing robot, with the child being the director
3. Interjecting thought-provoking questions
4. Altering pairings as needed for greater variety, challenge, compatability and so on
5. Identifying problem areas for discussion with a larger group
6. Identifying children who appear ready for working with larger numbers or more abstract procedures
7. Scheduling group meetings for introductory lessons and follow-up discussions as needed (since children make connections in their own time, many small groups are possible)
8. Keeping records of children's progress

Above all, the teacher regards the transition from concrete materials to formal symbolic representations as a long-term goal rather than a short-term objective. The apparent time lost in providing for this gradual transition ultimately results in a gain in both the children's understanding and their self-confidence and a reduction in the amount of time required for reteaching the standard addition algorithm throughout the grades.

These methods and activities have been offered as one possible approach. The next chapter describes similar methods and activities for a semidirect method of teaching the standard algorithm for subtraction. Before you decide to try this approach you are invited to read Chapter 14 on children's invention of

---

13. Until such time as the sequences for addition and subtraction algorithms overlap, number story problems are not likely to be very challenging. However, they do provide a real-life context and require the children to align the addends by themselves.

nonstandard algorithms and the indirect teaching methods that support this process. The comparison of approaches found in Chapter 15 also will help you make an informed decision. Whichever approach you may choose, feel free to adapt it in ways that make sense to you.

REFLECTIONS AND DIRECTIONS

1. Locate some third graders in your school or neighborhood and explore their understanding of the standard algorithm. Compare your findings with those reported here. Discuss them with at least one other teacher.

2. A common method teachers use to record children's progress is to check off each topic once they have achieved the computational objective with the standard algorithm.

	2-Digit Addition No Regrouping	2-Digit Addition Regrouping	3-Digit Addition No Regrouping	3-Digit Addition Regrouping	2-Digit Subtraction
Allison	✓	✓	✓	✓	
Billy	✓	✓	✓		
Charles	✓	✓	✓		
Donna	✓	✓	✓	✓	
Eileen	✓	✓	✓		
Fred	✓	✓	✓	✓	

Suggest an alternative method of recording learning progress based on what you've learned from children in this chapter.

3. Assuming that you choose to teach a standard addition algorithm to all children, reflect on an appropriate grade placement. Compare your ideas with those of at least one other person.

13

UNDERSTANDING OF THE STANDARD SUBTRACTION ALGORITHM

PREVIEW

Just as for addition, textbooks present a standard algorithm for subtraction, although alternative computational procedures are available. In nearly every classroom in the country, children are taught to take the smaller number from the larger one, column by column, starting from the right and *borrowing* as needed. Also, typically, second graders are expected to subtract from two-digit numbers, and third graders to subtract from three-digit numbers, like the following.

$$
\begin{array}{r} 43 \\ -26 \\ \hline \end{array}
\qquad
\begin{array}{r} 343 \\ -181 \\ \hline \end{array}
$$

As in the preceding chapter, the focus here is to explore whether third graders are not only capable of performing the computations with the standard algorithm but also of understanding and rationalizing the procedure. Particular attention will be paid to the place value and conservation concepts underlying the borrowing process. Also, special attention will be given to a specific problem area—subtraction from large numbers involving *borrowing across zero*. This chapter also describes an alternative, semidirect method for teaching the standard subtraction algorithm with a focus on developing meanings.

Based on what you've learned from children, predict the range of responses to the following subtraction task.

COMPUTATION	EXPLANATION	DEMONSTRATION	CONNECTION
1. READ THE QUESTION. 2. NOW DO IT. 43 −26	TELL ME HOW YOU WORKED IT OUT. ³⁄₄3 −26 17 *(Inaccurate responses are explained also.)*	NOW LET'S SEE HOW YOU DO IT WITH MATERIALS. ³⁄₄¹3 −26 *Answer is covered.*	*Pause immediately after regrouping.* TELL ME HOW YOU GOT THE 3 AND THE 13 *(pointing to the written form).* 3⁄₄¹3 −26

COMPARISON	VERIFICATION
For inaccurate computers I WONDER IF ONE OF THESE ANSWERS IS BETTER OR IF THEY'RE BOTH JUST AS GOOD. 43 −26 *Inaccurate answer*	IF YOU WEREN'T SURE ABOUT THE ANSWER, WHAT ELSE COULD YOU DO TO CHECK? OR SUPPOSE WE PUT THOSE CUBES THAT YOU TOOK AWAY BACK ON THE BOARD. HOW MANY CUBES WOULD THERE BE ON THE BOARD? *(Counting discouraged.)*

SUBTRACTION FROM A TWO-DIGIT NUMBER

Similar to the addition task described in the preceding chapter, the interview task for subtraction has phases exploring a child's computation, explanation, demonstration with grouped materials and connection back to the paper-and-pencil algorithm. Unlike the addition task, other phases of the subtraction interview explore the child's reaction when two phases of the task produce contradictory answers and assess his ability to verify the correctness of an answer. Again, the results from different phases of the interview will be discussed separately.

Computation

The same group of third graders as described in preceding chapters[1] found the subtraction computation to be more difficult than addition. While 23 of 29 children computed correctly for a two-digit addition task, only 15 computed correctly for a two-digit subtraction task. All but one of the accurate computers used the standard right-left procedure. The following illustration shows Chris working from the opposite direction.

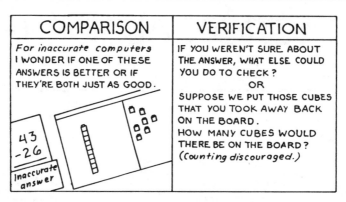

Took away 20 and had 23 left.
Then took away 23, 22, 21, 20, 19, 18.

43
−26
17

17.

Chris B.

---

1. This interview was the last in a series of three. Addition and subtraction tasks were alternated during the course of the interview.

The remaining children made one of two computational errors. Nearly half of the total group subtracted 3 from 6 regardless of vertical position. By using this *reverse subtraction* they ignored or circumvented the need to consider the borrowing process. Three of these children realized their error, either spontaneously or following a prompt. Although all three children real-

ized the need for borrowing, only one was able to complete the process with the aid of prompts. The regrouping/renaming involved in borrowing appears to be more difficult than that involved in carrying. The following children had difficulty in either recalling or reconstructing the process.

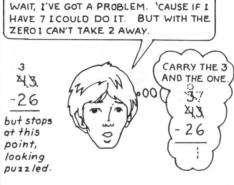

Explanation

The place value system allows taking away the whole bottom number from the whole top number by operating on digits in individual columns, as in the standard algorithm. When the top digit in any column is less than the bottom digit, you borrow from the top digit in the adjacent column to the left.[2] Here, you reduce the digit from which you are borrowing by 1 (ten) and increase the adjacent digit to the right by 10. Thus, the total value of the top number does not change although the parts are regrouped and renamed to facilitate taking away. Do these third graders have a place

value rationale for this procedure or do they merely see it as a mechanical "dance of the digits"?

Accurate Computers. In response to "Tell me how you worked it out," or "Tell me how you got your answer," the children tended toward abbreviated descriptions of the procedure that showed little or no place value awareness. In some instances place value awareness was indicated spontaneously in the context of describing an automated response. Probing questions served to clarify the children's understanding, encouraging them to step back and reflect on the underlying meaning.

---

2. Except in the case of a zero digit, where you continue left and borrow from the next nonzero column.

Reflect on how the children's understanding of borrowing might be probed further.

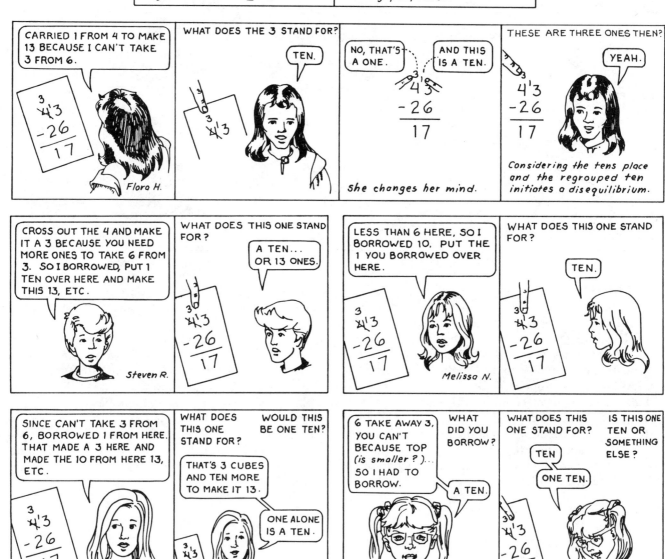

Gail appeared to be functioning at a level of manipulating digits in the algorithm, being unable (or unwilling) to undertake an explanation of borrowing. Most of the accurate computers, like Melissa, Paula, and Roberta, were able to think of 10 being borrowed from one column and placed in the adjacent column to the right. However, they gave no indication of being aware of 1 ten being regrouped/renamed as 10 ones based on the relationship of adjacent columns. Although this relationship is not stated explicitly, Steve appears to be at this level of understanding.

Inaccurate Computers. The children taking away the smaller digit from the larger one, regardless of top or bottom position, seemed quite confident about their procedure. They give the following range of explanations.

According to Piaget, children tend to assimilate new situations such as having the top digit less than the bottom digit to existing ideas—for example, "subtract the smaller number from the larger one"—regardless of fit. Joey illustrates such an assimilation.

Although this idea was once quite useful, it now needs to be reconstructed and integrated within a place value framework. Here, the whole quantity represented by the top numeral must be considered as well as its parts. Although the columns are processed separately, the individual digits are part of numerals that only make sense when considered together.

The discomforting awareness (disequilibrium) of the mismatch between the new situation and the child's existing ideas serves to motivate change. One child,

Hughie, spontaneously became aware of the mismatch and was able to rethink subtracting the smaller number from the larger one within a place value context. Some children may experience this unsettling awareness but are unable to bring place-value ideas to bear. Zoya's response reflects this continuing frustration, being unhappy with both attempts.

43
−26
20

43
−26
23

Zoya K.

Puzzled   *Still uncertain*

→ *Hesitates*

Although momentarily aware that 6 couldn't be subtracted from 3, she was unable to go any further.

Sometimes an interviewer's question can initiate awareness of a mismatch of ideas. Following such a question, Harold (p. 341) was aware of an error but had difficulty in reconstructing the regrouping/renaming process. Yet given an opportunity to work with materials in the next phases of the task, he not only corrected his procedure, but also explained it in place value terms.

Demonstration

Following the explanation, the written answer was covered and the child was asked to demonstrate the subtraction using Unifix cubes. With these materials, regrouping involves decomposing a ten-group into individual cubes. A variety of prompts for this regrouping was available as needed.

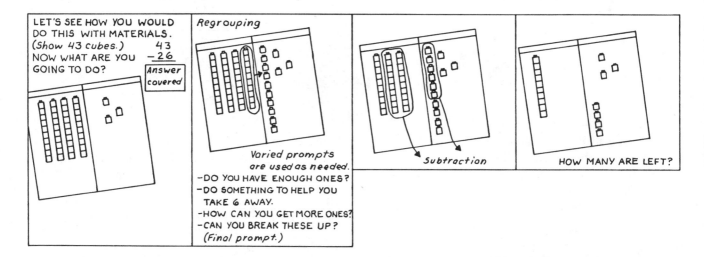

LET'S SEE HOW YOU WOULD DO THIS WITH MATERIALS. (Show 43 cubes.) NOW WHAT ARE YOU GOING TO DO?

43
−26
Answer covered

Regrouping

Varied prompts are used as needed.
−DO YOU HAVE ENOUGH ONES?
−DO SOMETHING TO HELP YOU TAKE 6 AWAY.
−HOW CAN YOU GET MORE ONES?
−CAN YOU BREAK THESE UP? (Final prompt.)

Subtraction

HOW MANY ARE LEFT?

Of the 29 children, 12 spontaneously decomposed a ten-group into ten individual cubes to complete the subtraction. Six more children regrouped following a single prompt, while three others required additional prompts. The remaining children responded in a variety of ways as illustrated below. Notice that the next four examples involve left-right procedures with materials.

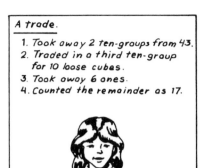

A trade.

1. Took away 2 ten-groups from 43.
2. Traded in a third ten-group for 10 loose cubes.
3. Took away 6 ones.
4. Counted the remainder as 17.

Sally T.

Short-cut with no regrouping

1. Removed 3 ten-groups from 43.
2. Returned 4 individual cubes to the board.
3. Counted remainder as 17.

Hughie C.

A two-step subtraction

1. Took away 2 ten-groups. Stopped.
2. Removed 3 loose cubes. (Took away all she could without regrouping.)
3. Decomposed a ten-group.
4. Removed 3 more loose cubes.

Brenda S.

Variation of no regrouping

1. Removed 2 ten-groups from 43.
2. Broke off 6 cubes from a ten-group and removed them.
3. Decomposed the four-group.

Chris B.

It is interesting to note that the children used left-to-right procedures with materials while the standard written algorithm requires the reverse.

Rather than displaying only the materials for the top numeral, Roberta displayed them for both numerals. After proceeding to decompose a ten-group from the smaller set, she appeared confused. Rather than attempting to help her build on the initial response, I suggested that she begin by showing only the larger quantity. In doing so, I may have missed observing another kind of subtraction.

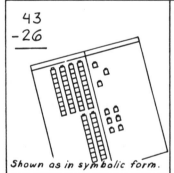

43
−26

Shown as in symbolic form.

NOW, WHAT ARE YOU GOING TO DO?

TAKE AWAY 26.

Roberta C.

Roberta regrouped a ten in the set of 26 cubes. Then she changed her mind and regrouped the set of 43 cubes.

I'M WONDERING WHY YOU PUT DOWN 26 CUBES.

I DON'T KNOW.

Shrugs.

SUPPOSE YOU PUT THESE 26 CUBES AWAY AND TAKE 26 AWAY FROM WHAT YOU'VE GOT ALREADY.

She successfully completed the regrouping and subtraction.

Could I have followed her lead and possibly facilitated her original attempts?

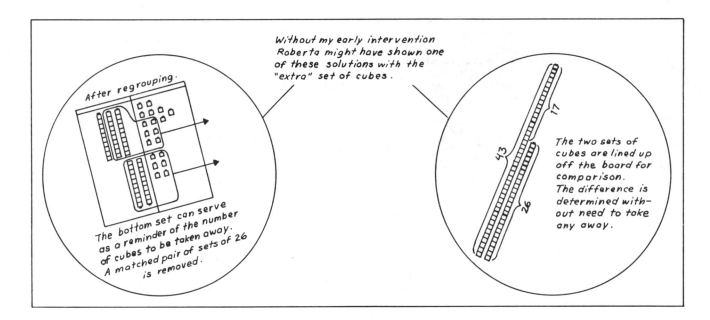

Without my early intervention Roberta might have shown one of these solutions with the "extra" set of cubes.

After regrouping.

The bottom set can serve as a reminder of the number of cubes to be taken away. A matched pair of sets of 26 is removed.

The two sets of cubes are lined up off the board for comparison. The difference is determined without need to take any away.

As noted in Chapter 8, the take-away approach to subtraction does not cover all possible contexts, although it is often presented in textbooks as doing so. The written algorithm often renders the different contexts indistinguishable.

Connection

Immediately following regrouping of materials and just prior to taking away, the children were asked to step back and reflect on their actions with objects and to connect them to the symbolism of the written algorithm. The accurate computers were asked either to explain the regrouping/renaming of the written algorithm in the presence of materials or to point to something on the board indicative of the regrouping/renaming process.

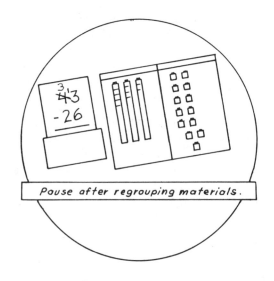

Pause after regrouping materials.

Accurate computer
NOW TELL ME HOW YOU GOT THE 3 AND THE 13 HERE *(pointing to written algorithm).*

IT'S COMPLICATED.

(No further response.)

Gail L.

Other accurate computers
YOU'VE JUST CHANGED THINGS AROUND ON THE BOARD AND YOU ALSO CHANGED SOMETHING AROUND ON PAPER. POINT TO SOMETHING ON THE BOARD THAT THIS 1 STANDS FOR.

Roberta C.

Paula L.

Incomplete Regroupers— Computation Phase
- YOU JUST MOVED SOME CUBES AROUND ON THE BOARD.
- WHAT DO YOU HAVE ON THE BOARD NOW?
- NOW CAN YOU SHOW THIS ON PAPER?
- (CAN YOU FIGURE OUT WHAT TO PUT HERE TO SHOW THE BORROWING?)

OH, I GET IT.

THERE'S A 4 HERE. I TOOK A TEN AND PUT IT HERE (13) AND NOW THERE'S ONLY 3 TENS LEFT.

THAT SHOULD BE 3 and 13. (*Referring to regrouped material.*)

Harold S.

Betty A.

He completes the written algorithm and explains it in place value terms (when asked HOW DID YOU GET 13?) During the demo phase he needed three prompts to complete it.

She completes the written algorithm. In the explanation that followed she referred to the top row as 33.

Harold and Betty, two children unable to complete the regrouping in the written algorithm, were asked to relate it to the regrouping action on materials and to complete the algorithm. Other inaccurate computers were not included in this phase of the task. Similarly, children producing inconsistent though accurate procedures in written and material formats were not asked a connection question. Here, the child could not be asked to make connections readily since the procedures were not parallel.[3]

Although the connection phase was not pursued with all children, it did provide some useful information. Some children are able to recognize that the position of the digits gives clues about the quantity of materials. They build on this understanding to complete the recording of the regrouping/renaming pro-

cess or to explain it. Others do not see the connection despite the parallel formats. It is interesting to note that although Harold and Betty (pp. 341, 347) were unable to finish the written algorithm, they did so with some understanding following their experience with materials. By contrast, Gail (pp. 342, 346) completed the written algorithm but was unable to explain it in terms of place value even in the presence of materials.

Paula's pointing to a ten-group as representative of the borrowed ten provides insight into my earlier difficulty in clarifying her understanding of regrouping/renaming during her explanation of the written algorithm (pp. 342, 346). Even following the regrouping of objects, she was only aware of transferring ten from one column to the next in borrowing. She was unaware that in the process of borrowing, 1 ten had been regrouped/renamed as 10 ones. This level of understanding was characteristic of most of the third graders capable of making place value references in explaining the algorithm, including Harold and Betty.

Comparison (Inaccurate Computers Only)

Manipulating the materials helped most of the inaccurate computers to find a remainder of 17. Then they had generated two different answers using different methods—17 with materials and 23 with the written algorithm. Did these children realize that both physical and written procedures should yield the same answer? If not, where did they place their trust, in physical or written procedures?

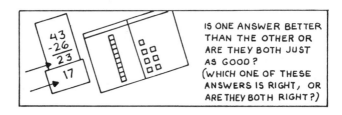

IS ONE ANSWER BETTER THAN THE OTHER OR ARE THEY BOTH JUST AS GOOD?
(WHICH ONE OF THESE ANSWERS IS RIGHT, OR ARE THEY BOTH RIGHT?)

---

3. For example, Hughie did the written algorithm showing regrouping/renaming and followed it up with a materials procedure requiring no regrouping.

Some children saw nothing contradictory in the two answers, indicating that both were right. For them, 17 was the answer when working with materials and 23 was the answer when working with paper and pencil, and there was no connection between the two.[4] Other children recognized a contradiction and placed their faith in only one of the answers. Some of these children believed in the result of the written procedure, while others were more confident in the reality of their physical experience. Nowhere is the gap between the world of things and written symbols more clearly demonstrated than in the preceding examples.

O'Brien and Casey (1982a) report that contradictory answers obtained mentally and on paper also may go unrecognized by children. For example, a child may

$$\begin{array}{r} 43 \\ -26 \\ \hline 23 \end{array}$$

obtain through the partially-learned school proce-

dure of the written algorithm, but obtain 43 − 26 = 17 through a self-learned mental procedure. Yet when challenged, the child may be satisfied with both answers. O'Brien and Casey point out that many children learn the "dance-of-the-digits" rules of the written algorithm as an isolated atom of knowledge, unrelated to any other knowledge of number. The preceding examples underscore the importance of helping children build bridges between actions on objects, existing ideas of number including mental computational strategies, and written algorithms.

Verification

The children were asked to check the results of their computational procedure to verify its accuracy.

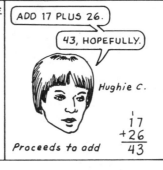

The preceding question proved too difficult for most of the accurate computers. This question was rephrased in a physical context and asked of most children. In the

presence of the completed written algorithm, the children were asked to identify the quantity resulting from recombining the separated objects.

> LET'S PUT BACK THE BLOCKS THAT YOU TOOK AWAY.
> HOW MANY BLOCKS DO YOU HAVE ON THE BOARD?
> HOW COME?
>
> (Counting was encouraged to check the prediction, but not before.)

---

4. Children who see the procedures as unrelated experience no disequilibrium. In other words, they see no problem. The children who realize the contradiction and show trust in their experience with materials recognize a problem and are ripe for experiences that help them build a bridge between the two procedures.

Ten children were able to predict and justify a response of forty-three cubes with reference to either the materials or to the completed algorithm. In some instances the children counted—an inference I made from observing both response delay and subvocalization.

BECAUSE I HAD 43 WHEN I STARTED. ...OH, LIKE 17 + 26 = 43.

Harold S.
He spontaneously referred to the written algorithm.

BECAUSE 26 + 17 IS 43.

Paula L.   Betty A.

While both girls referred to the written algorithm, Betty did so spontaneously, while Paula counted first before realizing a connection.

BECAUSE THAT'S HOW WE STARTED.

Chris B.   Joey R.
Roberta C.

However, they were unable to extend the idea to the written algorithm.

BECAUSE WHEN I TOOK THEM APART I REMEMBER I HAD 43.

Cal M.

(Silence).

Brenda S.

She was unable (or unwilling) to comment.

While three children dealt with the initial question directly, eight others needed the physical context to make sense of the question. At this point three of these children were able to make spontaneous connections to the written algorithm. Yet others were unable to do so. In my attempt to help another child, a communication problem resulted in an unexpected interpretation of my question.[5]

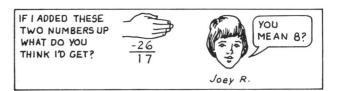

IF I ADDED THESE TWO NUMBERS UP WHAT DO YOU THINK I'D GET?

$$\begin{array}{r} -26 \\ \hline 17 \end{array}$$

YOU MEAN 8?

Joey R.

At the same time, nine other children gave a variety of unexpected predictions for the recombined quantity, while three were unable to respond.

Other predictions:	76	50	46	44	36	37	26	23	16

There is usually some hidden logic in children's responses. Yet this wide range of responses suggests that they may have lost sight of the question near the end of a lengthy task. After being encouraged to count the recombined cubes, some of the children expressed surprise at getting 43, while others did not. When asked why there were forty-three cubes on the board, two children gave logical responses. In the context of this task some children were unaware of the relation between addition and subtraction.

This extended interview task has revealed a range of levels at which third-grade children understand the written subtraction algorithm. Furthermore, it has provided evidence that the interaction of some children's existing ideas and their actions on objects result in a reconstruction of their understanding of the algorithm. Out of their initial confusion with the written algorithm, Harold and Betty reconstructed the regrouping/renaming of the algorithm within a place value framework while interacting with grouped materials. Although not achieving the most advanced level of place value understanding, they showed amazing progress. Out of the relative comfort of a completed written algorithm and an explanation with some place value references, Roberta faced the conflict of making sense of the materials. In doing so, she appeared to be in the process of elaborating her place value perspective of regrouping/renaming at the highest level of relations. The next section will allow us to follow the stability of the children's developing ideas as they will be asked to regroup/rename a hundred as 10 tens. If the children become disoriented on attempting the more challenging task, the earlier success may be attributed to one of

---

5. In this book a distinction has been made between number, numeral, and digit for the adult reader. However, no such distinction was made in the interviews. Perhaps the use of number and digit might suffice.

the vacillations of disequilibrium—understanding at one moment but not the next. However, if they are able to fit the new task into their newly-reorganized ideas of regrouping/renaming, we may have evidence of an approaching stable equilibrium. Sometimes the children who are the most confused make the most progress.

SUBTRACTION FROM A THREE-DIGIT NUMBER

Following an intervening task on the addition algorithm involving regrouping/renaming of 10 tens as a hundred, the children were presented with a subtraction task requiring regrouping/renaming of a hundred as 10 tens. The materials consisted of both individual tiles and arrays of tiles.

COMPUTATION	EXPLANATION	DEMONSTRATION	CONNECTION
1. READ THE QUESTION. 2. NOW DO IT. 243 - 61	TELL ME HOW YOU WORKED IT OUT. 243 - 61 182	NOW LET'S SEE YOU DO IT WITH MATERIALS. 243 - 61 *Answer is covered*	*(Pause immediately after regrouping.)* HOW DID YOU GET THE 1 AND THE 14 HERE? *(Pointing to the written form.)* 243 - 61

Based on what you've learned from the children about their understanding of addition and subtraction algorithms, anticipate the problem areas in this task.

Predict how Harold, (pp. 341, 347) Betty, (pp. 341, 347) and Roberta (pp. 342, 345, 346) will fare on this more challenging subtraction task.

Computation

Out of 24 children, 13 computed accurately using the standard algorithm. Again, Chris showed a successful left-right procedure.

243
- 61
182

200, TOOK 60 OF IT. HAD 140 LEFT. ADDED 40 → 180.
Chris B.

Two children read the problem as addition. After the operation sign was pointed out, one of them subtracted while the other persisted with addition and required an additional prompt. This time, 7 children ignored or circumvented the need to consider regrouping/renaming by reverse subtraction.

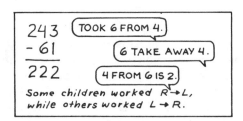

243
- 61
222
TOOK 6 FROM 4.
6 TAKE AWAY 4.
4 FROM 6 IS 2.
Some children worked R→L, while others worked L→R.

At the same time that more children were aware of the need to regroup/rename than on the preceding task, they weren't always able to do so. Betty was one of these children.

BORROW A HUNDRED. *She knows that 100 was borrowed but is unable to complete the regrouping/renaming.*

Betty A.

Teddy carries 3 from the ones column to make a 7. He then carries 2 from the hundreds column to the ones column. Each remaining column of digits can now be subtracted.

Teddy T.

CARRY THE 2 AND TAKE 6 FROM 6.

Vern L.

He gives up at this point and requests the tiles.

243
− 61
_____
2

Joey R.

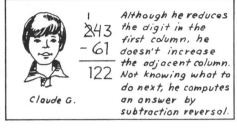
Although he reduces the digit in the first column, he doesn't increase the adjacent column. Not knowing what to do next, he computes an answer by subtraction reversal.

Claude G.

Explanation

Again, in response to "How did you get your answer?" most children gave abbreviated descriptions of the procedure. They either referred to the face value of the digits alone or combined mixed references to both face value and place value. Even this inconsistent place value reference indicated some improvement over preceding tasks on addition and subtraction algorithms. However, only one child spontaneously gave a detailed explanation of the standard algorithm.

The other children's responses were probed to encourage them to stand back and reflect on the regrouping/renaming process within a place value framework and to clarify their perspective. In particular, the probes attempted to find out if they could see the 1 in fourteen as representing 10 tens based on its position. The probes diagrammed were made following an uninterrupted opportunity to give a detailed explanation.

A single probe suggested that Gail was still viewing the standard subtraction algorithm in terms of manipulating digits. Further probes might have shown otherwise.

I KNEW I COULDN'T TAKE 6 FROM 4, I CROSSED OUT 200, MADE 100, THEN I ADDED 100 TO THE TENS...TO MAKE IT 14 TENS.

Frank M.

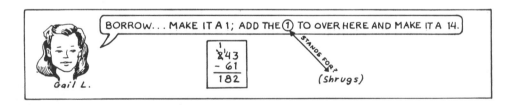
BORROW... MAKE IT A 1; ADD THE ① TO OVER HERE AND MAKE IT A 14.

STANDS FOR?

(Shrugs)

Gail L.

Bert and Cindy are aware that digits have values but think that a ten is borrowed.

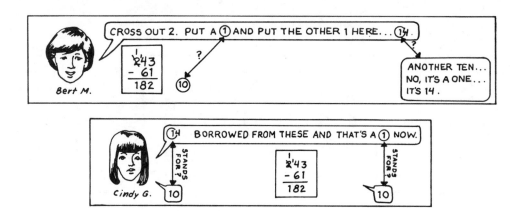

From the responses of Hughie and Paula, it is possible to infer an awareness of borrowing a hundred from one column and transferring it to another. The probes don't reveal any further awareness of regrouping/renaming a hundred as 10 tens. Yet artful questioning might have coaxed out evidence of this awareness.

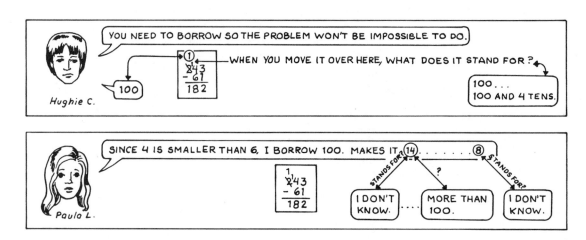

Although not always explicit, the responses of Harold, Steve, and Gordie provide some evidence that they understood the regrouping of a hundred as 10 tens.

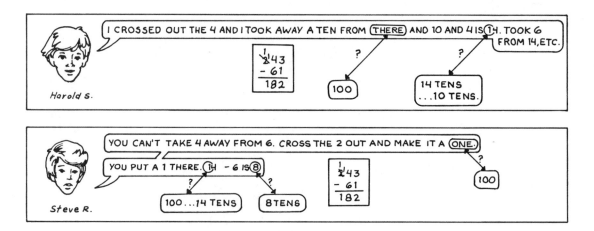

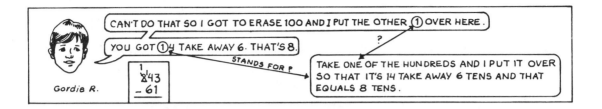

However, Charlie's response cautions us against making hasty conclusions.

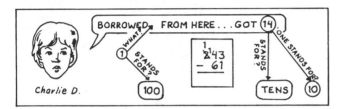

In answering the questions about 14, Charlie shifts perspective from the place value system of the three columns to the place value within the numeral 14 itself.[6] To deal with the value of "1" requires that he consider two interacting place value systems. Asking the right question at the right time may have helped him both become aware of the contradiction in his answers and resolve it in terms of regrouping/renaming. Further questioning may have begun with, "How did you get 14 tens?" and eventually returned to the value of the "1". Similarly, my inability to ask the right question at the right time obscured a view of Roberta's understanding of regrouping/renaming. Instead, my leading questions may have imposed my thinking on her.

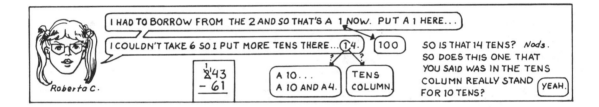

By not explicitly requesting an explanation at the start, I had to resort to questions focusing on the values of the regrouped numbers, and to infer understanding of the regrouping/renaming process from the responses. Clearly, asking for an explanation for a first grader may have encouraged the children with place value understanding to generate a more integrated reply having specific references to the regrouping/renaming process. Despite limitations in the questioning, four levels of understanding the subtraction algorithm have been identified.

1. Borrowing 1 from the adjacent column to the left regardless of position.

2. Borrowing 10 from the adjacent column to the left regardless of its position, even if the adjacent column is in the third position from the right.

3. Borrowing 100 from the third place from the right and moving it one place to the right without renaming the hundred.

4. Borrowing 100 from the appropriate place and moving it one place to the right while renaming it as 10 tens.

The highest level of understanding extends beyond borrowing an amount and returning it to the next column to regrouping and renaming it. Most of the third graders interviewed had not achieved this level of understanding.

---

6. A discussion of multiple meanings of "one" and systems within systems is found in Chapter 12, p. 331.

Demonstration

In a demonstration of the standard subtraction algorithm using individual and arrayed (fixed) tiles, the regrouping involves indirect decomposition by means of a trade. In this task an array of one hundred tiles is exchanged for ten arrays of ten tiles each.

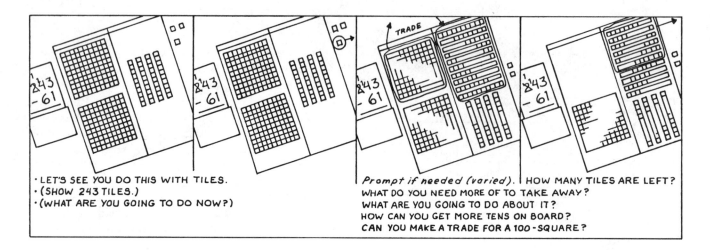

- LET'S SEE YOU DO THIS WITH TILES.
- (SHOW 243 TILES.)
- (WHAT ARE YOU GOING TO DO NOW?)

Prompt if needed (varied). HOW MANY TILES ARE LEFT?
WHAT DO YOU NEED MORE OF TO TAKE AWAY?
WHAT ARE YOU GOING TO DO ABOUT IT?
HOW CAN YOU GET MORE TENS ON BOARD?
CAN YOU MAKE A TRADE FOR A 100-SQUARE?

Eight of the children completed the demonstration, spontaneously regrouping (trading) a hundred-square for 10 ten-sticks, while others required prompts. Six of the children circumvented the need for a trade. Every child requiring prompts was successful in regrouping and finding the remainder. Examples of both correct and incorrect interpretations of the task are illustrated below.

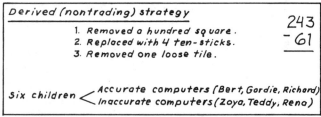

Derived (nontrading) strategy   243
1. Removed a hundred square.   − 61
2. Replaced with 4 ten-sticks.   _____
3. Removed one loose tile.

Six children ⟨ Accurate computers (Bert, Gordie, Richard)
                Inaccurate computers (Zoya, Teddy, Reno)

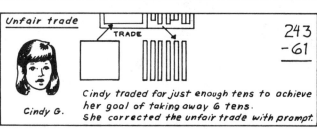

Unfair trade   243
TRADE   − 61
                                                      _____
Cindy traded for just enough tens to achieve
her goal of taking away 6 tens.
Cindy G.   She corrected the unfair trade with prompt.

Two-step subtraction   243
She removed as much as possible   − 61
before trading, then subtracted   _____
again as needed.
1. Removed 1 loose tile, then 4 ten sticks.
2. Traded a hundred square for 10 ten-sticks.
3. Removed two other ten-sticks.
Brenda S.

Redundant subtraction   243
She attempted a derived strategy and   − 61
lost track of what she was doing,   _____
thus subtracting a second time.
1. Removed a hundred square.
2. Replaced 4 ten-sticks.
3. Took away 6 ten-sticks and 1 loose tile.
Gail L.

It is interesting to note that the derived strategy requiring no regrouping was generated by both successful and unsuccessful computers with the written algorithm. For the accurate computers, this nonstandard procedure represented a second avenue for approaching the same problem. For the inaccurate computers, it may have been the only avenue to the problem, providing access at a time when the standard written algorithm presented a roadblock.

Connection

Immediately following the regrouping of materials, just prior to taking any away, some children were asked to reflect on their actions on objects and to connect them to the symbolism of the written algorithm. The varied questions used to provoke this connection are described for Zita, Joey, and Betty—all inaccurate computers in the first phase of this task. Their initial approaches to the task are included, allowing you to follow any progress being made.

While two of the children made no progress with the written algorithm, Betty was able to complete it as a recording of her actions on the tiles. Furthermore, she made reference to the regrouping/renaming process as "a hundred put back in tens" while referring to the written algorithm.

Other children—the accurate computers—were not asked a connecting question until demonstration of both regrouping and subtraction had been completed. Here, the question focused on the meaning of 14—the result of regrouping/renaming. Neither Gail nor Charlie was able to connect the symbols of the written algorithm to the experience with materials.

The evidence for resolving the contradiction in Charlie's original explanation was available in the process of regrouping materials, yet he ignored it. It appeared that he was unaware of this contradiction, since he didn't use the experience with materials to confront his own ideas. Stopping the action immediately following regrouping and encouraging reflection on that action (as done for inaccurate computers) may have aided both children. Furthermore, highlighting the contradiction of ideas may have provoked Charlie to resolve it.

While both Bert and Cindy showed clear, though not dramatic, progress, the extent of the progress made by Paula and Roberta was still unresolved following the connection question(s).

Both Roberta and Paula needed one more question to clarify their level of understanding, like "What is this '1' worth?" or "How did you get 14 tens?" It is clear to me now that my failure to tie up such loose ends in clarifying a child's understanding was due largely to my own level of understanding of the subtraction with regrouping process. My knowledge of the process evolved in working with the children, analyzing the videotaped interviews, and describing my experiences in writing. In the final stages of writing it has advanced further in studying the work of Resnick.[7] Despite these limitations in my work, I describe these experiences in the hope that your initial interaction with children will take place at a higher level of understanding.

During the explanation phase of this task, Harold appeared quite comfortable working at the level of written notation following a restructuring of the written algorithm in the preceding task with the aid of materials. He was able to transfer his understanding of the place value relations of regrouping/renaming to subtraction from larger numbers without the aid of materials. Meanwhile Betty became disoriented again when presented with the more challenging written algorithm task. Yet she demonstrated again that in the presence of materials she was able to reconstruct the algorithm meaningfully as a record of her actions. Both Harold and Betty showed remarkable progress from their earlier attempts at subtracting from a two-digit numeral in written notation. Yet other children, like Gail, were unable to make any connections between actions on objects and the standard subtraction algorithm. Again, we are reminded that children make connections in their own time.

Related Research

The 1978 National Assessment of Educational Progress (Carpenter et al., 1980) showed that only 66 percent of a large sample of the nation's nine-year-olds (fourth graders) could subtract from a two-digit number requiring regrouping. Also, only 50 percent of these children could subtract correctly from a three-digit number with regrouping. When the test items were presented horizontally rather than vertically, the success rate dropped by 8 percent. An earlier assess-

---

7. Resnick (1982); Omanson, Peled, and Resnick (1982).

ment in 1973 found that only 43 percent of the nine-year-olds could identify the two numbers that would be needed to check the result of a subtraction:

$$315$$
$$-179.$$

Since learning two- and three-digit subtraction are second- and third-grade objectives across the nation, the combined results of our interviews and the national assessment suggest that a reexamination of both teaching methods and grade placement is appropriate.[8]

REGROUPING, CONSERVING, AND RENAMING IN SUBTRACTION

We will now examine in further detail the regrouping of objects and the corresponding renaming of numbers in subtraction, together with the implied conservation of those numbers. Although presented here following the subtraction algorithm, in actuality these tasks were given to the children several weeks earlier.[9]

You are invited to anticipate children's responses to these tasks and to identify potential areas of difficulty.

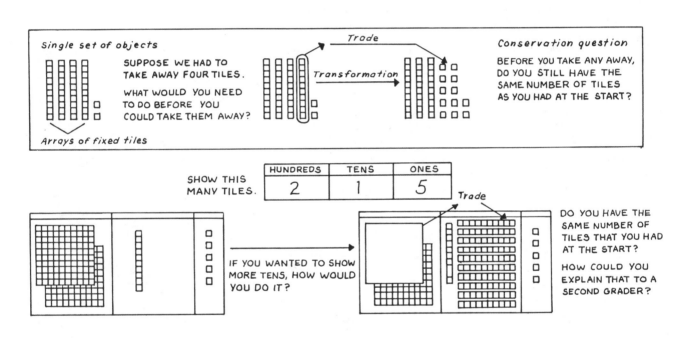

---

8. Bell and Burns (1981) report that only four of twenty-four beginning third graders could calculate 65 – 37 correctly with paper and pencil.

9. These tasks were part of the second interview. The tasks on standard addition and subtraction algorithms were given in the third, and final, interview of the series.

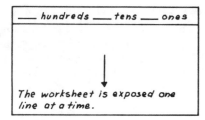

- WE'VE BEEN LOOKING AT DIFFERENT WAYS TO SHOW NUMBERS AND WE'VE LOOKED AT 215 AS ONE OF THOSE NUMBERS.

- WRITE 215 USING THESE BLANKS.

- NOW SHOW HOW YOU WOULD RENAME 215 USING NUMBERS IN THESE BLANKS.

Subtraction from Two-Digit Numbers

In the physical simulation of $\frac{42}{-\ 4}$, the children were asked to pause after regrouping to consider whether the number of objects had changed.

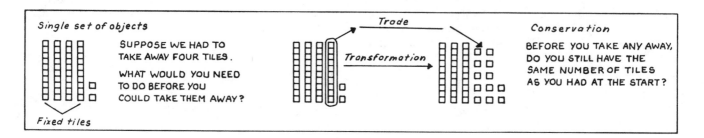

The task also provides a second look at the regrouping process of trading.[10]

Regrouping (Trading). The use of pregrouped objects glued on cardboard or sticks doesn't permit direct decomposition of the groups. Rather, a trade is necessary, in which materials are removed and replaced by others different in arrangement but equivalent in quantity. Twenty children traded a ten-stick for ten loose tiles. Nine children needed a prompt beyond the initial question. Prior to becoming familiar with the materials, seven children talked about breaking up the tiles in the ten-stick. One child traded a ten-stick for just enough tiles to achieve the goal of removing four. Six children traded a ten-stick for eight individual tiles, resulting in ten individual tiles on the board. Although making an unfair trade, they traded for "ten".

---

10. Again, the task was prompted by observations made of third graders by a classroom teacher—my wife, Shirley (Labinowicz, 1977).

From Tina's initial perspective, there was no need to regroup.

A child who didn't see the need for regrouping.

Tina J.: YOU COULD TAKE THESE (four) AWAY AND THERE WOULD BE TWO RIGHT HERE.

THERE'S FOUR RIGHT HERE.

With prompts, she counted 42 tiles.

These examples show the complexities that regrouping for subtraction present to a child. At the same time, it is interesting to note that the frequency of unfair trades was reduced several weeks later during interviews on the subtraction algorithm.

Conservation. Twenty-two children conserved during trading and justified their conservation responses. They gave logical reasons for the preservation of the initial quantity despite the removal and replacement of different arrangements of tiles.

SAME NUMBER?

Betty A.: YES, BECAUSE I HAVE 10 (ones) HERE AND TWO THERE. IF I PUT THESE BACK TOGETHER LIKE THAT (ten-stick) IT WOULD BE 42.

She mentally reverses the regrouping.
Two children reversed the process through physical demonstration.

Paula L.: ...JUST TOOK AWAY A TEN (stick) AND ADDED 10 (ones) TO THE OTHER SIDE.

Chris B.: ALL YOU DID WAS TRADE...

Steven R.: SAME AMOUNT, JUST PUT TOGETHER AND GLUED TO A STICK- JUST BUNCHED TOGETHER.

These children focused on equivalent quantities in the two cancelling steps- a form of compensation.

Seven children gave nonconserving responses initially. Being asked to justify their responses caused three of these children to rethink the task and offer adequate explanations for it. Obie vacillated between responses

before justifying the conservation of quantity. In his disequilibrium, he vacillated between a focus on the removal of the ten-stick and the compensating effect of removal and replacement of objects in the trade.

After a spontaneous trade

SAME NUMBER OF TILES?

Obie B.: NO...LESS.

HOW DO YOU KNOW?

I TOOK TEN AWAY.

AND WHAT DID YOU PUT IN ITS PLACE?... HOW MANY ONES?

TEN ONES.

WHAT DOES THAT DO TO THE NUMBER?

IT MAKES IT THE SAME.

Vacillating...

IT'S LESS.

Focusing on the ten-stick removal.

IT'S THE SAME STILL. ...YOU TOOK AWAY A TEN BUT THERE'S THE TEN (ones) OVER HERE.

These three children may account for a noticeable reduction in unfair trades several weeks later.

Related Research. Both the difficulty of conservation of number during regrouping/renaming and the gradual construction of the concept for larger quantities are supported by anecdotes reported over several weeks during Behr's extended teaching experiment with second graders (1976).

> Cubby is not always sure that the number of sticks in all remains the same. . . . Cubby finally understands that the total number of sticks remains the same after trading.

> Susan and David no longer (re)count the blocks after a trade. They conserve and remember how many there were before the trade. Celia counts the "new" blocks; if she makes an error in trading or counting she trusts the new number and assumes that the amount has changed.

> Kids readily accepted that 12 ones or 1 ten and 2 ones were the same value but on amounts such as 35, 87, or 77, where it ended up with 7 tens and 17 ones, children would have to count again to say 87. That is, 8 tens and 7 ones: kids say this is eighty-seven; trade 1 ten for 10 ones, 7 tens and 17 ones, all agree that it's the same value but must count to find 87. Yet this is not the case for 12, where the ones are not as numerous.

> They are (now) convinced that the total number does not change with trading. Have difficulty deciding whether to exchange 1 ten for 10 ones or vice versa.

> Phil made mistakes because he counted the ten he traded twice. . . . He knows how to trade 1 ten for 10 ones but sometimes forgets to remove his ten. Others understand when and how to trade and had no difficulties.

> Clair began trading a ten for the number of ones she needed to subtract instead of 10 ones.

> Kids were very excited to (re-)discover that trading did not change the value.

> Calvin and Karl sometimes get the wrong answer due to not returning the ten after trading.

> Most common errors are: putting 10 ones instead of the correct number after trading; trading when it isn't necessary.

> Cubbie and Carrie had difficulty understanding that the same number could be expressed in different ways.

Hendrikson (1980) reports that beginners also may be unwilling to decompose an existing group as opposed to forming a group. This may be due to either a lack of reversibility or to a resistance to showing a nonstandard display of that number.

Subtraction from a Three-Digit Number

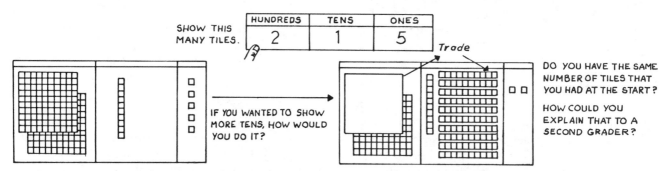

Regrouping (Trading). Only eight of the children traded a hundred-square for 10 ten-sticks without needing additional prompts, such as the following.

> Could you trade for something? What are you going to trade for?
>
> Suppose I took off this one hundred-square, what would you trade for it?

Eleven children initially traded a one hundred-square for 9 ten-sticks, leaving a total of 10 ten-sticks on the board. These unfair trades were adjusted once they were brought to the children's attention.

The location of the trade was also a problem for eight children as they initially traded a ten-stick for 10 ones. This error was corrected following a prompt. Following the trading, twenty children justified the preservation of the original number of tiles. Again, in the context of the standard algorithm several weeks later the frequency of the trading errors was reduced. No experiences with similar materials in the classroom had intervened.[11]

Renaming: Numbers have different names. Just as nonstandard arrangement of materials (with more than nine in a column) during regrouping shows the same number, nonstandard written notations of these arrangements represent the same number. The children were asked to write different nonstandard names for the same number.

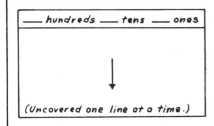

_____ hundreds _____ tens _____ ones

(Uncovered one line at a time.)

· WE'VE BEEN LOOKING AT DIFFERENT WAYS TO SHOW NUMBERS AND WE'VE LOOKED AT 215 AS ONE OF THOSE NUMBERS.

· WRITE 215 USING THESE BLANKS.

· NOW SHOW HOW YOU WOULD RENAME 215 WRITING NUMBERS IN THESE BLANKS.
  Prompts: · SUPPOSE WE HAD NO TENS, WHAT NUMBERS COULD YOU WRITE HERE AND STILL KEEP IT 215?
  · SUPPOSE WE NEEDED MORE TENS...
  · SUPPOSE WE HAD ELEVEN TENS...
  · SUPPOSE WE NEEDED MORE ONES...

11. These tasks were given in the second interview, several weeks prior to the presentation of tasks on standard addition and subtraction algorithms in the third interview. Another explanation for the apparent reduction in trading errors may be due to the lack of clarity of the above task. Although given in the context of conservation by being preceded by 3 tasks having such a focus, the key question for the above task is unclear when considered in isolation.

Six children were able to write two nonstandard notations for the number. These notations were ones the children would need in subtraction of tens or ones.

Richard was very comfortable with this task, generating several nonstandard names.

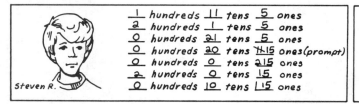

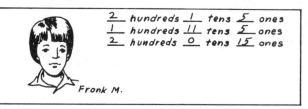

Six other children given the task were quite uncomfortable with renaming numbers in nonstandard ways, succeeding only in writing the standard name in this format. They juggled digits from column to column without regard for place value. Some examples follow.

The latter group needed to deal with the task as a recording of action on objects such as tile arrays. To understand the standard subtraction algorithm, children must be aware that just as the starting quantity of objects remains unchanged during regrouping/renaming, so does the top number that represents them. The notion of conservation is extended to the abstraction of written notation in terms of both standard and nonstandard names for the same number.

SUBTRACTION WITH BORROWING ACROSS ZERO

Our study of the standard subtraction algorithm would not be complete without a look at another problematic area for children—borrowing across zero.

You are invited to place yourself in the children's shoes.
    Complete the following problems using the standard algorithm.
    Explain your procedure in detail to another person.

$$\begin{array}{r} 702 \\ -368 \\ \hline \end{array} \qquad \begin{array}{r} 7002 \\ -\ \ 25 \\ \hline \end{array}$$

Now as a teacher, anticipate the different responses children might give.

Since our third graders were not given this supreme challenge, their limits of understanding were not fully explored. However, the research of others, particularly that of Resnick (1982), is available, both to fill the gap and to sharpen our focus on the earlier interviews.

In borrowing across zero, rather than an amount being borrowed and transferred to the adjacent column, this amount is distributed among two or more columns. Furthermore, this distribution involves a series of successive one-to-ten decompositions. Resnick (1982b) and Omanson, Peled, and Resnick (1982) illustrate these components of understanding with impressive explanations given by two children. The following drawings are adaptations of their transcripts.

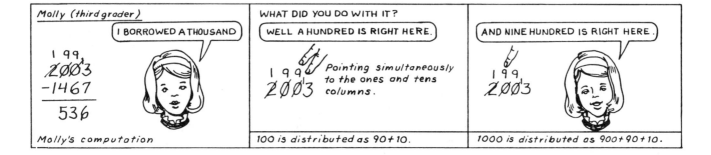

Using a compensation perspective, Resnick's Molly sees the borrowed thousand as redistributed over the hundreds, tens, and ones columns, leaving the top number represented as unchanged. Her method of recording the borrowing is consistent with her explanation. Jan shows us another perspective of the process.

Jan's detailed explanation filled in the gaps of the regrouping/renaming process in a way that suggested that her own record of the process would include all of the intermediate steps she described. Her ability to elaborate on the compensation of borrowing in terms of a series of successive one-to-ten decompositions indicates the highest level of understanding. Although both Molly and Jan constructed impressive rationales, there is a fine distinction in level of sophistication. Mol-

12. In this study, Jan was identified only as one of a group of fourth, fifth, and sixth graders undergoing remediation for systematic errors in subtraction. The remediation involved structured experiences with Dienes blocks.

ly's reasoning took a total-value perspective of the digits based on their position, while Jan's reasoning was from a place-value perspective based on not only the position of the digits, but also on the ten-to-one relations between positions.

Despite the high level of understanding shown by the two children, subtraction from "middle zeros" in the top number is notorious for its difficulty. In interviewing third- and fourth-grade children in three schools from three school districts, Davis and McKnight (1980) found no child was able to calculate the correct answer for:[13]

$$\begin{array}{r} 7003 \\ -\ 25 \\ \hline \end{array}.$$

Based on a large national sample of nine-year-olds (fourth graders), the National Assessment of Educational Progress (Carpenter et al., 1981) reports that only 28 percent responded correctly to the problem "Subtract 237 from 504." In another study, Resnick (1982) probed children's understanding of borrowing across zero. She asked third graders to identify the value of the "borrow digits" in problems such as

$$\begin{array}{r} 300 \\ -139 \\ \hline \end{array}.$$

She also asked them to pick out blocks that represented each of the "ones". Virtually every child interviewed named the "borrow digits" rather than assigning a total or place value, and selected the smallest block. In this type of subtraction, both procedure and meaning are problematic areas.

Many of the above children were able to subtract successfully with the standard algorithm when borrowing was required from nonzero digits, but they did so with little or incomplete understanding. If children learn the algorithm as a set of arbitrary rules, they have no basis for generating specific rules to extend the procedure to numbers with "middle zeros." If their understanding is incomplete at a time when a high level of integration of place value ideas is essential, again they have no solid conceptual basis for adapting to the situation. As a result, the children do the best they can with what they have, inventing error-laden procedures for dealing with these challenging subtraction problems. These "buggy algorithms" (Brown and Burton, 1978) are not merely random occurrences, but are systematically applied over several days or weeks. Five common "buggy algorithms" from Brown and Burton (in Resnick, 1982a) are described below. The typical responses given by third and fourth-graders interviewed by Davis and McKnight (1980) are given as well.

Can you go beyond these behavioral descriptions and infer specific gaps in the children's understandings for each "buggy algorithm"?

1. Borrow-From-Zero. When borrowing from a column whose top digit is 0, the student writes 9 but does not continue borrowing from the column to the left of the 0.

$$\begin{array}{r} 6\,\overset{9}{\cancel{0}}\,2 \\ -4\,3\,7 \\ \hline 2\,6\,5 \end{array} \qquad\qquad \begin{array}{r} 8\,\overset{9}{\cancel{0}}\,2 \\ -3\,9\,6 \\ \hline 5\,0\,6 \end{array}$$

2. Borrow-Across-Zero. When the student needs to borrow from a column whose top digit is 0, he skips that column and borrows from the next one.

$$\begin{array}{r} \overset{5}{\cancel{6}}\,0\,2 \\ -3\,2\,7 \\ \hline 2\,2\,5 \end{array} \qquad\qquad \begin{array}{r} \overset{7}{\cancel{8}}\,0\,4 \\ -4\,5\,6 \\ \hline 3\,0\,8 \end{array}$$

3. Stop-Borrow-At-Zero. The student fails to decrement 0, although he adds 10 correctly to the top digit of the active column.

$$\begin{array}{r} 7\,0\,3 \\ -6\,7\,8 \\ \hline 1\,7\,5 \end{array} \qquad\qquad \begin{array}{r} 6\,0\,4 \\ -3\,8\,7 \\ \hline 3\,0\,7 \end{array}$$

---

13. Although the children studied by Davis and McKnight (1980) had learned to represent numbers with Dienes blocks, they had not integrated this knowledge into their understanding of the subtraction algorithm.

4. Don't-Decrement-Zero. When borrowing from a column in which the top digit is 0, the student rewrites the 0 as 10 but does not change the 10 to 9 when incrementing the active column.

```
  6
 7 1012          1 2 0 5
- 368           -     9
 ----            -----
  344            1106
```

5. Zero-Instead-Of-Borrow. The student writes 0 as the answer in any column in which the bottom digit is larger than the top.

```
  326            542
- 117          - 389
 ----           ----
  210            200
```

6. Borrow-From-Bottom-Instead-Of-Zero. If the top digit in the column being borrowed from is 0, the student borrows from the bottom digit instead.

```
  7 0 12         5 0 8
- 3 5 8         - 4 8 9
   / 68            / 9
  ----            ----
  454             109
```
- -
```
   6               8 5
 1 0 0 12        1 0 0 12
- 25           - 25
 ----           -----
    7            5087
```

The answers typically given by the children in the Davis and McKnight study suggest that, not having an understanding of the algorithm, the children also have no basis for judging the reasonableness of what they are doing.[14]

Resnick (1982b) has identified major components of understanding the standard subtraction algorithm. Although these major components were discussed in general terms in the context of earlier interviews, an explicit statement at this point may serve to sharpen your focus as it did mine.

1. Knowing the goals and subgoals giving structure to the algorithm—for example, the primary goal is to take away the whole bottom number from the whole top number.

2. Knowing the correct values of all the quantities involved in borrowing, including the value of each digit and the value of each increment and decrement.

3. Recognizing the compensation of borrowing, saying that the amount added to one column must equal the amount taken from another column.

4. Explaining the compensation of borrowing as a one-to-ten decomposition (or trade).

5. Recognizing that borrowing preserves the value of the whole top number.

Having such understanding would provide a child with an inner critic that would prohibit the "buggy algorithms" described above.

In contrast to the bleak picture painted above, there is also evidence that given appropriate instruction, children can not only eliminate buggy algorithms and explain their limitation, but also explain the correct algorithm. In an extended pilot study involving what Resnick (1982) calls "mapping instruction" (similar to our sequence for the addition algorithm in Chapter 12), a small group of second and third-grade children linked their understanding of action on objects to the written notation of the algorithm. Molly's explanation of borrowing a thousand (p. 363) is taken from that study. A second study of Omanson, Peled, and Resnick (1982) applied mapping instruction to

---

14. The Davis and McKnight study explored other avenues to the problem and to checking the reasonableness of the answer in followup interviews. During this time, a few students solved this problem correctly.

fourth, fifth, and sixth graders with identified buggy algorithms. Based on an orientation session for Dienes blocks, a pretest interview, and a single forty-minute intensive session of mapping instruction, the written performance of some of the children did not improve, despite evidence of improvement in specific areas of understanding.[15] The researchers conclude that successful borrowing across zero requires an integrated, rather than piecemeal, understanding. Another phenomenon observed was that of unstable understanding—that is, understanding and explaining the source of the 9 in borrowing across zero at one moment, and reverting back to a buggy algorithm at the next.

Even the best of teaching methods are not successful if they do not respect the time requirements of learning. Adequate time is needed for children to overcome the resistance of previously learned buggy algorithms and to integrate new understandings. In the conclusion of their report, Omanson, Peled, and Resnick (1982) acknowledge that children need more time. Another observation was that the children who verbal-ized the values of digits most often during mapping appeared to make the most progress. Thus, the researchers recommend maximizing the verbalization of values during instruction. One success story resulting from this short-term study has already been demonstrated (p. 363). Not only did Jan eliminate her systematically applied buggy algorithm, but replaced it with the standard algorithm at the highest level of understanding. An interesting sidelight in Jan's success is that despite exclusive exposure to Dienes blocks during the remediation study, she explained regrouping/-renaming in terms of successive one-to-ten decompositions rather than trades. It was as if she had been working with discontinuous materials rather than the continuous Dienes blocks. Once again we are reminded of two characteristics of children's learning (Wheeler, 1977):

> Place value ideas evolve slowly in children's minds. They impose their own structures on experience and make their own connections in their own time. (p. 101)

AN ALTERNATIVE APPROACH TO TEACHING THE STANDARD SUBTRACTION ALGORITHM

The preceding interviews with third graders on subtraction with regrouping/renaming have provided evidence of their difficulties in both carrying out the standard procedure and in understanding it. Furthermore, many children showed a lack of both association of actions on objects with the paper-and-pencil algorithm, and of integration of place value ideas needed to understand the algorithm. This information suggests strongly that if we choose to teach the standard paper-and-pencil algorithm, we should do so gradually, beginning with discussion of action on objects. The approach described here will provide for a gradual transition from concrete activities and verbalization of values to written notation of the algorithm.

This approach is adapted from the work of Fuys and Tischler (1979), Irons and Jones (1979), Labinowicz (1977), Swart (1980), Resnick (1982), and Thompson and Van de Walle (1981), and parallels the one presented for addition in the last chapter.

The interviews also provided evidence that at least some of the children were capable of developing their own alternative computational procedures. Further examples of child-constructed, non-standard methods are described in the next chapter. Meanwhile, this chapter will close with a discussion of a semidirect approach to teaching the standard subtraction algorithm in the context of meaning.

Level One: Action on Objects

As for the addition algorithm, it is possible to identify a set of steps with materials that correspond with those of the written algorithm. Yet unlike addition, the dynamics of subtraction are more difficult to represent in a two-dimensional, static manner. Insisting on a one-to-one correspondence in concrete and written notation can be somewhat restricting for sub-traction. In activities to teach subtraction, it seems more natural to take objects off the transition board entirely, rather than to move them down to the next row. Unless there is some special place designated for locating these objects off the board, like a take-away box, these objects can no longer be used for verification when taken off the board. By the same token, it is more natural to leave the remainder where it is. However, if the transition board is large enough, the remaining pieces can be pushed down without disturbing those in the middle row. Doing so reserves a particular location of the board for the starting quantity, thus mapping the written algorithm closely. A more flexible alternative is to provide a board with only

---

15. Swart (1980b) reports that 20 percent of a first-grade class were able to complete three-digit subtraction in written notation following extensive work with Dienes blocks. However, no materials were available during the test.

place value columns and do what comes naturally.

Small group lessons in which each child and the teacher have their own materials and transition board initiate the sequence. The teacher presents a subtraction task in the context of a number story problem to be worked out as a group. Although the children are invited to show the starting quantity, regroup as needed, take away, and find the remainder, they do so within certain constraints. Initially, the process is stopped at the end of each step to identify these constraints of the standard procedure. In addition to right-left direction of subtraction, the teacher may identify particular cells on the board for location of the quantities at each stage of the developing process. To avoid blind imitation, the teacher invites the children to think about the values and locations of the materials, as well as their regrouping actions on them.

Introductory Lesson. Here is an example of what might happen during an introductory activity/discussion in response to a number story problem.

> We have thirty-three children in our class. If sixteen of you will be practicing for the concert this afternoon, how many will be left in the room for art?

ORAL LANGUAGE (DIALOGUE) CONCRETE MODELING

Exhibiting the whole/representing the problem

Put out the number of tiles (or cubes) that stands for the whole class.

How many tiles stand for the part of the class going to concert practice?   (16) Are they on the board now?   (part of the 33 tiles)

Just so we remember how many to take away, I'll draw 16 tiles in the middle row.   (Invite input.)

Subtracting ones/regrouping tens as ones (if needed)

We're going to take them away column by column. How many ones do we take away?   (Where do we start?)

Do you have enough loose ones to take away?   (No)
Where can we get more ones?   (trade/break up a ten)

How many ones do we have now?   (13)
(Do we still have 33 tiles? How come?)
Now take away 6 ones and push them to the next row and cover the dots.

How many loose tiles are left?   (7) Push these down to the answer space.

Subtracting tens/regrouping a hundred as tens (if needed)

What do we take away next?   (a ten)
Do we have enough tens to take away?   (yes)
   (If not, the questioning cycle repeats for different levels of regrouping—
   hundred for tens, thousand for hundreds.)

To which space do we take away the 1 ten? (down to middle row/cover up the ten-mark)

How many tens are left?   (1)
Where do we move it?   (down to answer space)

Restating the problem

So, how many children will be left in the room during concert practice?   (17)

How can you say the problem we just solved as an equality sentence?   (16 + 17 = 33)

Verifying

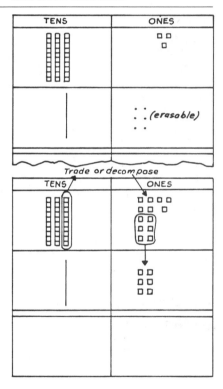

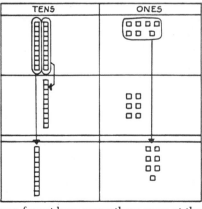

When we take away the tiles we lose track of how many we had at the start. Suppose you forgot how many there were at the beginning; how could you find out?

Suppose we pushed these two parts together; how many tiles/cubes do we have on the board? How come?

Following the introductory lessons, pairs or small groups of children engage in related activities in a variety of contexts.

Breakdown (Trade) Game. Each player starts the game with the same quantity of grouped and ungrouped objects on his board and the goal is to be the first one to discard all the objects. Each child rolls a pair of dice on his turn and discards the quantity indicated, regrouping (trading) as needed. The competition of the game encourages the children to check each other's regroupings. The size of the starting quantity can be increased with experience. A variation of the game can use pregrouped materials such as Dienes blocks for trading.

Detective Game. After showing a display of grouped objects on the board and naming the number represented, a child closes his eyes. The partner then rearranges the materials, by regrouping, removing, or adding more objects. The first child looks over the display and decides whether the number of objects has remained the same or not. A variation of the task requires one child to identify whether trades demonstrated by his partner are fair or unfair, while observing the demonstration.

Number Story Problem Cards. One child in the pair takes the top card from the deck and reads the number story aloud. The second child displays the initial quantity, regroups as needed, takes away, and identifies the remainder, restating the problem in the context of an oral number sentence. Variations of this activity are found in the last chapter (p. 334). Addition story problems should be mixed in the deck. The *Robot Game* can also be played, beginning with a number story problem.

Level Two: Pictorial Representation of Action on Objects

As the teacher demonstrates subtraction on the transition board with materials, the children are asked to represent each step of the procedure on outlines of the board, without use of numerals or written words. The children's pictorial representations are then compared in terms of ease of communication. For a follow-up activity, the Robot Game is extended to include the robot director's pictorial record of the robot's activity with materials.

Level Three: Written Notation of Action on Objects

Once the children are able to subtract by regrouping and taking away materials and can explain the process in terms of the value of the groups, the transition to the written algorithm can be undertaken. It can be approached as a formal record of what is already understood at the preceding levels.

Initially, the teacher keeps a symbolic record as the children subtract with materials. Pausing at each step, the teacher focuses questions on the correspondence between the values of the groups, their position and regrouping, and the numerals that represent them. The questions initiate a dialogue in which connections between existing ideas and new methods of representation begin to be constructed. Later, the children can record concurrently beside the materials at each step. An alternative is to have children pair up, with one child doing the subtraction with materials and the other doing a simultaneous written recording on an outline of the transition board as they talk their way through the problems. In either case, it is essential that the representation alternate from materials to written notation at each step of the procedure. Without this synchronization, there may be no conscious association between the parallel procedures.

Following this encoding, children are involved in interpreting the written record in terms of values of objects represented and actions on these objects (decoding). Both specific and broad questions can be used during this reconstruction. Specific questions such as, "What does this ( ) stand for?" and "Where did it come from?" help children to focus on values represented by specific symbols and on specific regrouping/ renaming. Broad questions such as "Explain all the changes that are recorded in the top row," encourage them to integrate their understanding of the process over a succession of regroupings/renamings.

Introductory Lesson. The teacher says, "This time I need you to slow down and work each step together so I can keep a record."

There are 223 pages in Sally's library book. She's already read 158 pages. How many pages are left to read?[16]

16. Ordinarily a two-digit subtraction would be used to introduce each level.

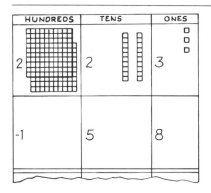

Recording the problem

If I write a 2 here, what does it stand for on the board?

I'm going to write down 158 so we remember how many things to take away.
(What does the 5 stand for?)
Read the problem.

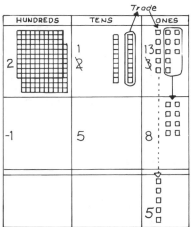

Recording subtraction of ones/regrouping of 10 as ones

How many ones do we have now? (following regrouping)
How can we show the change in my record?

How many tens do we have now? How can we show the change in the record?

Is the number in the top row still the same? How come?

How many ones do we have left? Where do I record this?

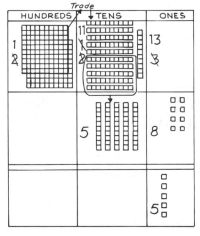

Recording subtraction of tens/regrouping of 100 as tens

(Recycle the questions for the next level of base groupings.)

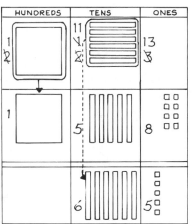

Recording subtraction of hundreds

(Recycle the questions as needed.)

(continued)

CONCRETE MODELING/
FORMAL SYMBOLIC RECORDING   ORAL LANGUAGE

HUNDREDS	TENS	ONES
1 ~~2~~	11 ~~X~~ ~~2~~	13 ~~3~~
1 ◻	5 ⦀⦀⦀	8 ⊡⊡
	6 ⦀⦀⦀	5 ⊡

Restating the problem

Read the problem from my record.

Breakdown (Trade) Game. While the children are playing the game with materials, they can also keep a record of their scores.

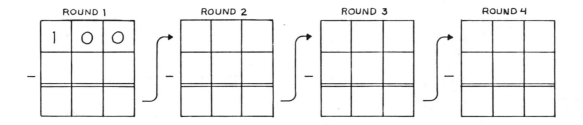

ROUND 1   ROUND 2   ROUND 3   ROUND 4

Robot Game. In this extension of the game, the robot director not only gives explicit directions for the robot to follow with materials, but keeps a written record of the activity. The robot is still programmed to respond to values of the materials.[17]

Teacher Game. Given completed subtraction records of other children (and the teacher), including some with errors, the children use the materials to check the answers and mark them right or wrong, and determine the correct answer as needed. A variation of the task involves having the "teacher" explaining either the correct procedure or the error to another child.

Level Four: Written Notation of the Standard Algorithm in the Absence of Materials

Eventually the children are given subtraction problems without any materials available. They are encouraged to visualize the materials and the regrouping process to bridge the gap to the final stage of symbolic recording in absence of guiding transition board outlines. A variety of activities are available for pairs of children.

Task Cards. Subtraction situations are presented on cards in a variety of ways. These include number story problems, and both horizontal and vertical notations. In the former situations, the children must set up the written algorithm in the absence of any guiding outlines. Addition situations should be mixed in to increase the challenge.

---

17. Swart (1980a) reports that the Robot Game is effective for eliminating buggy algorithms for borrowing across zero.

Teacher Cards. Two types of tasks can be presented, one requiring error detection and the other an explanation of procedures to another child. No materials are made available for reference except in the case of serious disagreement.

Just as in addition with regrouping/renaming, the pace of the transition from concrete action to written notation of the standard algorithm requires careful monitoring to accommodate the progress of individual children. Different groups of children can be working on the same problems at different levels. The high abstractors of place value ideas and the good visualizers of the concrete modeling process may progress rapidly to the written algorithm. By contrast, those children having difficulty in abstracting place value ideas, visualizing, and verbalizing values and regrouping procedures, will make the transition more gradually. Continued listening to children in individual interviews and small group discussions will allow teachers to guard against being seduced by empty symbolism and thus adopting unrealistic expectations for the latter group.

REFLECTIONS AND DIRECTIONS

1. Locate some third graders in your school or neighborhood and explore their understanding of the standard subtraction algorithm. Compare your findings with those reported here. Discuss them with at least one other teacher.

2. If you are unable to locate any children for interviewing, select two of the following children for closer study of their understanding of place value numeration both in the context of computational algorithms and independent of them.

   Zoya   Harold   Betty   Roberta
   Rena   Gail   Paula

   See the index for specific page references.

3. Assuming that you choose to teach the standard subtraction algorithm to all children, reflect on an appropriate grade placement. Compare your ideas with those of at least one other person.

4. Decide how you would reply to the child in the following situation.

Based on what you've learned from the illustrated interactions with children, select those responses that are appropriate to the situation.[18]

Discuss your selection with at least one other person.

18. These response options have been generated by members of the Association of Teachers of Mathematics (Wheeler., et al, 1977).

14

INVENTION
OF
NONSTANDARD ALGORITHMS

PREVIEW

The standard algorithms for addition and subtraction are generally accepted, concise procedures that are universally applicable regardless of the size of the numbers involved. Despite their long history of development and codification in which cumbersome methods were distilled to a small number of precise steps, these algorithms are presented to children stripped of any clues to their origins. Often, the methods are so concise that novices have to struggle hard to follow what's happening. Exposed to singular methods, many children are unaware of how these algorithms were developed, why they work, and that alternative methods are possible for a given task. For example, in the Davis and McKnight study (1980), third and fourth graders were unable not only to borrow across zeros to subtract 7002 – 25 according to the standard algorithm, but also to find another avenue for approaching the task. They appeared trapped by a dependence on an adult convention of which they had incomplete understanding. At the same time, there is ample evidence described in Chapters 12 and 13 and elsewhere (Lankford, 1974; Ginsburg, 1977) that despite being taught standard methods of computation, some children do generate their own alternative methods. These are methods that, from my observations of classrooms and those of others, go unnoticed—or worse, are discouraged by teachers. For some of these children, self-constructed algorithms provide alternative avenues to computational tasks, while for others the self-created methods provide the only avenue to the task in which they have confidence.

This chapter will examine available information on the ability of all children to construct algorithms of their own in problem-solving environments that encourage them to do so. Since my own interviews did not explore this aspect of children's numerical thinking directly, the chapter will rely on various sources: a philosophical position taken by members of the Association of Teachers of Mathematics (Wheeler et al., 1977), a teaching experiment conducted by Hatfield (1976, 1978, 1981), and a continuing program at the Village Community School initiated by Madell (1978, 1980). Then some teaching methods that support these creative efforts of children will be described.

Having the capacity to consider numbers simultaneously as wholes and parts (usually in terms of hundreds, tens, and ones) facilitates computation and opens up the possibility of generating multiple algorithms for the same task. By focusing on the parts in different ways and combining or separating them in any order (direction), children can attain an answer in different ways. The standard regrouping algorithms for addition and subtraction tasks are not the only algorithms possible.

You are invited to experiment with numbers and to generate nonstandard procedures for the following computational tasks on a separate sheet of paper. Feel free to use Unifix cubes, Dienes blocks, or other materials if you like.

$$\begin{array}{r} 248 \\ 375 \\ +197 \\ \hline \end{array} \qquad \begin{array}{r} 657 \\ -389 \\ \hline \end{array}$$

Since a structured written format may suggest a single learned procedure to both adults and children, the preceding tasks are rewritten below in another format. This new format may allow you to approach the tasks from a different perspective and thus generate alternative procedures.

$$248 + 375 + 197 = \rule{2cm}{0.4pt}$$

$$657 - 389 = \rule{2cm}{0.4pt}$$

Similarly, the disembodiment of computational tasks from real-life situations tends to narrow our perspectives. To generate other alternative procedures, try to envision these tasks in a practical context where you have no pencil and paper for figuring—for example, making change, checking the restaurant bill, computing mileage on a trip.

A PHILOSOPHICAL POSITION ON NONSTANDARD ALGORITHMS

In their publication *Notes on Mathematics for Children* (Wheeler et al., 1977) members of the Association of Teachers of Mathematics outline a philosophical position in which the construction of computational procedures is a problem-solving area for children. They encourage teachers to break free from teaching standard algorithms and instead to emphasize a central theme of mathematics—change or transformation.

> Running through all the examples we discussed we can see transformations of one kind or another. Change is at the heart of doing mathematics. We can substitute numbers for other numbers, change numbers into others, change the shape or position of an object, change our point of view with respect to it, change the form of a problem so that we can now solve it, change to solving another problem which we see in some way related to the first, change the representation of an idea using objects, speech, written words, symbols and diagrams. (p. 13)

For these writers, computation and calculation are the labels given to the processes of turning numbers into other numbers. Their perspective of mathematics is illustrated by the following examples (Wheeler et al., 1977).

STANDARD ALGORITHM FOR ADDITION	ALTERNATIVE VERTICAL AGORITHM
$\begin{array}{r} 22 \\ 248 \\ 375 \\ +197 \\ \hline 820 \end{array}$ This procedure produces an immediate answer. Interviews have shown that answers can be obtained by blind adherence to the rules and manipulation of independent digits.	$\begin{array}{r} 248 \\ 375 \\ +197 \\ \hline 600 \\ 20 \\ 200 \\ \hline 820 \end{array}$ Since numbers can be decomposed into multiples of hundreds, tens, and ones, the order of recombination is insignificant and may be varied. Although offered here as a hypothetical example, children have invented such methods when given the opportunity. The intermediate steps reveal thinking at a level above focusing on digits.

Three procedures for addition invented by a nine-year-old: $(248 + 375 + 197)$[1]		
$248 + 375 + 197$	$248 + 375 + 197$	$248 + 375 + 197$
$= 648 + 75 + 97$	$= 223 + 400 + 197$	$= 250 + 370 + 200$
$= 648 + 72 + 100$	$= 220 + 400 + 200$	$= 450 + 370$
$= 748 + 72$	$= 820$	$= 820$
$= 750 + 70$		
$= 800 + 20$		
$= 820$		

Four subtraction procedures worked out by a nine-year-old			
$657 - 389$	$657 - 389$	$657 - 389$	$657 - 389$
$= 658 - 390$	$= 358 - 90$	$= 668 - 400$	$= 650 - 382$
$= 668 - 400$	$= 308 - 40$	$= 268$	$= 600 - 332$
$= 268$	$= 268$		$= 300 - 32$
			$= 298 - 30$
			$= 268$

These examples illustrate that computation can be viewed as transformation of number situations into other situations having already-known or more accessible answers. Further, for this nine-year-old there are many ways to make such transformations.

For these members of the Association of Teachers of Mathematics, studying mathematics from the perspective of transformation can offer children the opportunity of taking maximum responsibility for their own learning (Wheeler et al., 1977).

> The need is to make the construction of arithmetical procedures a problem area for children. Telling them how to do their sums, or so to structure their learning that they are inevitably led to a set of algorithms of which the teacher approves, short-circuits a rich field of mathematical experience. Decision-making, creating problem-solving strategies, and the refining of home-made algorithms, are integral parts of mathematics and it is precisely in these areas that children will gain experience if they are allowed to generate their own computational procedures. (p. 95)

Furthermore, studying mathematics from the perspective of transformation not only encourages children's taking responsibility for their own learning, it also widens a teacher's response-ability to the children's construction of their own mathematics.

The writers of *Notes on Mathematics for Children* are quick to point out that their examples of

children's creative inventions were produced in a nonrestrictive classroom environment. The teacher's acceptance and encouragement were essential for such productivity. Rather than viewing the standard algorithms as something handed down from on high, the teacher considered them as merely "a sequence of transformations chosen frequently by one person, and, over a period of time, by several people." From this broad view of algorithms, it follows that children should be encouraged to construct their own and to share them. Furthermore, the teacher is careful not to restrict the children's view of different possibilities by disembodying computational tasks from real-life contexts or by presenting them only in a particular format. Just as a mathematician's algorithm is refined and condensed prior to becoming a well-known pathway, so are children provided opportunities to refine and condense their own rather than handed ready-made, polished methods. An important characteristic of child-constructed algorithms illustrated in this chapter is that they often reflect early stages of development. Intermediate steps in a child's algorithm provide valuable clues to his numerical thinking, indicating which aspect of the numerical information he is paying attention to in making the transformations. Such understanding is essential if the teacher is to guide the child in making refinements of his algorithm. On the other hand, sometimes a child's self-constructed algorithm may be something far in advance of what his teacher has seen. Such experiences are not uncommon.

Children develop new abilities in response to the challenge of problem solving. They adapt existing

---

1. The child's extended horizontal written solutions have been organized to make the transformations easier to follow.

methods to cope with novel situations. Presenting challenges to children, such as "In how many different ways can you solve this problem?" and "Which is the best way?," encourages them to be both inventive and evaluative. Although there are many ways to solve a problem, some ways are better than others, depending on what materials are available (blocks, paper and pencil, neither), the size of the numbers involved, the particular combination of numbers involved, the priority given to speed or accuracy, the context in which the numbers arise, and so on.

Wheeler and his colleagues (1977) also stress the potential of this radical approach to teaching computation to create an environment in which more children will feel confident in dealing with mathematics. The main sources of children's negative attitudes toward mathematics are identified as countless hours of mindless, uninteresting practice to maintain or increase skills of computation, the lack of sufficient time for children to develop their own methods, or the outright discouragement of methods other than the standard ones. Expanding on the latter source of negative attitudes, the ATM members write:

Many people have been made fearful of arithmetic by being forced to use what, to them, is a blindly incomprehensible juggling of symbols associated with "carry one", "borrow one," "pay it back" . . . This has often led to people being characterized as without number ability simply because the self-same methods have been acceptable to a sufficiently large number of others. (p. 15)

Piaget (1973) identified children's forced acceptance, often without understanding, of "an entirely organized discipline" with its polished, ready-made methods as a major source of feelings of inferiority and inhibition in mathematics. Children's construction of algorithms through autonomous activity allows them to develop a sense of control over numbers. Furthermore, involving them in the creative process of generating their own procedures not only increases their interest, but also reduces the need for external rewards to maintain involvement. The resultant development of both an interest in and a sense of control over numbers makes an important contribution towards the goal of learning-how-to-learn that extends far beyond the classroom. On this goal of education, Piaget (1974) wrote:

The ideal of education is not to teach the maximum, to maximize the results, but above all to learn to learn, to learn to develop, and to learn to continue to develop after leaving school. (p. 30)

A TEACHING EXPERIMENT: PROJECT FOR THE MATHEMATICAL DEVELOPMENT OF CHILDREN

An exploration of second grader's algorithmic thinking within a context of learning-how-to-learn was conducted under the direction of Larry Hatfield for the Project for the Mathematical Development of Children. The study aimed to facilitate the development of child-created algorithms. It then capitalized on the children's constructions and modifications of methods to examine their thinking independent of the dictates of adult conventions. The unadulterated thinking of different groups of second graders was studied during two successive school years.[2] The teaching experiments focused on child-created algorithms not only for addition and subtraction, but also for multiplication and division.

For his studies, Hatfield (1976) redefined an algorithm in psychological terms. Traditionally, the computational algorithm is thought to be synonymous with its written record of recognizable steps. Thus, chil-

$$\begin{array}{r} \overset{3}{\cancel{4}}3 \\ -26 \\ \hline 17 \end{array}$$

dren who complete tasks such as $\begin{array}{r}\overset{3}{\cancel{4}}3\\-26\\\hline17\end{array}$ are considered as "knowing" or "having" the standard algorithm for addition. Yet interviews in Chapter 12 have indicated that children's underlying thinking differs considerably despite identical written records. Some of these children interviewed moved from left to right, while others moved from right to left. Some understood the regrouping process, while others were merely juggling digits. Thus, according to Hatfield (1976), our interest must go beyond the recorded steps to the sequence of actions in which the child engages while solving the task. These actions extend beyond the external to internalized activity such as construction of mental images and performance of mental operations or transformations. Psychologically, a child's algorithm is his thought process during the computational task. Written records give only traces of this activity. Further, Hatfield attributes an algorithm to a child only after it is demonstrated to be reproducible—that is, once the same action sequence has been applied successfully in more than one instance, with the same reasons given. Examples of four child-created algorithms are illustrated next. (Hatfield, 1981).

---

2. The separate studies were conducted in 1974-75 and 1975-76.

Although the children's records of intermediate steps provide detailed clues to their thinking, listening to children can contribute additional information and test preliminary inferences.

In Hatfield's exploration, children's algorithms had their beginnings in their responses to computational tasks presented in physical contexts similar to those described in Chapter 10.

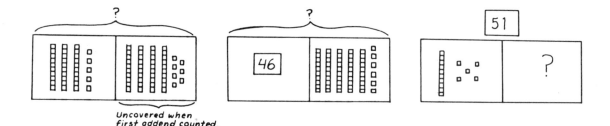

Uncovered when first addend counted.

Both the tasks and the children's algorithms went through different levels of representation from materials to pictorial and then written notation. As these algorithms evolved they underwent successive revisions and refinements. The children were encouraged to keep records of their reasoning, both as an aid to memory and a reflection on the progress made to that point. Hatfield found that counting was a natural and powerful response by children even when they were specifically asked not to count. Thus, a child's earliest record may have consisted only of tally marks of the tens and ones counted-on. Alternative procedures invented by children included those requiring no counting. With continued encouragement to keep a written record, traces of intermediate steps in their thinking appeared. In addition to encouraging detailed records, Hatfield interprets the teacher's role as guiding chil-

dren toward more conventional written formats for their algorithms.

The teaching was conducted in small groups by Hatfield and other mathematics educators at the University of Georgia. Time was provided for children to invent, think through, and talk through their procedures with reference to their materials and records. These discussions were aimed at helping children bring intuitive methods into conscious awareness and to understand why they work. Children's methods were shared, with some of them being adapted, refined, or extended by other children. For example, some children not only adopted "Adam's Greedy Way" of subtraction, but made it their own by generalizing it to subtraction with larger numbers. By taking time to compare different algorithms, the children not only considered which way was best, but also for what purpose. Some children were able to identify when certain methods worked best (Hatfield, 1981).

Hatfield concluded that within problem-solving contexts, second graders have the capacity to construct computational algorithms, not only for addition and subtraction but also for multiplication and division. Furthermore, some of the second graders constructed a variety of algorithms and even knew when they were best used. The teaching experiment also provided evidence of continuous restructuring on the part of children in modifying and refining methods, as well as evidence of children learning through interaction with their peers.

A CONTINUING PROGRAM: THE VILLAGE COMMUNITY SCHOOL

Instead of receiving explicit instructions in the standard algorithms, children at the Village Community School in New York City are encouraged to construct their own computational algorithms in problem-solving contexts. This program has provided Rob Madell, a mathematics specialist at the school and initiator of the program, an opportunity to study children's unadulterated algorithmic thinking over several years. He has observed children constructing progressively more sophisticated procedures beginning in the first grade and has noticed patterns in their behavior. Madell (1978, 1980) identifies three natural tendencies of children that do not match up well with the demands of standard algorithms:

1. Their early reasoning doesn't proceed column by column.

2. Even when it does so, it is universally from left to right.

3. Their method of carrying and borrowing avoids numbers larger than ten.

The following paragraphs will describe examples of children's natural approaches and contrast them to those underlying the standard algorithm.

Grouping for Column-by-Column Work. The standard procedures require an understanding of systematized grouping by tens—serving as the basis of breaking up the numbers into parts and operating on like parts separately in a column-by-column procedure. Meanwhile, Madell (1978, 1980) observes that this fundamental requirement of the standard algorithm is not accessible to many children prior to the third grade. First and second graders show an inability or reluctance to join tens to tens (count-on by tens).

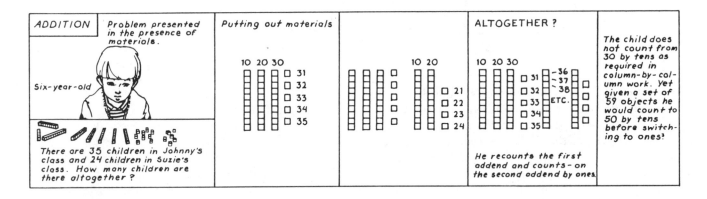

---

3. We have observed similar behaviors with some third graders (Chapter 10).

They are also reluctant or unable to take away -teen quantities as tens and ones.

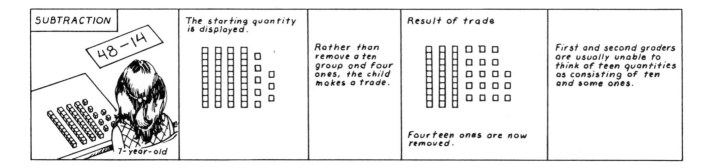

Prior to considerable progress in children's ability to count-on by tens and hundreds and in their development of numeration as systematized grouping by tens, column-by-column procedures such as the standard algorithms may be inaccessible to them, except at a level of manipulating independent digits. Although column-by-column thinking is perhaps most advanced, it is not always most appropriate. For example, in

$$\begin{array}{r} 39 \\ + \ 3 \end{array}$$, counting 39, 40, 41, 42, is a more natural response

than to think of 39 as 3 tens and 9 ones.

Directionality. The standard algorithms for addition and subtraction proceed from right to left. Our observations of third graders in Chapters 12 and 13 indicate that some of them work in the opposite direction despite specific instruction. Madell's observations reveal how children proceed in the absence of adult influence. Left to their own devices, these children universally work from left to right, even when working column by column in the third grade. Madell (1980) identifies three levels of efficiency in seven-to-eight-year-old children's processing in the left-right direction.

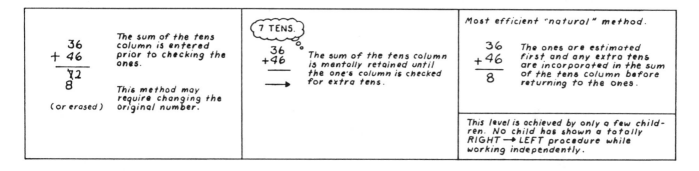

The highest level is achieved by only a few of the most sophisticated children. This left-to-right procedure is applied uniformly even for addition of three- and four-digit numerals. At the end of the third grade when the standard algorithm is introduced, the children are quite resistant to its right-left directionality.[4] Many argue

that their own methods are comparable in speed and accuracy.

In subtraction, the children develop a variety of strategies consistent with their left-to-right approaches to addition. These procedures may vary with the numbers involved.

---

4. According to Levin (1981), the existing techniques for standard paper-and-pencil computation, such as the right-to-left nature, the way of recording carrying and borrowing, were developed to allow reliable operation without erasure.

Same child shown responding to different tasks.

Records 29.

In the last example, the child takes the absolute difference between 3 and 4, compensating for it by taking more away from the remaining quantity. Other children, such as Cochrane's Kye (1970) and Davies' Mark (1978) invent an extension of this procedure to negative numbers—for example, 3 − 4 = − 1. Left to their own devices, children construct sophisticated procedures. This, in turn, may present a problem for the teacher. On the difficulty of interpreting these procedures, Madell (1980) writes:

> You should not be disappointed if you had a little difficulty following the reasoning in these subtraction arguments. Nearly everyone does. There is a lesson in that. It is hard to follow the reasoning of others. No wonder so many children ignore the best of explanations of why a particular algorithm works and just follow the rules. I believe the understanding comes best through re-creation, (p. 5)

In searching for an explanation of children's natural directionality, it is interesting to note that numerals are read from left to right and that multiple groupings of objects are usually counted by starting with the largest groupings. Further, children are taught to read from left to right. Thus, a "front-end" focus in children's algorithms appears to be a natural extension of earlier abilities.

Regrouping. Another critical aspect of the standard algorithms for addition and subtraction is the process of regrouping. We have already seen the difficulties that third-grade children experience in understanding the regrouping process. Madell (1980) observed that children's successful alternative algorithms avoid regrouping in the usual manner. Furthermore, children's natural methods tend to eliminate the need for memorizing addition facts beyond ten. In the latter half of the third grade, the Village Community School children are asked to compute the ones first. Here's how they do it.

For subtractions such as the last example, Madell reports that he has never seen a child deal with them by trading a ten for 10 ones and saying, "four from thirteen is nine" when working on his own. Children's successful methods circumvent difficulties of regrouping and the need for knowing combinations between 11 and 19. Nuttall's Stephen (1980), a second grader, invented an algorithm showing both these characteristics and applied it successfully to large numbers in a left-to-right direction. His method allowed him to keep a running total in his mind without need for a written record.

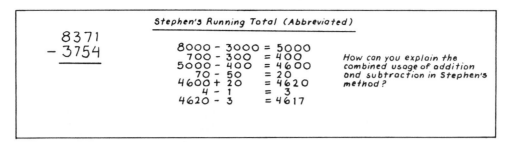

Stephen's Running Total (Abbreviated)

| 8371 |
| −   3754 |

8000 − 3000	= 5000
700 − 300	= 400
5000 − 400	= 4600
70 − 50	= 20
4600 + 20	= 4620
4 − 1	= 3
4620 − 3	= 4617

How can you explain the combined usage of addition and subtraction in Stephen's method?

A noteworthy aspect of child-constructed algorithms is that they are often applicable as mental strategies in the absence of paper-and-pencil contexts, to which standard algorithms are restricted.

Children's natural procedures that avoid the need to know sums greater than 10 allow the teachers at the Village Community School to spend less time on repetition and more on meaningful learning. Madell (1980) cautions that the traditional early focus on memorization obscures the view of mathematics as reasoning—a distortion that may be impossible to reverse at a later time.

As in Hatfield's study, the children's algorithms at the Village Community School had their beginnings in their responses to computational tasks presented in physical contexts with Dienes blocks. Also, both the tasks and the children's algorithms went through different levels of representation from materials to pictorial and then written representation. Although children develop rather sophisticated algorithms without explicit instruction, Madell emphasizes that these algorithms do not emerge from children's heads full blown. Rather, they may go through different levels of development.

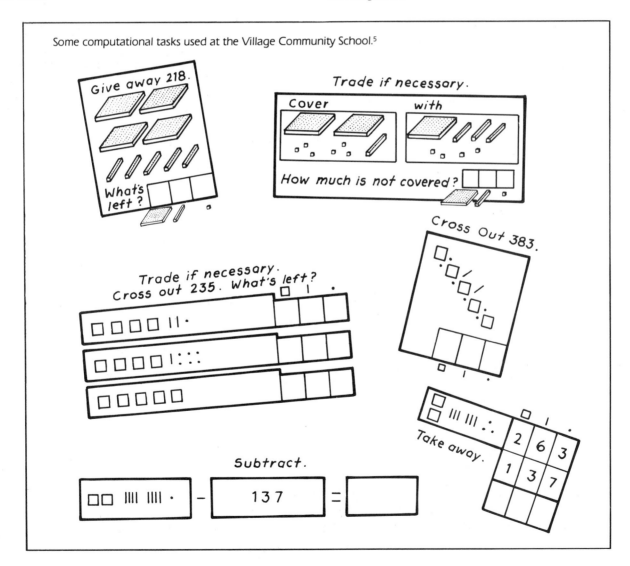

Some computational tasks used at the Village Community School.[5]

5. Taken from Madell and Stahl (1977), pp. 94, 96, 103, 108, 109, 115.

Since the class size in this private school was below twenty and since teachers were often assisted by both an aide and frequent visits by Madell, the challenge of understanding each child's computational procedures through careful observation and requests for clarification during individual and small group interactions was easier than it might be for many teachers. The teachers were in a good position to provide provoking questions or suggestions as needed, or to encourage children with different viewpoints to interact by juxtaposing the differences in discussion, without providing any explicit teaching. Following a long-term experience with this continuing program at the Village Community School, Madell argues that "children not only *can* but *should* create their own computational algorithms, and the teacher's role is 'merely' to help (p. 1)." To make a start in this direction could be a reasonable goal for teachers in general.

Rather than urging children to avoid errors, teachers can accept errors and treat them as opportunities for learning. During a visit to the Village Community School I observed Madell interacting with a second grader. In approaching an addition task by setting out and counting Dienes blocks, she started counting by tens, but failed to switch counting units when all the tens had been exhausted. Rather, she continued to count the smallest cubes as tens. Madell showed no disappointment at the child's answer; nor did he label it as wrong. Rather, he gently asked the child to try the task again. She did so, this time switching units as needed. At this point, Madell asked the child which answer was preferred. When she chose the second answer, she was then asked to explain what had happened earlier to get a different answer. The child was able to identify the problematic area in her counting following a brief pause. Reflecting on the actions bringing about her error likely prevented future errors of this type. Rather than suppressing errors, teachers can encourage children to deal with the discrepancies involved.

Madell's examples of children's natural approaches to computational tasks argue simultaneously for child-created algorithms and the postponement of any instruction on the standard algorithm. He contends that while not all children's algorithms are quite as efficient as the standard procedure, the loss in efficiency is more than compensated by understanding and confidence. Madell (1980) summarizes the effect of this radical approach to teaching arithmetic on everyone involved.

> For all the apparent delay, we are very happy with the results. Our children score well on standardized tests. THEY ENJOY MATH AND SO DO THEIR TEACHERS. All involved (parents too!) have been educated to see arithmetic as thinking, and not as just a collection of meaningless rules. (p. 1)

According to its initiator, this radical program not only had a positive impact on the children's computational abilities and understanding, but also on the attitudes of all concerned.

Madell's decision to provide primary-grade children with opportunities to create their own computational algorithms while simultaneously delaying their exposure to the standard paper-and-pencil algorithms is supported in the work of Resnick (1982b). Based on her research, she has proposed a model for children's development of decimal knowledge, with three levels of ability.

Levels of Decimal Knowledge (Resnick, 1982)

Level One: Unique Partitioning of Multidigit Numbers
- *Recognizing relative value of different parts making up whole numbers*

- *Comparison of numbers*

- *Adding and subtracting 10 from any quantity more quickly than adding or subtracting numbers such as 3, 4, 5, 6, 7, 8, or 9*

- *Counting up or down by tens from any starting numbers*

- *Constructing mental addition and subtraction algorithms*

Level Two: Multiple Partitioning of Multidigit Numbers
- *Recognizing that numbers can have more than one representation with blocks—for example, there can be more than nine blocks in a column.*

- *Preservation of quantities by exchanges that maintain equivalence—the whole is conserved during rearrangement of the parts*

Level Three: Application of Part-Whole to the Standard Written Algorithm
- Identifying the value of carry and borrow marks

- Explaining the written algorithm based on a place value rationale
Equivalent changes of the parts conserve the whole:
A decrement of 1 hundred = an increment of 9 tens and 10 ones.

The intellectual demands of understanding the standard algorithms require construction of ten-to-one relations between adjacent positions based on systematized grouping by tens. Considerable time is needed for this construction. Prior to this advanced level, children are capable of computations based on total values of individual digits in numerals rather than on their place value. Madell's children demonstrated growing capacity to generate their own nonstandard algorithms leading to good computation. Yet according to Resnick's proposed levels, a later focus on the standard algorithm in conjunction with grouped materials may provoke a deeper understanding of the decimal number system as a systematized grouping by tens.

CLASSROOM TEACHING METHODS AND ACTIVITIES

The richness of children constructing their own algorithms, as noted by Easley (1980b), defies encapsulation by textbook writers.

> Sagaciously, we note that elementary arithmetic is not limited to being traversed in a strict linear order, but is interrelated and interconnected in a structure something like a tree, in which teachers and children can climb around like monkeys, easily leaping from branch to branch. (p. 164)

With this approach, there are no textbooks for teachers to follow page by page; rather, they are left to their own devices in providing a variety of situations that motivate children to construct progressively more elaborate and integrated structures, while constructing and refining their own algorithms. Guided by their own structures of interrelated number ideas, teachers can initiate a rich variety of activities while remaining open to the possibilities of a spontaneous change in direction fueled by children's interests and abilities. Rather than model computational procedures or transmit specific algorithms, teachers model problem-solving processes and attitudes, and maximize children's opportunities to interact with a variety of problems, materials, and alternative ideas and methods of their peers in an atmosphere of friendly acceptance.

Easley (1980b) points out that there is no easy recipe to follow in encouraging problem solving such as child-constructed algorithms.

> There is no rule, no procedure, for making one's own discoveries. You must keep encouraging, begging, providing clues that preserve the dignity of the discoverer, and then it happens. Some children discover several things in an hour—others only in a month. But discovery is such fun, and it builds one's confidence in facing unknowns. (p. 165)

There are, however, general guidelines for indirect teaching methods. Illustrations of such methods mentioned earlier are applicable for many areas of problem solving. In particular, methods for facilitating children's construction of alternative counting methods (Chapter 4) and construction of bi-directional equality (Chapter 7) are likely to be helpful. This discussion will continue with a consideration of the teacher's role as provider of rich contexts for problem solving, modeler of problem-solving processes and attitudes, and facilitator of interaction.

The Teacher as Provider of Varied Contexts for Problem Solving

By allowing children to work with grouped objects in an unrestricted manner while solving computational tasks, the teacher can provoke the emergence of a variety of computational methods. Most of the child-constructed algorithms in this chapter had their basis in action on objects. To further illustrate the potential for creativity in such an unrestricted environment, the following list tells of different ways in which children might vary the standard algorithm for addition and subtraction with regrouping:

1. For addition, regroup only after all like groups are joined, instead of joining and regrouping as needed, one column at a time.

2. For subtraction, complete all regrouping prior to taking away any objects, instead of regrouping and taking away one column at a time.

3. For the preceding variations, proceed from left to right rather than right to left.

	3	7
+	8	5
	11	12
1	2	2

Since children develop algorithms that tend to avoid regrouping, at some point it may be appropriate to restrict the manipulation of materials to methods involving regrouping without use of other shortcuts. This constraint requires children to focus on the regrouping process and encourages them to extend their thinking about ten-to-one relations involved, while still having the opportunity to be creative. When children are allowed to record their procedures in their own ways, interesting variations of the standard written algorithm are also possible, depending on how and where the regrouping is recorded. Different contexts and formats of presentation encourage generation of varied methods, which when recorded provide rich springboards for comparison and discussion.

Another particularly productive context for encouraging the generation of varied methods is the hundreds board. The following activities suggested by O'Brien (1975) illustrate its potential.

Forget for the moment what you know about addition and subtraction algorithms and generate alternative ones by referring to the hundreds board.

What is 7 + 2?   7 + 4?   17 + 2?   17 + 24?   47 + 24?

What is 9 – 3?   19 – 3?   19 – 13?   14 – 6?   24 – 6?   84 – 36?

1	2	3	4	5	6	7	8	9	10
11	12	13	14	15	16	17	18	19	20
21	22	23	24	25	26	27	28	29	30
31	32	33	34	35	36	37	38	39	40
41	42	43	44	45	46	47	48	49	50
51	52	53	54	55	56	57	58	59	60
61	62	63	64	65	66	67	68	69	70
71	72	73	74	75	76	77	78	79	80
81	82	83	84	85	86	87	88	89	90
91	92	93	94	95	96	97	98	99	100

In how many ways can you do these?

Are some ways better for some problems and not others?

What about 15 + 19 versus 15 + 26?

And 92 – 29 versus 92 – 26?

The use of two or more hundreds boards in combination can extend the size of the numbers being considered.[6] Furthermore, children's interaction with these boards can serve to integrate their numeration ideas.

The presentation of number story problems in the absence of any aids (objects, hundreds boards, paper and pencil) encourages the further development of mental computation strategies such as initiated with the hundreds board.

> Sally's aunt is having a birthday today. She was born in 1895. How old is she today?

The following suggestion for the development of children's ability in problem solving comes from Brownell (1942).

> Practice in problem solving should not consist in repeated experiences in solving the same problems with the same techniques but should consist in the solution of different problems by the same techniques and in the application of different techniques to the same problem. (p. 439)

Ideally, providing for such a balance should result in sufficient practice with a single self-constructed method in a variety of situations to refine and to consolidate it. At the same time, the experience of developing alternative methods should establish the value of cross-checking and develop an appreciation of the limitations and advantages of different methods.

The Teacher as Model of Problem-Solving Processes and Attitudes

Rather than model specific computational procedures, teachers can model problem-solving processes and attitudes in dealing with children's constructions. When interpreting these constructions on the spot, the teacher needs to call upon abilities developed in other situations, such as the ability to infer recurring patterns of thinking from observation of other children. Some situations demand abilities not developed previously

---

6. Davis and McKnight's fourth graders (1980) were unable to generate alternative methods for

$$\begin{array}{r} 7002 \\ -\ 25 \end{array}$$

when their understanding of the standard subtraction algorithm failed them. It is unlikely that these children were without such alternative computational methods; rather, they had not extended these methods to numbers in the thousands.

or ones that are rusty from limited practice. Burns (1982) identifies critical problem-solving attitudes teachers can model to encourage children's own problem solving.

> You must be willing to learn along with your students, and to share your own striving with your students when some problem is difficult, so they can see you struggle. Most importantly, children need to see you, their teacher, as a learner, willing to explore, and even being wrong at times. Not knowing does not need to be perceived as failing; it can be thought of as not having found out yet. . . . They (teachers) must be seekers who are willing to plunge into new situations, not always knowing the answer or what the outcome will be. (p. 47)

Spontaneous interpretation of children's creative constructions—the basis for the next step in teaching—requires problem solving on the part of the teacher. In observing their teacher in action, the children need to notice attitudes of curiosity, flexibility, reflection, willingness to ask clarifying questions, resistance to premature closure, and willingness to risk making an error.

The Teacher as Facilitator of Interaction

To provoke children to construct, evaluate, and refine computational algorithms, the teacher must present problems not only in a variety of novel and challenging contexts, but in an atmosphere that is conducive to exploring and sharing ideas. The teacher's curiosity and acceptance of error as part of the problem-solving process make an important contribution to an attitude of psychological safety. Three mathematics educators with considerable experience with problem solving in the elementary classroom, Jacobson, Lester, and Stengel (1980), conclude that children need time to reflect and work on problems more than they need specific instruction in problem solving. Jacobson et al. also make the following proposition for the improvement of problem-solving ability:

> Teachers should provide experiences that encourage students to try different approaches, talk with their classmates while working on problems, and discuss the relative merits of different approaches. Such an atmosphere will not only motivate students but also lead to more mature problem-solving procedures. (p. 134)

In observing teachers and children becoming involved in this vital, productive, and enjoyable process of making mathematics, these mathematics educators noted that the process evolved uniquely in each classroom.

The following examples are offered as suggestions for starting points in the teacher's development of a personal style that works with the children in his/her classroom.

To set the stage for the children's activities in the construction and refinement of algorithms, the teacher calls the class together, presents the problems, encourages multiple methods, gives an overview of activities to take place, and provides guidelines for the children's interaction.[7]

Introduction			
Here are some different problems to work on. When you find a way to get an answer, try another way to get the same answer. See if you can find some interesting shortcuts.	After about twenty minutes we'll get together and give you a chance to show and compare the different ways you approached each problem.	You can work by yourself or with someone else. If you work with someone else, just make sure that your talking doesn't disturb anyone around you.	Here's a way to work together and still get a chance to do your own thinking. If you're working with someone else, give your partner(s) enough time to work out their own method before comparing and explaining. If you're waiting and keeping your method a secret, try checking your answer with another method.

Presenting problems in the context of searching for shortcut methods of computation motivates the continued refinement of methods and the construction of increasingly powerful methods. The guidelines encourage reflection while providing for different learning styles.

During the next twenty minutes or so, while the children are working out different methods of computation, the teacher circulates and observes both individual and group constructions and interactions—ensuring that the process is reasonably orderly—makes on-the-spot interpretations, and deals with questions and obstacles that arise. An effective method of keep-ing the children's thinking going is to accept their efforts in the absence of adult authority and judgment, at the same time that feedback from other sources is maximized. Piaget (1973) suggests that teachers provide children with counterexamples that compel re-thinking of over-hasty conclusions, and encourage their collaboration as a means of advancing intellectual development through continual mutual stimulation, exchange of ideas, and exercise of critical thinking. The possible responses to a child's error shown next illustrate the teacher's role in provoking rethinking of incomplete or faulty procedures through individual or group interaction.

---

7. Although problems and guidelines may be presented to the total group, most of the ensuing interactions are within smaller groups. The impressive work of Hatfield and Madell was conducted with small classes and/or small groups within the classes.

CHILD'S FAULTY PROCEDURE	POSSIBLE TEACHER RESPONSES AND FOLLOW-UP QUESTIONS	
IS THIS RIGHT? 347 −182 — 245	Is there another way to check your answer? Try working it out with materials to see if you get the same answer.	What happened? Which one of these answers is right? . . . How come you got the other answer the first time?
	Is there a particular part you're unsure of? . . . I noticed that in the ones you took the bottom number from the top one. In the tens you took the top number away from the bottom one. Is that alright?	How could you fix it? . . . How come you got the other answer the first time?
	Try working out this problem. 347 − 102	What happened? Which one of these is right? . . . How come you got the first one to come out to the same answer?
	(Referral to another child.) I noticed something interesting. Both of you got different answers when working out the problems in similar ways. I wonder how come. (Leave the children to their own devices at this point.)	

These responses have the potential of provoking a child's awareness of an error, rethinking of the procedure, and/or rethinking of underlying ideas. Notice that children also are encouraged to look back at their errors and analyze the underlying faulty reasoning. A further response to a child's difficulty might be to provide alternative related activities that build essential ideas—for example, regrouping and trading games—prior to returning to computational problems days or weeks later. In the case of pairing of children while juxtaposing conflicting answers and inviting interaction, it is critical initially to find children using dissimilar procedures before matching them from different parts of the classroom. Another criterion in pairing is whether a child is capable of being helpful; that is, being able to accept another's errors without dismissing them as merely "dumb." Encouraging children's collaboration is important in advancing both their intellectual development and their independence from the teacher.

At the end of the designated time, or after the children have made considerable progress with a set of problems, the children are called back to a large group seating arrangement. Here, the teacher encourages children's presentation, clarification, comparison, and evaluation of methods, as well as a continuation of the process.

PRESENTATION
- Would anyone like to show their methods (on the board)?

- Gloria, I noticed you were doing something different. Would you mind sharing it?

- Bill and Cindy, are you willing to share your methods? (Selected for contrast.)

CLARIFICATION
- Would you explain how your method works?

- Let me see if I understand you. (Teacher carefully verbalizes own understanding.)

- Does anyone want to ask any questions about Gloria's method?

COMPARISON
- How many different methods do we have on the board?

- Are any methods close enough to be called the same? (Different representation of the same procedure.)

- What's the same/different about these two methods?

EVALUATION
- Is one method better than the other? What makes you think so?

- Suppose the problem was changed to ____. Would this method still be better?

- What kind of problem would this method be best for?

- Is there a method that's best for all problems we've tried so far?

EXTENSION
- Have you got all the possible methods on the board?

- Would someone be willing to record these methods in a problem book for our class library? We'll leave some blank pages so that you can add other methods to the book as you think of them. (Each method is credited to its inventor.)

During these follow-up process discussions, children's preverbal experience is consciously provoked into awareness in attempts to explain and clarify. Children who are mentally engaged during the discussion, although not heard from, also can benefit from the experience. Children who appear to be daydreaming actually may be engaged in thinking through their own unique methods. The awareness that other children have viewpoints different from their own plays an important role in getting them to rethink ideas underlying computation, rethink and refine their procedures, as well as expand their repertoire. This complex process of classroom problem solving has been stated simply in terms of goals for the teacher and the child by Jacobson, Lester, and Stengel (1980):

> The teacher's goal is to keep the students thinking and generating ideas, and the students' goal is to find satisfying solutions. (p. 131)

In search of satisfying algorithms, some children become actively engaged in raising their own questions of clarification, comparison, evaluation, or discussion.

Discussion of the teacher's role as facilitator of problem-solving interaction would not be complete without mention of the potential conflicts arising in making the transition to this approach. There is likely to be conflict between your professional goals for the children and the goals of parents and administrators. Imagine how classroom visitors having a perspective of computation as isolated skills might interpret the following classroom interaction (Jacobson, Lester, and Stengel, 1980):

> . . . The students would gather in their groups and begin working on the problem. One group would meet in the corner of the room and use the chalkboard to sketch out a solution. Another group would meet under a table. Others would work in small groups at their desks, and a few would work alone. Often the members of the group would work quietly on their own until they found a possible method of solution, and they would share it with the group. This sharing often generated some discussion and eventually led to a solution. The students who preferred to work alone frequently checked with their classmates once they had attempted the problems on their own. (p. 133)

Sometimes children's algorithms are incomplete and simply wrong. Also, in comparison to the customary orderly worksheets in which children fill in the blanks, their own written algorithms are messy. Thus, for the uninitiated, children's written nonstandard algorithms are difficult to interpret and, therefore, to appreciate. Similarly, children working out procedures in the context of materials may be considered to be messing about. Since discoveries don't take place on cue, but only after considerable thought, discussion, or trial and error, children's efforts may again be misinterpreted as daydreaming and more messing about. Also, the consultation and excitement of discovery may be regarded as generating excessive movement and noise.[8]

What has been productive activity proceeding with reasonable order may suddenly feel uncomfortable in the presence of parents and administrators. As a defensive reaction, you may find yourself making sudden demands for neatness, attention, or quiet. In making the transition to this radical approach, feelings of discomfort are natural. Yet it is important to reach a level at which you do not feel the need to distort your direction in order to meet the expectation of others, but rather to follow your professional conscience. Communication with parents and administrators must be opened up to clarify your intentions and to convince them that you are working in the children's best interests before inviting them to visit your classroom.[9] Communicating your intentions requires the ability to justify them in terms of the goals of education and the changing needs of society, and in terms of what is known about children's learning abilities. The next chapter will help you clarify, integrate, and consolidate your thoughts in these areas. The final chapter also suggests ways to open up communication with parents, administrators, and other teachers.

REFLECTIONS AND DIRECTIONS

1. Many years ago a teacher in need of an extended period of uninterrupted time assigned his class a computational task designed to occupy the children for that duration. The task was to add the numbers from 1 to 100. Much to the teacher's dismay a child brought up his slate almost immediately. On it was recorded the correct answer.[10] How can you account for the child's rapid processing? Try the computational task yourself and do it in ways that avoid the drudgery of numerous separate additions. Compare your methods with at least two other people.

    $$1 + 2 + 3 + \ldots\ldots\ldots\ldots\ldots\ldots\ldots\ldots\ldots\ldots\ldots\ldots + 98 + 99 + 100 = \underline{\quad\quad}$$

2. Locate some children to interview and find ways to coax out nonstandard computational methods of their own invention. Compare your observations with those reported here.

3. If you can't locate any children, you might locate and examine the examples of third graders' nonstandard computational methods illustrated in earlier chapters.

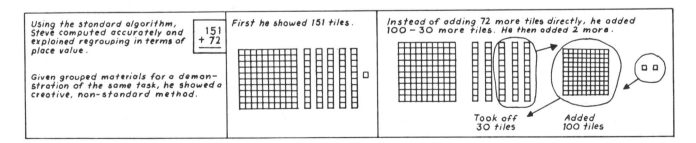

Using the standard algorithm, Steve computed accurately and explained regrouping in terms of place value.

$$\begin{array}{r} 151 \\ + 72 \end{array}$$

Given grouped materials for a demonstration of the same task, he showed a creative, non-standard method.

First he showed 151 tiles.

Instead of adding 72 more tiles directly, he added 100 − 30 more tiles. He then added 2 more.

Took off 30 tiles   Added 100 tiles

---

8. Furthermore, by not providing children with the standard procedure and correcting their answers directly, nor "speeding up the process" by praise, the teacher may find his ability to "teach" being questioned.

9. In his continuing program at the Village Community School, Rob Madell made regular presentations at parents' nights and curriculum nights.

10. This child later became a well known mathematician. His name was Frederich Gauss.

These alternative computational methods were generated despite a classroom emphasis on standard procedures. Reflect on the implications of this information for your classroom.

PART FOUR

CONCLUSION

15

THE THREE FACES OF CLASSROOM COMPUTATION: AN ANALYSIS

PREVIEW

Preceding chapters have focused on two faces of classroom computation—teaching standard algorithms in the context of meaning and facilitating children's invention of nonstandard algorithms in the context of problem solving. These approaches to teaching computation will be contrasted with the predominant classroom practice of teaching standard algorithms as isolated skills devoid of meaning. A detailed analysis of these three approaches will be presented side by side to facilitate comparison. The chapter will begin with a historical perspective and end with a forward look at society's changing needs for computation with the advent of calculators and computers. Also, the chapter considers a major roadblock to innovations in teaching computation—the standardized achievement test.

A HISTORICAL PERSPECTIVE

Many factors influence curricular change. These include advances in our knowledge of how children learn, in content, and in available technology, together with social, economic, and political pressures. The teaching of computational algorithms through rote memorization dominated classroom practice for decades. Despite some dissatisfaction with the methods in the educational community, the approach appeared to the general public to meet a societal need for proficiency. In the 1940s Brownell (1944, 1945, 1947, 1948) initiated a movement to make school mathematics meaningful by adapting the teaching to ways in which children learn best. In this approach, the standard algorithms were developed gradually in the context of place value ideas. This movement had barely gotten off the ground when it was quickly overshadowed in the late 1950s by a major revolution—New Math. This alternative movement had its beginnings in the Soviet Union's launching of Sputnik and the subsequent political pressure for improved American mathematics and science education. The availability of funding attracted mathematicians to development of curriculum that focused on expanding the content and meaning of elementary mathematics beyond computation to number theory, geometry, probability, graphing, function, measurement, and so on. The elegant reasoning presented by the mathematicians was difficult for many people to follow, especially teachers, whose exposure to New Math often amounted to brief crash courses in which they were asked to swallow whole the mathematicians' confections. Lacking adequate preparation for the teaching of New Math, teachers too often focused on its precise vocabulary rather than on its concepts. Some children continued to receive traditional computational lessons as many teachers saved sets of discontinued textbooks and conducted business as usual. Lacking the time or the background to help their children with homework difficulties in New Math, many parents became frustrated. Following an initial period of considerable public support came a time of increasing disenchantment, including public outcries against New Math. The New Math revolution ground to a virtual halt following fifteen years of curriculum development and implementation. Perhaps the most critical shortcoming of the movement was its lack of awareness that children's thinking is qualitatively different from that of adult mathematicians.

Although interest in child-constructed algorithms appears recent, that is not the case. In 1952 Hickerson expressed a commitment to this approach in his book *Guiding Children's Arithmetic*:

> There is no correct technique of performing calculations, but there is a best way for a particular person at a particular stage of their development. (p. 26)

Furthermore, he offered the following guidelines for helping children to develop diverse computational methods:

> To help each child develop to the fullest his ability to compute,
>   a. he should be encouraged to visualize a concrete situation represented by the symbolism;
>   b. he should be encouraged to think through for himself how each new type of computation should be performed;
>   c. he should be encouraged to discover computational shortcuts;
>   d. he should be encouraged to perform computations mentally as often as possible;
>   e. he should learn, according to his capacity, to make up for himself and his classmates computational exercises for out-of-context practice;
>   f. he should gain facility in estimating answers; and
>   g. he should gain facility in obtaining exact answers. (p. 67)

In this scheme, each child would be encouraged to develop his abilities to the utmost, and "discover for himself why the numbers behave as they do in each new computation". Unlike in the New Math movement, teachers were encouraged to adjust the content and the teaching method to meet the needs of the child, not the child to meet the "logic" of the math content, no matter how elegant. Furthermore, since the 1960s a comprehensive theory of learning to support the restructuring of teaching and learning in the classroom has been available in America—Piaget's theory of intellectual development. Piaget's writings in education (1973) stress the importance of developing critical thinking (problem-solving abilities) and reveal individuals as capable of educating themselves. Yet outside of a few classrooms, the need for developing meaningful and varied computational methods and related problem-solving abilities in the context of listening to children has been ignored.

Educational reform often takes the form of overreaction, in the pendulum effect. Overreaction to New Math came in the form of the Back-to-Basics movement—a retreat to traditional methods focusing on computational proficiency through drill and practice. The broad view of mathematics was narrowed again to mere arithmetic. This was not arithmetic in a rich sense, but arithmetic in a disembodied form—stripped of ideas, relationships, and applications that give meaning to computational methods. Teaching computation as isolated skills once again dominates classroom practice. Although textbooks give lip service to broader goals of mathematics education—developing meanings, applications, and relationships—they continue to ignore children's development of ideas, thereby fostering bold, raw, rote memorization. Despite rapid advances since the turn of the decade in our understanding of how children learn mathematics, existing social and economic pressures contribute to a climate that appears strongly resistant to change.

The teaching of computation in isolation from meaning and application has been widely accepted by educational decision makers and the general public. It is also comfortable for many teachers who themselves tend to view mathematics as a collection of isolated, arbitrary rules. However, computation in isolation has been disavowed by most mathematics educators, who have pointed to a litany of serious limitations over a number of decades. Most of the limitations were first discussed at length in the series of papers by Brownell (1944, 1945, 1947, 1948, 1956). A more recent warning of the potential harm of computation in isolation has been sounded by Davis and McKnight (1980):

> . . . these "back to basics" type movements can easily prove harmful. They seek to improve student performance by simplification—but the simplification they seek may leave large numbers of students with impoverished resources that handicap them in the long run. (p. 78)

With the advances in our understanding of how children learn and our changing societal needs in the face of recent technological advances, such as the availability of calculators and computers, mathematics educators are making a strong case for meaningful learning of computation and other areas of mathematics. The 1978 yearbook of the National Council of Teachers of Mathematics (Suydam and Reys, 1978) was devoted entirely to issues relating to the teaching of computation and its changing role in society.

Despite repeated indictments of Back-to-Basics methods by many mathematics educators and local attempts to initiate innovation, the teaching of computation as isolated skills still dominates classroom practice. A major barrier to implementation of innovative teaching is the measuring stick applied to most programs—the standardized achievement test.

STANDARDIZED ACHIEVEMENT TESTS —A MAJOR INFLUENCE ON THE FACE OF CLASSROOM COMPUTATION

Standardized achievement tests rate and rank children according to their competence on a narrow range of tasks *that can be scored readily.* At the same time that standardized achievement tests are unable to determine the quality of children's thinking underlying their computational responses, they fail to sample adequately the different areas of mathematics beyond computation. As a result, test scores are at best only a crude estimate of children's mathematical competence. The fact is that they rarely provide teachers with new information. Teachers, by and large, know their children much more fully than standardized tests can. Moreover, such tests underestimate the abilities of children because of their paper-and-pencil format. Although several modes may have been used in learning computational algorithms in the classroom, the standardized achievement tests restrict children to a single mode. Children who are working at place value concepts with physical materials and oral discussion may be progressing satisfactorily, yet the paper-and-pencil tests do not attend to this progress. Rather than providing needed information to the teacher to support future learning, such tests often merely provide children with frustration and failure. The tests give credit only for computational skills, thereby making little distinction between children who have mastered these skills in the context of arithmetic meanings and those who have mastered only the isolated skills. Thus the abilities of the latter group are grossly mismeasured by standardized achievement tests. In short, these test results communicate little practical information to the teacher. Most teachers know much more than the high-cost computer print-out of results can tell them.

In spite of their failure to provide practical diagnostic information for the teacher, the dangers of mislabeling children, the cost in learning time lost during extended periods of testing, and the cost in terms of damaged classroom learning atmosphere, not to say anything of the cost of the tests themselves, standardized achievement tests proliferate in the schools. They serve the American concern for a quantitative measuring stick that, despite gauging only the veneer of school learning, satisfies a political need. The test results allow administrators to compare classrooms and teachers, teachers to compare children, and parents to boast about their child's 80th percentile ranking on national norms. This focus on test score comparison puts pressure on teachers to look better on the next year's test results. As a consequence, standardized achievement tests exert a major influence not only on what is being taught, but also on how it is taught. Concerns for maximizing children's learning and looking good on the test results are often incompatible. Occasionally, a delay in symbolic emphasis coupled with conceptual development with materials in the first half of the year can prepare the children for understanding the paper-and-pencil format of workbooks and achievement tests. However, since numerical concepts evolve gradually and teaching from concrete to abstract is a long-range goal, the pacing dictated by textbooks and achievement tests is unreasonable for many, if not most, children.

Under such pressures, most teachers resort to traditional teaching methods encouraging superficial learning that shows up well on the tests. Since innovative programs based on concerns for developmental learning are unable to show immediate gains on these tests, but have the potential of doing so over a longer period, these programs are quickly phased out. Many administrators are more concerned with looking good in the community with apparent quick gains than with taking a risk that might foster potentially valuable innovations in teaching meaningful mathematics. By discouraging innovation, standardized achievement tests have tended to standardize the curriculum across the country (DeRivera, 1973; Project Torque, 1976). Nothing short of a moratorium on standardized achievement testing will free educators to listen to children and expand their perspective of teaching and learning.

REFLECTIONS AND DIRECTIONS

1. Below are thumbnail sketches of interviews conducted with three third graders. You are invited to study the interview data and rank the children from two different perspectives:
   - the perspective of standardized achievement tests—isolated computational skills
   - the perspective of the total interview—computational skills in the context of meanings and ability to learn.

	Gail L.	Betty A.	Harold S.

COMPUTATION
 	
A. 37   B.   151
  +28   + 72
   65   223
	Able to compute quickly and accurately.	Able to compute accurately, though slowly, using a counting method.	Able to compute quickly and accurately.

EXPLANATION "Tell me how you worked it out. . . . What does the 1 stand for?	Explained in terms of a rule, "12 carry 1." Unable to discuss procedure in terms of place value.	Explained in terms of place value.	Explained in terms of place value.
DEMONSTRATION	Able to demonstrate addition with regrouping of materials.	Able to demonstrate addition with regrouping of materials..	Able to demonstrate addition with regrouping. Regroups spontaneously.
CONNECTION "Point to something on the board that this 1 stands for."	A. Pointed to a ten-group saying, "I think." B. Points to a single tile.	A. Pointed to a ten-group, then to the traded ten-group. B. Spontaneously pointed to the traded one hundred-group.	A. Pointed to a ten-group. B. Spontaneously pointed to the traded one hundred-group.

2. You are invited to continue the comparison of children's abilities from the perspective of standardized achievement tests and individual interviews.

   One of these third-grade children was ranked at the 83rd percentile (above 83 percent of the children in the nation) on general mathematics ability at the beginning of the third grade. From your continued analysis of the data, predict which child received this ranking.

	 Gail L.	 Betty A.	 Harold S.
COMPUTATION $\overset{3}{\cancel{4}}3$ -26 $\overline{17}$	Computed accurately after hesitation, asking the interviewer, "Do you remember how to borrow?"	Computed with subtraction reversal. Became spontaneously aware of error but unable to complete regrouping.	Computed with subtraction reversal. Became aware of error with prompt. Unable to complete regrouping.
EXPLANATION	Gave rule without place value reasons despite probes.	——————	——————
DEMONSTRATION	Able to demonstrate.	Able to demonstrate with one prompt.	Able to demonstrate after three prompts.
CONNECTION	Unable to make connection between physical and written formats, saying, "It's complicated."	Able to record regrouping of materials, complete the computation, and to give a place value explanation for algorithm with one slip.	Able to record regrouping of materials, complete the computation, and to give a place value explanation.
COMPUTATION $\overset{1}{\cancel{2}}43$ $-\ 61$ $\overline{82}$	Computed accurately.	Unable to complete the regrouping/renaming process.	Computes quickly and accurately.
EXPLANATION	Gave rule without place value reasons.	——————	Explained the renaming in terms of a hundred and tens.
DEMONSTRATION	Has trouble demonstrating with materials.	Able to demonstrate with materials. Trade of a hundred for 10 tens demonstrated spontaneously.	Able to demonstrate with materials. Trade of a hundred for 10 tens demonstrated spontaneously.
CONNECTION	Unable to make connection between physical and written formats.	Completed the written algorithm as a recording of the physical modeling with one prompt. Algorithm discussed in place value context. Realized a hundred was borrowed and replaced as tens.	——————

3. A good test should verify the children's competence.
   Can these rankings be accepted at face value?

   Gail: 83rd percentile
   Harold: 76th percentile
   Betty: 53rd percentile

4. Imagine what classroom life would be like without regularly scheduled standardized achievement tests each spring and/or fall.

5. Imagine how you might work toward the suspension of standardized testing in your classroom, school, or district.

THE GOALS OF EDUCATION—ANOTHER INFLUENCE ON THE FACE OF CLASSROOM COMPUTATION

It is generally agreed that education should prepare children in terms of the needs of society. Beyond that point of agreement there are two major perspectives of how best to accomplish this. If the goal of education is to pass "tried and true" knowledge to the next generation, the teaching of selected algorithms as standard methods follows naturally. Depending on the range of this goal, the algorithms may be taught as isolated skills or in the context of meaning. However, if the knowledge essential for one generation is no longer adequate for the next generation because of rapid advances in knowledge, the goals of education should reflect this shift in societal needs. In this case, an appropriate goal would be to prepare individuals capable of educating themselves and adapting to new situations as they arise. This learning-how-to-learn would extend far beyond the classroom throughout life.

Piaget identified intellectual autonomy as the primary goal of education (in Duckworth, 1964):

> The principal goal of education is to create [humans] who are capable of doing new things, not simply repeating what other generations have done—[humans] who are creative, inventive, and discoverers. The second goal of education is to form minds that can be critical, can verify, and not accept everything that is offered. (p. 5)

More recently, Kamii (1981) has expanded on this view of intellectual autonomy as the aim of education.[1] From this goal of education another face of classroom computation emerges—child-constructed algorithms in the context of computation as an area of problem solving. Although this view of computation is far removed from existing classroom practices, it deserves closer examination in the light of current societal demand for ever-more quantitative thinking and problem-solving abilities.

THE THREE FACES OF CLASSROOM COMPUTATION: A COMPARISON

Prior to further consideration of society's changing needs for computation and how the three faces of classroom computation adapt to these needs, a detailed comparison of these approaches is in order. In addition to being discussed in relation to goals of education, they can be discussed in terms of the type of knowledge generated. Teaching computation as isolated skills can be characterized as passing on static knowledge, whereas facilitating child-constructed algorithms can be viewed as developing dynamic knowledge. O'Brien (1977a) offers the following concise comparison that speaks to their potential effectiveness:

> . . . I have characterized static knowledge as knowledge gained by copying. You have it or you don't. You cannot reconstruct it. It goes nowhere. Dynamic knowledge is gained by action, both physical and mental. It grows by construction and elaboration. It is susceptible to reconstruction. It generates new knowledge. (p. 1)

The teaching of meaningful computation with standard algorithms can be considered to contribute to the development of dynamic knowledge but not to the extent possible through child-constructed algorithms.

In the following table, the three faces of classroom computation are described side by side to facilitate further comparison. Understanding a particular approach to teaching involves not only knowing what it is but also knowing what it is not.

---

1. Kamii takes the position that education is to enable children to adapt to their society *and* to construct (ideals about) better forms of society. This position goes beyond mere adaptation, thus avoiding situations similar to that experienced by children in Germany during Hitler's Nazi regime.

COMPARISON OF THE THREE FACES OF CLASSROOM COMPUTATION[2]

	COMPUTATION AS ISOLATED SKILLS	COMPUTATION IN THE CONTEXT OF MEANINGS OF ARITHMETIC	COMPUTATION AS AN AREA OF PROBLEM SOLVING
Goals and Objectives	Cultural transmission. Discrete, mechanical skills are basic. Products and attitudes: Accuracy and speed of computation; obedience.	Cultural transmission. Understanding of standard procedures is basic. Products and process: Accuracy and understanding of computation.	Intellectual autonomy—ability to adapt to a changing world. Thinking—learning-how-to-learn beyond schooling—is basic. Process, attitudes, and products: Generation of multiple algorithms, flexible computational abilities, problem-solving attitudes, accuracy through cross-checking.
Psychological Foundations	Behaviorism Knowledge is structured externally. Computational procedures exist externally, independent of people. Knowledge is a copy of reality. Identify stimulus for child: $\begin{array}{r}17\\ +16\\ \hline\end{array}$ Identify response to be given: $\begin{array}{r}{}^{1}17\\ 16\\ \hline 33\end{array}$ Provide precise practice, rewarding successes until child reproduces desired response consistently. Repeat practice as needed.	Piaget's developmental constructivism (emphasis on levels or stages) Knowledge is structured internally. Computational procedures are given meanings by people through physical and mental activity. The meanings do not exist independent of people. Knowledge is a construction/interpretation of reality. What is taught is screened through a child's framework of ideas; different children may have different interpretations. The learner actively changes the content to fit his framework or reorganizes his framework to deal with new input.	Piaget's developmental constructivism (emphasis on dynamic features) Knowledge is structured internally. Computational procedures are both constructed and given meaning by people through coordinated physical and mental activity as a way of adapting to new situations. Knowledge is a construction/interpretation of reality that undergoes progressive reorganization at higher levels of understanding. The child is an active knower. He attends selectively and anticipates in his interaction with the world around him. He transforms input in terms of existing patterns of thought and reorganizes those patterns to incorporate new input. This organizing activity tends toward coherence, stability, economy and generalizability. At the same time the organizing tends toward equilibrium it is developing the seeds within it for new disequilibrium. The child doesn't merely organize, he quests. He constantly recognizes, engages, resolves—and generates—novelty (O'Brien, 1977).

2. Notice that two of the approaches to classroom computation are based on different interpretations of Piaget's theory. Teaching computation (with standard algorithms) in the context of meanings takes a conservative interpretation. It fails to acknowledge the capacity of children to invent their own methods.

COMPARISON OF THE THREE FACES OF CLASSROOM COMPUTATION (continued)

	COMPUTATION AS ISOLATED SKILLS	COMPUTATION IN THE CONTEXT OF MEANINGS OF ARITHMETIC	COMPUTATION AS AN AREA OF PROBLEM SOLVING

Psychological Foundations (continued)	What is taught is learned in a predictable manner. Learning is in control of the teacher.
	
	Learning is linear/cumulative; no restructuring required at different stages or levels because stages do not exist.
	    Activities are sequenced so that the computational procedure is applied to tasks with gradually increasing numbers, e.g., two-digit addition precedes three-digit addition. Regardless of the size of the numbers involved, no restructuring is required. The carried "one" in addition is treated the same regardless of the column. Since this "one" has no meaning attached to it, there is no reason to delay exposure to three- or four-digit addition, except that the same procedure applied to larger numbers requires more steps of the same kind.
		What is taught is not always learned in a predictable manner. Learning is in the control of the child.
		
		Learning is nonlinear; restructuring of ideas is necessary at different stages or levels; restructuring requires additional time.
		    Understanding in action is restructured at a higher level of abstract relations capable of representation in written notation. Understanding large numbers takes longer because further structuring/elaboration of place value relations is needed to incorporate progressively more systematized groupings by ten. The end result is a complex network of ideas.
			Learning is not always predictable since it is in the control of the child.

Teaching General Practice	Show/tell and practice. Explicit modeling of standard procedure. The ultimate form of the expected response, that is, written algorithm, is demonstrated at the outset. Any reference to materials is brief, usually part of teacher demonstration. Two-digit addition precedes three-digit addition.	Initial modeling followed by group discussion posing moderate confrontations to stimulate construction of meanings prior to practice. Modeling is not explicit. Rather, it serves to place constraints on the activity, such as right-left direction, so that it parallels the standard written algorithm. There is a gradual exposure to the final written form of the algorithm over several weeks based on children's capacities and constraints. The activities begin with extended experiences with concrete materials. The next level of activities is begun only after understanding at preceding level is demonstrated. The written algorithm becomes a record of previously understood actions with grouped objects. This sequencing implies progressive restructuring of number ideas.	Modeling of problem-solving processes and attitudes while facilitating children's construction and refinement of own procedures. Any modeling of procedures is restricted to providing a suggestion (not explicit). A variety of novel and challenging problems are presented. Children construct, modify, and evaluate procedures in the context of generating multiple procedures and searching for shortcuts and elegant methods. Initial methods are generated from action on objects. Children are encouraged to share their methods and to extend them in collaboration. Written algorithms are encouraged as an aid to memory, a basis for reflection on progress to that point and as a basis for sharing methods. Children's algorithms undergo different levels of refinement and representation.

	Column 1	Column 2	Column 3
	Large-group lessons. All children are at the same page in the same textbook.	Small-group lessons and follow-up discussions. Different groups at different levels based on observed capacities and constraints of children.	No lessons. Considerable small-group interaction. Different groups at different levels and topics based on their abilities and interests. Follow-up process where group discussion is primary source of "teaching." Here, the teacher facilitates children's presentation, clarification, comparison, and evaluation of multiple procedures.
	Individual practice in standard vertical format of written algorithm on prepared worksheets devoid of applications.	Individual, paired, and small-group practice in computation, using the standard algorithm, and discussion of meaning in a variety of formats and applications.	Individual, paired, and small-group problem solving and practice in a variety of formats and applications. Children are encouraged to make their own written representations of alternative, nonstandard procedures that they have generated.
Treatment of Errors	Errors are reduced by simplification of computational procedures—by ignoring place value and focusing on digits alone.	Errors reduced by careful sequencing and pacing of activities based on children's developing abilities. Yet errors are almost unavoidable when dealing with complex meanings of place value.	Exposure to error is accepted as part of problem solving. Opportunities are provided for children to analyze the faulty thinking underlying the errors.
	Errors are not accepted and are marked wrong. An answer key is available and marking is often delegated to the teacher's aide.	Errors are accepted as potential indicators of child's underlying thinking. No answer key is available for place value rationales of the algorithm. Teacher tries to understand the child's perspective, whether the answer is correct or not, by observation and discussion. Children become increasingly involved in correcting and justifying their own work in interaction with peers.	
	Errors are corrected by teacher providing modeling of accepted procedure and further practice.	Errors may be approached in the following ways: Direct the child to a simpler form of the same task, allowing the child to discover the error—for example, correcting a written algorithm by repeating the task with materials and making connections between the two levels. Ask provoking questions to help the child become aware of inconsistencies in reasoning. Do nothing at this time. Provide alternative activities and return to similar tasks in a few weeks. Encourage children with contrasting perspectives to interact.	

COMPARISON OF THE THREE FACES OF CLASSROOM COMPUTATION (continued)

	COMPUTATION AS ISOLATED SKILLS	COMPUTATION IN THE CONTEXT OF MEANINGS OF ARITHMETIC	COMPUTATION AS AN AREA OF PROBLEM SOLVING
Treatment of Errors (continued)	The locus of evaluation is external, as is the locus of new ideas. Teachers and textbooks are the only sources of authority.	The locus of evaluation is increasingly internal. One's own reasoning and materials are alternative sources of authority to teachers and textbooks.	Locus of evaluation is internal. Own reasoning is primary basis for evaluation—internal consistency of cross-checking using alternative methods of computation.
Motivation	Extrinsic motivation predominates in the form of praise, token rewards, competition, grades.	Promotes development of intrinsic motivation to some extent. There is less dependency on external rewards.	The generation of alternative methods and the search for satisfying solutions promotes intrinsic motivation and cooperation.
Testing	Evidence of learning is an abundance of correct answers (mastery level at 80 percent at a high rate of speed (built up through practice with timed tests).	Evidence of learning is the observation of correct answers accompanied by increasingly more elaborate rationales for the procedure.	Evidence of learning is the construction and justification of increasingly powerful methods of computation.
	Ease of testing—the answer is either correct or it isn't. Standardized paper-and-pencil tests are valid measures of computational skills.	Difficulty of testing—there are different levels of meanings. They cannot be evaluated on an all-or-nothing basis. Standardized achievement tests are invalid measures of meanings. They need to be supplemented by regular observations and periodic interviews of children.	Difficulty of testing—multiple methods at different levels and degrees of meaning. Standardized achievement tests are invalid measures of nonstandard methods and of meaning. Evaluation needs to be conducted through anecdotal records of day-to-day observation of children and periodic interviews.
Scheduling	Scheduling is based on textbook guidelines—originally determined by arbitrary decisions.	Although the content may appear similar, the scheduling may be widely out of synchrony with textbook guidelines since it is based on children's demonstrated capacities and constraints and provides for their gradual development of meanings.	Since content is dissimilar, scheduling is totally out of synchrony with textbook guidelines. It is based on children's interests as well as their capacities and constraints.
Category	Direct teaching.	A blend of direct and indirect teaching.	Indirect teaching.
Outcomes	A veneer of superficial skills with limited application beyond the specific content practiced. Meaningless manipulation of digits taught by endless, boring repetition. The rule-bound procedures are easily forgotten and require further practice at repeated intervals over different grade levels.	The child's developing internal organization of number ideas supports: - accurate computation with internal safeguards against absurd answers, - further learning of arithmetic—for example, decimal fractions, estimation,	The child's construction and refinement of progressively more powerful alternative computational methods supports: - accurate computation based on internal consistency of cross-checking with alternative procedures, - further learning of mathematics in school and beyond through learning how to learn and an expectation of finding satisfying solutions,

- understanding alternative procedures, so that the most efficient method can be applied in a given situation, or so that a method can be reconstructed if forgotten,

- confidence in dealing with the unknown and problem-solving attitudes of curiosity, persistence, tolerance for error, and so on.

- intellectual autonomy.

- understanding the algorithmic procedure, so that it can be reconstructed if it is forgotten,

- application of the standard algorithm in new situations and construction of alternative algorithms as needed.

The isolated rules provide a child with no basis for reconstruction of the procedure once it is forgotten, or for further learning of arithmetic—for example, estimation methods are not possible without an understanding of the numeration system.

SOCIETY'S CHANGING NEEDS

The widespread use of electronic calculators and computers in our society has been accompanied by a demand for ever-more quantitative thinking and problem-solving abilities. This forces us to reexamine our priorities in educating children for tomorrow's world.

Weighing the classroom time spent on learning the standard paper-and-pencil algorithms against their utility outside the classroom provides some insight into the existing problem for today's schools. Although most of the elementary school years are spent mastering these algorithms, their utility outside of school is limited, according to research conducted by Burns (1982) and Lave (1979). Burns gathered data from teachers attending her workshops to determine the everyday situations outside their classrooms in which computation was required, the methods of computation used, and their relative frequency. A summary of the survey data follows.

EVERYDAY APPLICATIONS OF COMPUTATION
• When I balance my checkbook.
• When I'm keeping track of spending at the supermarket.
• When I need to know how much wallpaper or floor covering.
• When I want to check the mileage on the car.
• When I'm deciding how long I need to bake a roast or turkey.

METHODS OF COMPUTATION
• mental strategies
• calculator
• paper-and-pencil (standard) algorithms
• manipulative strategy

Using a calculator or doing computations mentally were the two most consistently used methods. *Mental strategies* are nonstandard computational algorithms done without paper and pencil. Although universally applicable to numbers of any size, the standard paper-and-pencil algorithms are not always the most direct or the most efficient—for example, in the case of 7002 – 25. Other reasons for the infrequent use of paper-and-pencil algorithms may be the speed of the calculator for large numbers and the lack of either paper or confidence.[3] Lave (1979) confirms the rare use of paper-and-pencil computation in nonwork settings. He found that most of the mental strategies led to rough approximations or estimates of the needed results. In the business world, the advent of calculators and computers has almost eliminated the need for paper-and-pencil methods also. Thus, rather than the singular focus of computation in today's classrooms, societal needs would seem to require diverse methods of computation.

---

3. Burns suggests that on parents' night or curriculum night, teachers can survey the parents on their everyday application of computation as a way of provoking awareness of the need for change and opening the door to communication of the teachers' new program.

Would you pay this restaurant bill?

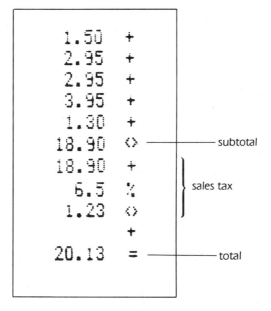

1.50	+	
2.95	+	
2.95	+	
3.95	+	
1.30	+	
18.90	⟨⟩	— subtotal
18.90	+	
6.5	%	} sales tax
1.23	⟨⟩	
	+	
20.13	=	— total

Following a delicious lunch with a friend, I received the written bill accompanied by this calculator strip. This was to assure me that the charges had been totaled correctly. After "eyeballing" the numbers I called the waitress and asked that the bill be recalculated. A few minutes later, the manager arrived with both an apology and an explanation. The waitress had failed to clear the calculator of the previous transaction prior to punching in my charges.

One method of computation having wide application in everyday life is *estimation*. In the absence of a convenient means of computing the exact subtotals of a mounting bill while shopping or ordering from a restaurant menu, another form of computation is needed to guard against overspending. As noted above, this form of computation—estimation—is also invaluable in checking the reasonableness of answers obtained by other means—even by calculator. Two common methods of estimation are described below (Trafton, 1978).

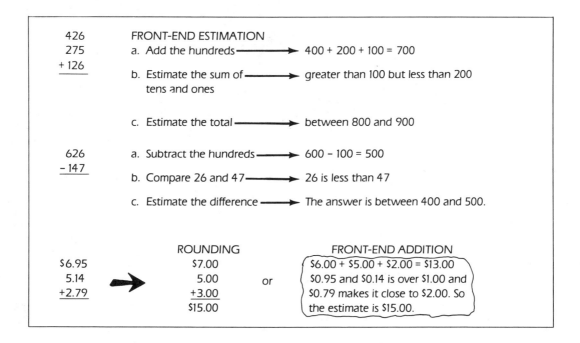

FRONT-END ESTIMATION

426
275
+ 126

a. Add the hundreds ⟶ 400 + 200 + 100 = 700

b. Estimate the sum of tens and ones ⟶ greater than 100 but less than 200

c. Estimate the total ⟶ between 800 and 900

626
– 147

a. Subtract the hundreds ⟶ 600 – 100 = 500

b. Compare 26 and 47 ⟶ 26 is less than 47

c. Estimate the difference ⟶ The answer is between 400 and 500.

	ROUNDING		FRONT-END ADDITION
$6.95	$7.00		$6.00 + $5.00 + $2.00 = $13.00
5.14	5.00	or	$0.95 and $0.14 is over $1.00 and
+2.79	+3.00		$0.79 makes it close to $2.00. So
	$15.00		the estimate is $15.00.

Despite the speed and sophistication of electronic calculators and computers, these devices are still prone to human error in keyboarding and in programming. The need to check the reasonableness of answers, therefore, has not been eliminated. A prior mental estimate is essential to guard against punching the wrong key and accepting the wrong answer to a problem. Yet estimation ability may not be enough. A study of high school students and adult professionals with demonstrated capacities in estimation by Reys (in Timnick, 1982) revealed a shocking fact in their interactions with a calculator: many showed more confidence in the machine-made result than they did in their earlier estimates, even when the machine answer was off by as much as 50 percent. Even subjects making good estimates were reluctant to challenge the machine. This overreliance on the calculator allowed good estimators to become intimidated and to fail to deal with the discrepancy between the answers. One example is illustrated below.

Many subjects, otherwise competent and confident in their abilities, automatically assumed that any error was theirs—that the machine was infallible. Although puzzled by the discrepancy they did not attribute the error to the machine. Some of the subjects did begin to challenge the accuracy of the calculator as the discrepancies got progressively larger. One seventh-grader made his challenge on the very first task.

Having developed an awareness of the fallibility of this tool, this youngster was not intimidated by the discrepancy and confronted it directly. He was sure enough to challenge the calculator. In a society where nearly every adult is already faced with computerized accounting and billing, Rey's research is most timely. It not only points out the importance of estimation abilities, but also the dire need for critical thinking to guard against an overreliance on the tools of the electronic age.[4]

Other needs already in existence, though brought to the forefront by the advent of calculators and com-

4. Esty (in Timnick, 1982) reports a caution expressed on this study: the calculator (HP-65) was a complex programmable model considered to be special and accurate, and the research involved subjects sophisticated in estimation but not in the tricks of wily investigators.

puters, are for problem-solving abilities that apply understanding of computation, measurement, geometry, and other areas of the curriculum to real-life situations. Recognizing this need, the National Council of Teachers of Mathematics launched the 1980s with a commitment to their development, reflected in their 1980 yearbook, *Problem Solving in School Mathematics* (Krulik and Reys, 1980). Although problem-solving abilities extend into areas beyond computation, many

are relevant for the computational arena. They include deciding which information is relevant, what operation is required, which computational method is most suitable, and whether the answer is reasonable. Deciding on the most suitable computational method requires a sensible choice among the calculator, variations in mental strategies, estimations, and paper-and-pencil algorithms (Hill, 1980).

ADAPTABILITY OF THE THREE FACES OF CLASSROOM COMPUTATION

In the area of computation, a rapidly changing society is demanding that its citizenry be able to apply selectively a diversity of methods, with accompanying problem-solving abilities. This section will examine the potential of the three approaches to computation for meeting these demands.

Computation as Isolated Skills

In terms of the changing needs of computation in our society, the teaching of computation as isolated skills leads nowhere. Year after year, children are required to use memorized algorithms to get correct answers. All errors are treated equally in the classroom, whether close to the answer or far off the mark. Meanwhile, everyday applications by adults in the real world tolerate minor deviations while avoiding the serious consequences of gross errors (Levin, 1981). The years spent focusing only on correct answers contribute to the development of an intolerance for anything but the right answer, thereby interfering with learning essential estimation skills. Although the ability to estimate requires a "feel for numbers"—a flexible understanding of the numeration system and operations—the development of such meanings have been avoided in teaching isolated algorithms, thereby providing no foundation for further learning. Furthermore, the teaching of estimation and other mental strategies' by rote is impractical because of their variety. A position paper prepared by the National Advisory Committee on Mathematics Education (1975) contained the following indictment of teaching computation in isolation:

> In isolation, computational skills contribute little to one's ability to participate in mainstream society. (p. 148)

> These methods contribute nothing to a confused child's understanding, retention, or ability to apply specific mathematical knowledge. Furthermore, such instruction has a stultifying effect on student interest in mathematics, in school, and in learning itself. (p. 24)

That a small number of children are able to develop some understanding and flexibility in their computation in spite of the limited methods of teaching is a testament to the power of the basic human tendency of making sense of experience. With all of the serious limitations of teaching memorized algorithms, it is surprising that this approach persists in any school.

Extending this narrow perspective to the use of calculators and computers in the classroom adds nothing to the child's experience. Rather, the child just experiences more of the same kind of practice, already shown to have limited effectiveness. Children memorize the procedure for operating the calculator, as demonstrated by the teacher. They use the device to check their answers to paper-and-pencil algorithms. The classroom computer serves to program the children to follow the steps of the standard algorithms, provides another context for drill and practice (although the images on the video monitor are likely to resemble existing worksheets), provides the child with feedback on the correctness of answers, and provides the teacher with a record of the child's progress. In this narrow view of teaching and learning, the teacher has essentially been replaced by the computer. These examples point out that despite a revolution in technology, classroom life can still be unproductive. What is needed is a complete rethinking of teaching and learning to approach the full potential of children as problem solvers and of computers as problem-solving tools.

Computation (Standard Algorithms) in the Context of Meaning

Teaching meaningful computation involves allowing time for children to develop the place value concepts underlying the standard algorithms. The resulting internal organization of ideas goes somewhere—both supporting and generating new learning. It supports a further study of the numeration system—of, for example, decimal fractions. It also supports children's incidental generation of alternative algorithms when paper-and-pencil algorithms are impractical. However, the basic philosophy of passing on knowledge from a broad perspective is to teach selected, efficient methods as needed, using a blend of direct and indirect methods. Once children understand the standard algorithms, classroom time can be devoted to some tried and true mental strategies and estimation procedures as suggested by Trafton (1978) and Musser (1982).

Developing Mental Arithmetic Skills[5]

Four alternatives to the standard algorithm for addition and one for subtraction follow. These algorithms are as systematic as the conventional ones; yet their front-end nature permits easier remembering of partial answers and more meaningful interpretation of the numbers themselves.

$(4)8 + (3)5$

1. Add the tens. (40 and 30 is 70)
2. Add the ones. (8 and 5 is 13)
3. Add the two sums. (70 + 13 = 83)

$(48) + (3)5$

1. Look at the whole first number. Add to it the tens in the second number. (48 and 30 is 78)
2. Add the ones in the second number to the sum. (78 + 5 = 83)

$(48)^{+2} + (35)^{-2}$

1. Add a number to one addend to make it a multiple of ten. (48 and 2 is 50)
2. Subtract the same number from the other addend. (35 − 2 is 33)
3. Add the two numbers. (50 + 33 = 83)

$(48) + 19$ (for use in adding 19, 29, etc.)

1. To the first number add 20 (the next higher multiple of ten). (48 and 20 is 68)
2. Subtract 1 from the sum (68 − 1 = 67)

$(62) - (2)5$

1. From the first number subtract the tens in the second number. (62 minus 20 is 42)
2. Now subtract the ones in the second number. (42 − 5 = 37)

---

5. From Trafton, 1978; pp. 207–208.

You may have recognized these mental strategies and realized an overlap between two faces of classroom computation. The major distinction is that in *meaningful computation* the children are taught mental strategies after the standard algorithm,[6] whereas in *computation as an area of problem solving*, the children construct mental strategies and record them on paper prior to exposure to the standard algorithm. Despite the potential of meaningful computation, one cannot help but wonder whether exposing children continually to someone else's algorithms over a number of years does not create an attitude of dependence on the teacher as the source of methods to think about and apply. Despite having an adequate conceptual background, children may be reluctant to tackle the unknown, to act independently and flexibly. Thus many children may restrict themselves to constructing meanings for methods selected by adults, while others construct original methods as alternatives.

When the cultural transmission of selected methods with meaning is extended to the use of calculators and computers, some unique opportunities for problem solving are created for children. However, these are likely to be restricted to classical problems or to areas otherwise considered important. The calculator provides occasions for using estimation procedures. Furthermore, it provides opportunities for searching for number patterns and extending problem solving so that large numbers are no longer a deterrent to computation. Attention previously given to computation can now be focused on observation and analysis. Rather than just providing more drill and practice, computers can be used to create opportunities for problem solving. One computer program called Buggy involves users (children and teachers) in the detection and interpretation of errors made by children on standard addition and subtraction algorithms. Besides providing practice in computational skills in an interesting context, this program engages users in experimentation—gathering relevant data, analysis, hypothesis formation, and testing. Furthermore, the experience provides users with insight into the value of errors. Originally developed for teachers by Brown and VanLehn (1980), this program has been used with sixth- and seventh-grade children by Barclay (1982). Below is a description of the computer simulation and its outcomes in the classroom (Barclay, 1982).

Buggy—The Great Error Hunt

Buggy has the computer simulate a child's consistent computational errors or bugs in addition or subtraction. The task of the user is to discover the bug in the child's thinking. After displaying an example including this bug, the computer waits for the user to generate additional tasks to gather information on what kinds of number combinations or number sizes trigger this "buggy" reasoning. For each task provided by the user the computer provides an answer, using the same buggy thinking when it normally would be triggered. Once the user thinks he/she knows what the bug is, the computer is notified by punching a particular key. Now the computer requests the user's explanation of the bug to be typed in (to articulate the idea). The user then tests his/her understanding of the bug by placing his/herself in the simulated child's shoes and using the "buggy" thinking to compute tasks presented by the computer. The computer provides feedback on whether the bug has been identified. If not, the user is notified, "That's not the bug I have," and the computer solves the same task with "buggy" thinking. The user is now required to rethink the bug and test it with further tasks and feedback from the computer until he is satisfied that he understands it.

Figuring out what the bug is means posing questions in the form of computational tasks that might reveal its nature. Following some aimless questioning in the initial exposure, the children become more systematic with experience. For addition they soon realize they should check for carrying in one or more columns. Also, they try problems with a different amount of digits in the addends. Although they may arrive at a hypothesis for a bug after only the first example, they eventually learn more than one hypothesis is possible from such limited information and that alternate hypotheses may need to be tested with several examples before eliminating one of them. Rather than "teach wrong answers" the "Buggy" simulation involves participants in analyzing the algorithms, disassembling them and manipulating the steps, and experimenting with "what happens if."

---

6. This reflects current scheduling. What we've learned from children (Madell, 1981; Resnick, 1982) indicates that mental strategies should be taught *prior* to exposure to the standard algorithm.

> Some of the reasons for using Buggy in the classroom are:
>
> To realize the positive value of error as a basis from which to make a next step. Students using Buggy progress from seeing bugs as "stupid and dumb" to realizing they are wrong, but interesting. Usually my students are always erasing wrong answers rather than seeing them as useful data;
>
> 2. To do problem solving. Students are involved in the powerful process of hypothesis formation, based on analysing the answers provided by the program to problems they have posed;
>
> 3. One final reason for using Buggy is that children enjoy it, both the math-able and the math-anxious. For the latter, the program offers a nonthreatening privacy in which they can continue to explore at their own pace. For someone who feels negative about mathematics, this is a rare experience. (p. 27)

In Buggy we have another partial overlap with computation as an area of problem solving, except here the focus is on analysis and interpretation of only the standard algorithms, rather than on multiple methods of children.

We have seen that teaching computation with meaning can meet societal needs for multiple methods of computation and for problem-solving abilities, if extended in those directions. However, it does so with a possible restriction—its structured methods of teaching may create an attitude to dependence. A second restriction is that exemplary methods and classical problems are selected by adults, leaving little room for children to spontaneously follow the direction of their interests.

Computation as an Area of Problem Solving

Children's construction of algorithms, followed by refinement and evaluation of alternative procedures, results in both dynamic knowledge and positive attitudes toward problem solving conducive to the generation of new knowledge. In the words of Wheeler (1975), children (and teachers) experience mathematics as a "constructive, or dynamic, even creative discipline." Steffe (1978) compares such activity to that of a mathematician:

Yet the young child, in his state of mathematical naivete but innate curiosity, in many respects resembles the creative mathematician working at the frontiers of his discipline. (p. 1)

If the goal of education is to prepare individuals to educate themselves and to adapt to the ever-changing needs of society, then approaching computation as an area of problem solving has much to contribute to this education.

We have seen that mental strategies used by adults (Trafton, 1978) are similar to those constructed by primary-grade children in problem-solving environments (Hatfield, 1981; Madell, 1980). The left-to-right direction, assumed naturally by children, facilitates memory of partial answers during mental processing.[7] In the work of Hatfield and Madell, the children's written work was a record of their thinking, a way of keeping track of their progress, and a means of communicating with others. Thus mental strategies can be formalized into nonstandard paper-and-pencil algorithms. Alternative activities include solution of written or orally-presented problems in the absence of paper and pencil. Children also are encouraged to make a mental check of answers resulting from paper-and-pencil methods using an alternative approach.

Research by Ginsburg, Baroody, and Russell (1982) reveals that primary-grade children invent methods of estimation when faced with the need to do so.[8] Rather than being taught estimation procedures directly, the children are placed in a variety of situations requiring quick, rough approximations in the absence of convenient means of computing exact answers. For example, the children experience shopping simulations requiring the purchase of several items in a store with a limited amount of money. Or they play the Estimation Game. Here, children give an estimated answer to an addition or subtraction problem

---

7. See Nutall's Stephen (Chapter 14, p. 381).

8. In comparing the estimation abilities of first, second, and third graders, Ginsburg, Baroody, and Russell (1982) report an improvement in accuracy of estimation, coherence of explanations, and sophistication of estimation strategies with age. The researchers attribute the superiority of the latter group to more extensive arithmetic experiences and greater development of general arithmetic knowledge.

prior to computing the answer with paper and pencil or calculator. The difference between the estimate and the computed answer becomes the child's score. The child with the lowest score at the end of the game is the winner. In adapting to these situations, children build on what they already know from earlier algorithm-invention experiences—left-right processing, a "feel" for numbers, the flexibility of working with multiple methods, and a tolerance of their own errors. Periodically, following these game experiences, the children's estimation methods are shared for comparison, evaluation, and refinement.

Once children have the opportunity to explore the functioning of calculators and use them for computation, the evaluation process of selecting the most appropriate method for computation and verification can be extended further. Selection can be made after consideration of time constraints, need for accuracy, and availability of tools. Here, children get involved in testing the speed of computation with existing methods for different-sized numbers. Thus, we see that a program facilitating child-constructed algorithms is consonant with the societal need for development of flexible, diverse computational methods.

Extending this dynamic view of education to work with electronic calculators and computers serves to enhance children's intellectual functioning further. As mentioned earlier, the calculator provides opportunities for searching for number patterns and extend-

ing problem-solving where large numbers are no longer a deterrent to computation. Rather than restricting children's experiences to classical examples, this approach allows children to branch out to study problems of their own choosing. Also, rather than the computer programming the child to learn isolated computational skills, the child programs the computer to solve problems of interest. According to Papert (1980), primary-grade children already have demonstrated the ability to do so. A computer program is just another algorithm. Through computers, children can transform numbers in a greater variety of ways. The kinds of problems for which children can use computers extends beyond number to other areas of mathematics, to science, and further into the unknown. No one can envision all the possible problem-solving applications of computers. Yet we do know that children experiencing extensive construction of computational algorithms are likely to bring crucial problem-solving abilities and attitudes to yet-unknown challenges. Children who have constructed their own algorithms have kept their innate curiosity alive and have developed confidence in dealing with the unknown by the creation of solutions, by task persistence, by flexibility in dealing with diverse methods, and by critical thinking abilities. They are well on their way to becoming intellectually autonomous individuals.

In summary, the highlights of the preceding discussion are placed side by side for ease of comparison.

SUMMARY COMPARISON OF THE THREE FACES OF CLASSROOM COMPUTATION

	COMPUTATION AS ISOLATED SKILLS	COMPUTATION IN THE CONTEXT OF MEANINGS	COMPUTATION AS AN AREA OF PROBLEM SOLVING
Adaptability to Society's Changing Needs for Computation	Methods of computation needed are too varied for memorization. Yet this approach provides no conceptual base for further development of mental and estimation strategies. Sole focus on correct answers develops an intolerance for error of any kind—another obstacle to acquiring estimation methods.	Provides a conceptual basis for further study of mental and estimation strategies.	Child-constructed algorithms have their beginnings as mental strategies. Their front-end (left-to-right) focus allows a running total to be kept in mind. Both this front-end focus and children's tolerance for error in the process of constructing their own algorithms prepare children for further generation of estimation methods.
Utilization of Available Technology	Application of the calculator is restricted to computation—serving to check answers obtained by the standard algorithm.	Calculator provides occasions for practicing estimation and extending problem-solving capabilities—for example, exploring number patterns, dealing with problems having large numbers. Application restricted to classical problems.	The calculator provides occasions for practicing estimation extending problem-solving capabilities and evaluating a broader range of computational methods for different situations.

(continued)

	COMPUTATION AS ISOLATED SKILLS	COMPUTATION IN THE CONTEXT OF MEANINGS	COMPUTATION AS AN AREA OF PROBLEM SOLVING
Utilization of Available Technology (continued)	The computer programs children to follow the steps of the standard algorithm, provides for more drill of isolated skills. Also provides feedback on correctness of answers and recordkeeping.	Computer creates opportunities for extending problem-solving capabilities. Applications restricted to accepted procedures—for example, the standard algorithm in the Buggy program or to selected problems. Although problem solving is enhanced, it is in the context of adult-constructed programs.	The computer creates opportunities for extending problem-solving capabilities. The child programs the computer to solve problems. The computer program is just another algorithm to be constructed. Applications are unlimited.

FINAL THOUGHTS ON THE FACES OF CLASSROOM COMPUTATION

The advent of inexpensive calculators and computers in the 1980s forces us to reexamine our priorities in educating children for tomorrow's world. The presence of high technology has intensified already-existing needs for diversified methods of computation and problem-solving abilities to the extent that they can no longer be overlooked. This advance in technology coincides with major advances in our understanding of how children learn mathematics.

It is impossible to foresee the total impact of calculators and computers on classroom learning. Although I do not propose that quality learning is restricted to interaction with electronic devices, any educational planning must give serious consideration to their inclusion. With the spread of home computers expected to match that of color TV sets by the turn of the century, schools may find an unexpected ally in support of innovation. For the first time, a large segment of the general public (including teachers and administrators) will have an awareness of the need for change and the importance of children's previously unnoticed problem-solving capacities. This growing awareness is likely to translate into a greater acceptance of the importance of meanings and self-constructed knowledge, as well as a particular interest in computer applications in the classroom. Furthermore, this climate of openness to change may lead to a moratorium on standardized achievement testing in mathematics. The moratorium would remove the pressure of looking good, thereby allowing administrators and teachers to work together on the important issue of improving the quality of teaching and learning.

Another outcome of the moratorium would be to allow test makers time to work on the difficult task of developing alternative tests capable of assessing meanings and diverse abilities in computation and problem-solving. With this convergence of technology, advances in knowledge of how children learn, and growing public interest, change may now be imminent.

Two goals of education have been discussed and illustrated—passing on tried-and-true knowledge to the next generaton, and preparing individuals capable of educating themselves and adapting to new situations in an evolving society. The first goal was interpreted from the perspectives of skills and meanings. In contrasting the interpretations of the two goals from the broad perspective of meaning, we found some areas of overlap in classroom practice for *meaningful computation* and *computation as an area of problem solving*. Selecting one as a major goal does not preclude the other from being a minor goal, thereby allowing a search for a useful blend of the two. Although in Piaget's view, the primary goal of education should be intellectual autonomy, his writings on education suggest a blend of methods. He often uses as a metaphor of an ideal situation the image of two classrooms—one with a teacher and one without a teacher (in Duckworth, 1964). He expands on this metaphor in the following way:

> We need pupils who are active, who learn early to find out for themselves, partly by their own spontaneous activity and partly through the materials we set up for them. (p. 5)

In stressing the importance of presenting problems in physical contexts and paying attention to the child's interests and attitude towards mathematics, Piaget wrote (1973):

> . . . Every normal student is capable of good mathematical reasoning if attention is directed to activities of his interest, and if by this method the emotional inhibitions that too often give him a feeling of inferiority in lessons in this area are removed. In most mathematical lessons the whole difference lies in the fact that the student is asked to accept from outside an already entirely organized intellectual discipline which he may or may not understand, while in the context of autonomous activity he is called upon to discover the relationships and ideas by himself, and to re-create them until the time when he will be happy to be guided and taught. (pp. 98–99)

It is important to find a blend of emphasis that is not only suitable for your children, but that also matches your own needs and abilities as a teacher.

REFLECTIONS AND DIRECTIONS

No grand curriculum design is available for teaching place value concepts to all children in every school situation. Nor is one possible, for various reasons. However, the preceding chapters have left you with resources for developing your own curriculum:

- Information on children's perspectives on number ideas, revealing both the power and the limitations of their thinking

- Theoretical framework for explaining how children's numerical thinking develops
- Framework for classroom implementation that provides alternative teaching methods and starting points for classroom activities that are congruent with children's numerical thinking
- Interviews tasks to learn about children's entry levels and progress.

Combine forces with at least one other person and design your own curriculum outline for teaching numeration and place value concepts, along with addition and subtraction algorithms, to the children in your class/school.

16

TEACHERS AS ACTIVE AGENTS OF CHANGE

PREVIEW

In recent years, the increasing federal influence on school programs and public disenchantment with the quality of education have been accompanied by the phenomenon of numerous people outside the classroom exercising power over teaching and learning in the classroom. Rather than being recognized as informed professionals, capable of independent decisions in the best interests of children's learning, teachers have been designated as intermediaries between legislated programs, extremely vocal parent groups, textbooks and standardized tests, and the children.

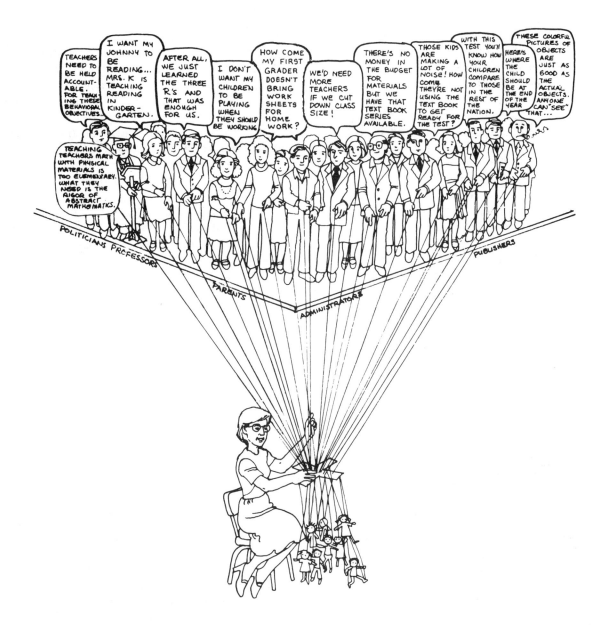

Is teaching a profession?

What role can individual teachers, and groups of teachers working in collaboration, play in 1). furthering the development of teaching as a profession, and 2). initiating change in the classroom based on both caring for children and knowledge of their numerical thinking?

In the traditional approach to education, children are expected to be passive and obedient learners—regurgitating information on call. Teachers are viewed as passive recipients of curricula and directives from on high with the expectation that they will be obedient intermediaries in the transfer of this curriculum to children as given. Yet in our study of children's numerical thinking, considerable evidence has been provided that children are not only active selectors and interpreters of what is passed on to them in terms of existing patterns of thought, but also reorganizers of these patterns in dealing with new input. Furthermore, children are not merely reactors to what is given; they are also initiators of interactions and generators of knowledge. This self-constructed knowledge often is generated despite what is being taught in the classroom.

Like children, teachers are both selective interpreters and initiators. Many innovative curricula have been imposed on teachers with no long-term in-service education to support the rethinking of teaching and learning needed to implement them successfully. As a result, ideas basic to these curricula have not "taken." Rather, they have been selectively screened out and interpreted in terms of each teacher's existing view of teaching and learning. On the other hand, some teachers have independently restructured their perspectives of teaching and learning, based on readings and classroom experiences. These teachers have been ready to implement innovative curricula despite inadequate in-service preparation. At the same time, other teachers have constructed their own innovative curricula or have made creative adaptations of available resources to fit their local conditions.

When innovative curriculum packages have failed to influence classroom teaching and learning in many classes, there is a swing of the pendulum in the opposite direction. The New Math movement was succeeded by a Back-to-Basics movement stressing the teaching of computation as isolated skills. Despite these new expectations, defined by long lists of objectives, textbooks, and standardized achievement tests, there are teachers who teach in ways that are contrary. These teachers are risking administrative sanctions by basing their teaching on a commitment to children and a deeper understanding of their numerical thinking. Teachers can and do act as autonomous professionals and as agents of change.

A profession is characterized by a body of knowledge that is not available to the general public and by the autonomy of its members to make independent decisions in applying this knowledge. Obviously, teaching is an underdeveloped profession. Yet rather than focusing on the inadequacies of professional preparation offered by universities and the lack of support provided by school systems and the general public, the next section will highlight ways in which individuals and groups of teachers can further their professional development and initiate positive, informed change in the schools. Rather than viewing the teacher as a pawn, it will focus on the teacher as an active agent of change.

SOME DIRECTIONS FOR CONTINUING PROFESSIONAL DEVELOPMENT AND INITIATING SCHOOL CHANGE

Schools themselves don't change; it's the people within them that do. Change, whether it relates to improving your professional status or introducing innovative curricula, is a complex and gradual process. There is no single way to approach the initiation of change. The following pages describe a number of possibilities for different teachers at different levels of involvement. You are invited to select those that are most appropriate for you at this time in your professional development and in your current school situation.

Be Selective About Courses and Workshops You Take

Search out child development, psychology, and education courses taught by people spending considerable time studying individual, and groups of, children. Also look for courses providing direct experiences with children rather than just lectures.

Children have a lot to teach us.

Find math content or methods courses that allow you to explore materials, to study your own process of learning—paying attention to your confusion and comparing your solutions with the approaches of others—and to improve your problem-solving abilities.

You'll only be able to allow children freedom to explore materials to the extent that you feel free.

You'll respect children's thinking once you respect your own.

You'll be comfortable with children's multiple solutions to problems and feel free to join them in solving problems having outcomes unknown to you to the extent that you've had similar experiences.

Seek out methods courses taught in the context of theory and research in thinking, learning, and teaching. Attend a theory and research session at the next mathematics teaching conference.

Justifying innovative methods to colleagues, principals, and parents requires that you be conversant with the underlying theory and research.

Even as this book goes into publication, other researchers are listening to children and are elaborating and refining the explanations offered here.

Professional growth is a continuous process.

Participate in a class for developing communication skills—one providing practice in active listening, asserting needs responsibly, and resolving conflicts.

Communication skills are a key factor in initiating school change, as well as in successful day-to-day living with a classroom of children.

Help to Make the Education of Children a Collaborative Effort

Team up with another teacher to schedule time for intensive interaction with individual children and small groups.

The views of number that you will uncover, although they may shock test makers, will allow you to begin adapting your teaching to the capacities of children.

Organize a study-and-support group of teachers for exploring children's numerical thinking. In it, you might:

- try some of the interview tasks described earlier with children and compare experiences
- use some of the interview tasks for multiplication, division, and fractions found in Appendix F
- devise your own tasks for studying numerical thinking

- bring in examples of children's numerical thinking from the classroom for discussion
- consider the implications of your findings for the classroom
- establish a research partnership with a university professor—not only collecting data in studying children, learning, and teaching, but also identifying research questions for study.

Social interaction is an effective way of expanding your understanding.

Listening to children and to other teachers is important in helping them work through their evolution.

Professional study need not be restricted to universities and course credits.

Teachers as researchers are an untapped natural resource.

Share portions of this book with fellow teachers and your administrator. Take the time to talk about them. Offer workshops to teachers and administrators for the viewing and discussion of selected videotaped interviews with children.

Examples of children's numerical thinking can serve as a discrepant event for provoking a rethinking of teaching and learning.

Dealing with the complexities of children's thinking when answers are not intact can clarify the need for smaller class size and for thoughtful reexamination of the mathematics curriculum. This consideration might include what is to be taught, when it is to be taught, and how progress is to be determined.

Organize parent workshops in which they:

- examine their own restricted use of school paper-and-pencil computational methods
- explore materials for developing understanding of alternative computational methods and for developing a network of relationships
- view and discuss videotapes of children's numerical thinking
- receive an invitation to visit, observe, and even participate in classroom math activities

Parents can be convinced of the value of innovative teaching practices through direct experiences that are insightful, challenging, and fun.

Parents can be valuable partners in the education of children.

Initiate Changes in the School Curriculum

Organize a study-and-support group of teachers for exploring changes in the school curriculum based on what is known about children's numerical thinking.

A concentration of teachers committed to this task provides both the intellectual stimulation and emotional support necessary for extending themselves, for withstanding external pressures to conform to traditional methods, and for initiating challenging teaching methods and innovative curriculum suited to the needs of children.

Be wary of prepared lists of behavioral objectives as a definition of the curriculum for all children. Develop your own objectives once you get to know your children. As you get to know them better, feel free to revise your objectives as needed.

Lists of objectives have the most meaning for those who prepare them when designed with specific children in mind.

Lists of objectives, however carefully designed, can still close the door on the unexpected.[1]

Reject using a single textbook as a definition of the mathematics curriculum indicating what is to be taught and when it is to be taught for all the children in your class. Develop your own unit of activities based on interviews and observations of children, together with the starting points found throughout this book. Supplement this unit by using available textbooks and other resources selectively. Although it is unreasonable to expect each teacher to prepare all learning materials for all topics in mathematics given the limitations of time and specialized background, it is important that he/she be aware of available resources. While experimenting with these activities, once again look to the children for feedback on difficulty, pacing, and so on.

Many resources other than textbooks—such as materials—are now available through state adoptions.

Children are an untapped natural resource for curriculum planning.

---

1. Wheeler et al. (1978).

Help parents and administrators become aware of the limited educational value and potential harm of standardized achievement tests. Suggest their elimination while providing alternative indicators of children's progress that are congruent with developing numerical thinking. Include selected interview tasks from this book as checkpoints in the teaching/learning process.

> *Testing should grow from what is taught, and what is taught should grow from who is taught. If the test fits, then it might be useful. Otherwise, the tail is wagging the dog.[2]*

Consult other resources for informed change found in Appendix G.

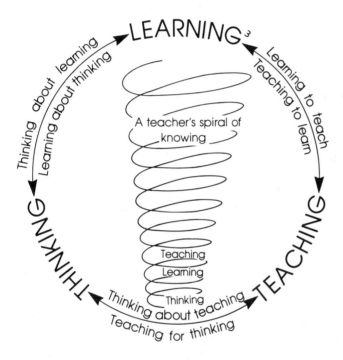

Develop a Coherent View of Education as a Basis for a Principled Stand

Reflect on the following declaration of professional conscience. Discuss it with other teachers. Revise it to fit your personal framework. Decide how you would be willing to discuss it with the school faculty, administrator, or parents.

2. Postman and Weingartner (1973), p. 36.

3. Labinowicz (1980).

Just as listening to children is important in helping them work through their evolution, listening to your professional conscience is important in identifying directions for your own evolution.

A public statement is a test of your professional commitment.

A Declaration of Professional Conscience for Teachers[4]

Our schools can be no better than the people working within them. These same people can determine the direction of change through a self-examination of their professional conscience and taking a principled stand.

> *As teachers we hereby declare ourselves to be in opposition to the industrialization of our schools (and the resulting dehumanization of teachers and children.) We pledge ourselves to become advocates on behalf of our children. We make the following declaration of professional conscience.*

> *We are professionals. We are prepared and will continue to prepare ourselves for teaching by constructing knowledge of human learning and development, pedagogy and curriculum.*

> *We continually update our knowledge of education, of our fields of instruction, and of the real world because of our professional dedication to use all means to improve our effectiveness as teachers.*

> *We expect school authorities to support us in our professionalism and self-improvement. And we will oppose all policies that restrict our professional autonomy to use new knowledge or new pedagogical practices on behalf of our students.*

> *We have no greater accountability than we owe our students. We will make the welfare of our students our most basic criterion for professional judgment. We will work with parents and policy makers to formulate programs that are in the best interests of our pupils. We will work with the children to personalize these programs.*

> *We pledge ourselves to make school a warm, friendly, supportive place in which all children are welcome. Our classrooms will be theirs. We will provide guidance and leadership to support our students in development of problem solving, decision making, and self-discipline. We will help them to build a sense of respect and support for each other.*

---

4. The declaration of professional conscience is an abbreviated form of a position paper circulated by Kenneth Goodman of the University of Arizona (1980). It has been adapted further with the addition of the paragraph on computational proficiency prepared with reference to O'Brien (1977b).

We acknowledge the importance of children's proficiency in computation, but within the broad context of understanding, flexibility of thinking, and problem solving. Research has shown that drill and practice alone, stressing speed and accuracy without understanding, is not an effective approach to teaching. To limit children to memory alone is to cheat them and to mislead them, if not to cripple them. The changing needs of our society demand going beyond right answers obtainable by single paper-and-pencil methods. They demand the flexibility of choosing between available resources—calculator, paper and pencil, or mental calculation methods—and the power of judging the reasonableness of answers through estimation or cross-checking with alternative methods. Furthermore, they demand the decision-making ability to apply correct procedures to problems not previously experienced. We therefore commit ourselves to teaching computational methods that encourage children's further intellectual development and survival in a changing world.

We use our knowledge base to support our students in their quest for knowledge. The real curriculum is what happens to each learner. We as teachers, are the curriculum planners and facilitators. We will not yield that professional responsibility to publishers of texts or management systems. We will select and use the best educational resources we can find but we will not permit ourselves or our pupils to be controlled by them.

We accept the responsibility for evaluation of our pupils' growth and we will make no long- or short-range decisions that affect the future education of our pupils on the basis of single examinations, no matter what the legal status of the examination. We will evaluate through ongoing monitoring of our pupils during our interactions with them. We will strive to know each pupil personally using all available professional tools to increase our understanding of each and every one.

Organize a study-and-support group of interested teachers, administrators, and parents for discussing the preceding statement of professional conscience and adapting it as the basis for planning an alternative school. Such an alternative school would offer the emotional support and the intellectual stimulation for teachers to extend themselves within a coherent program developed through continuing collaboration and study of children.

Teachers have the right to teach according to their professional conscience.

Parents have the right to influence the kind of education received by their children.

Children have the right to a quality education.

REFLECTIONS AND DIRECTIONS

Consider your teaching in terms of what you want to change, what you want to change it to, and what you are going to do to make that change. The following table[5] may help you to organize your thoughts and make some important decisions. By weighing the anticipated risks against the potential payoffs, you may decide the result is worth the risk. In considering the worst thing that could happen to you in risking change, you may discover that tenure laws give you considerable power.

WHAT I WANT TO CHANGE	WHAT I WANT IT TO BE/ PAYOFFS	HOW I'LL GET THERE	RISKS/RESISTANCES/ CATASTROPHIC EXPECTATIONS
1.			
2.			
3.			

---

5. Hunter (1972).

REFERENCES

Allardice, Barbara. "The Development of Representational Skills for Some Mathematical Concepts." Unpublished dissertation, Cornell University, 1977.

Alward, Keith, and Saxe, Geoffrey. *Exploring Children's Thinking: Part 3, Conservation.* San Francisco: Far West Laboratory for Educational Research and Development, 1975.

Aviv, Cherie. "Place Value Objectives." Paper presented at annual conference of the National Council of Teachers of Mathematics, Seattle, 1980.

Baratta-Lorton, Mary. *Mathematics Their Way.* Menlo Park, Calif.: Addison-Wesley, 1976.

————. *Workjobs II: Number Activities for Early Childhood.* Menlo Park, Calif.: Addison-Wesley, 1978.

Barclay, Tim. "Buggy—Outfitting for the Great Error Hunt." *Classroom Computer News,* March/April 1982, pp. 26-27.

Baroody, Arthur, and Ginsburg, Herbert. "Effects of Instruction on Children's Understanding of the 'Equals' Sign." Paper presented at the national AERA meeting, 1983.

Bednarz, Nadine, and Janvier, Bernadette. "The Understanding of Numeration in Primary School." *Educational Studies in Mathematics,* February 1982, pp. 33-57.

Behr, Merlyn. *Teaching Experiment: The Effect of Manipulatives in Second Graders' Learning of Mathematics.* PMDC Technical Report #11. Tallahassee: Florida State University, 1976 (ERIC ED 144809).

Behr, Merlyn; Erlwanger, Stanley; and Nichols, Eugene. *How Children View Equality Sentences,* PMDC Technical Report #8, Tallahassee: Florida State University, 1976.

Bell, Max, and Burns, Jean. "Counting and Numeration Capabilities of Primary School Children: A Preliminary Report." In T. Post and M. P. Roberts (Eds.), *Proceedings of the Third Annual Meeting of the North American Chapter of the International Group for the Psychology of Mathematics Education.* Minneapolis: University of Minnesota, 1981, pp. 17-23.

Brown, J. S., and Burton, R. B. "Diagnostic Models for Procedural Bugs in Basic Mathematical Skills." *Cognitive Science,* 1978, 2, pp. 155-92.

Brownell, William. "Psychological Considerations in the Learning and Teaching of Arithmetic." In W. D. Reeve (Ed.), *The Teaching of Arithmetic,* Tenth Yearbook of the National Council of Teachers of Mathematics. New York: Bureau of Publications, Teachers' College, Columbia University, 1935.

————. "Problem Solving." In N. Henry (Ed.), *The Psychology of Learning,* Forty-first Yearbook of the NSSE, Part Two. Chicago: University of Chicago Press, 1942, pp. 438-40.

————. "The Progressive Nature of Learning in Mathematics." *The Mathematics Teacher,* April 1944, pp. 147-57.

————. "When is Arithmetic Meaningful?" *Journal of Educational Research,* March 1945, pp. 491-98.

————. "The Place of Meaning in the Teaching of Arithmetic." *Elementary School Journal,* January 1947, pp. 256-65.

————. "Learning Theory and Educational Practice." *Journal of Educational Research,* March 1948, pp. 481-97.

Brownell, William, and Chazal, Charlotte. "The Effects of Premature Drill in Third Grade Arithmetic." *Journal of Educational Research,* 29 (1935): 17-28.

Burns, Marilyn. "Organizing the Learners." *Meeting the Math Challenge.* Palo Alto, Calif. The Learning Institute, 1980.

————. "How to Teach Problem-Solving." *Arithmetic Teacher,* February 1982, pp. 46-49.

Burns, Marilyn, and Richardson, Kathy. "Making Sense Out of Word Problems." *Learning,* January 1981, pp. 26-32.

Carpenter, Thomas. Personal communication, September 1981 (a).

————. "Heuristic Strategies Used to Solve Addition and Subtraction Problems." Paper presented at the annual Psychology of Mathematics Education conference, Minneapolis, September 1981 (b).

Carpenter, Thomas; Corbitt, Mary Kay; Kepner, Henry; Lindquist, Mary Montgomery; and Reys, Robert. *Results from the Second Mathematics Assessment of the National Assessment of Educational Progress.* Washington: National Council of Teachers of Mathematics, 1981.

Carpenter, Tom; Hiebert, James; and Moser, James. "The Effect of Problem Structure on First Graders' Initial Solution Processes for Simple Addition and Subtraction Problems." Wisconsin R&D Center for Individual Schooling Technical Report #516. Madison: University of Wisconsin, 1979.

Carpenter, Tom, and Moser, James. "The Development of Addition and Subtraction Problem-Solving Skills." Paper prepared for the Wingspread Conference on the Developmental Theory of Addition and Subtraction Skills, Racine, Wisconsin, November 1979.

Casey, Shirley. Personal communication, 1981.

Cochran, Beryl, Barson, Alan, and Davis, Robert. "Child-created Mathematics." *Arithmetic Teacher*, March 1970, pp. 211–215.

Curwin, Richard, and Fuhrmann, Barbara Schneider. *Discovering Your Teaching Self: Humanistic Approaches to Effective Teaching.* Englewood Cliffs, N. J.: Prentice-Hall, 1975.

Davidson, Patricia. Personal communication during the National Council of Teachers of Mathematics national meeting, Toronto, 1982.

Davies, H. B. "A Second Grader's Subtraction Techniques." *Mathematics Teaching*, 83 (1978): pp. 15–16.

Davis, Robert, and McKnight, Curtis. "The Influence of Semantic Content on Algorithmic Behavior." *The Journal of Mathematical Behavior*, Autumn 1980, pp. 39–87.

deBono, Edward. *The Mechanism of Mind.* Baltimore: Penguin, 1969.

DeRivera, Margaret. "Academic Testing and the Survival of Open Education." *EDC News*, Spring 1979, pp. 7–9.

Duckworth, Eleanor. "Piaget Rediscovered." In R. Ripple and V. Rockcastle (Eds.), *Piaget Rediscovered.* Ithaca: Cornell University, 1964, pp. 1–5.

_____. "The Virtues of Not Knowing." *Elementary Principal* 54 (1975): pp. 64–66.

_____. "An Introductory Note About Piaget." *Journal of Education*, Winter 1979, pp. 5–12.

_____. "Understanding Children's Understanding." Unpublished paper, DSRE Working Paper Series, Massachusetts Institute of Technology, Cambridge, 1981.

Easley, Jack. *On Clinical Studies in Mathematics Education.* Columbus, Ohio: ERIC Center for Science, Mathematics and Environmental Studies, June 1977.

_____. "Alternative Research Metaphors and the Social Context of Mathematics Teaching and Learning." *For the Learning of Mathematics*, July 1980, pp. 32–40. (a)

_____. "On Becoming a Resource Person." In J. Easley (Ed.), *Pedagogical Dialogues in Primary School Mathematics.* Urbana: University of Illinois Bureau of Educational Research, August 1980, pp. 159–170. (b)

Easley, Jack, and Zwoyer, Russell. "Teaching by Listening—Toward a New Day in the Math Classroom. *Contemporary Education*, Fall 1975, pp. 19–25.

Forman, George. Personal communication at the annual meeting of the Jean Piaget Society, 1980.

Fuson, Karen. "Counting Solution Procedures in Addition and Subtraction." Paper prepared for the Wingspread Conference on the Developmental Theory of Addition and Subtraction Skills, Racine, Wisconsin, November 1979.

_____. "The Development of Counting Words and of the Counting Act." Paper presented at the Fourth International Congress on Mathematical Education, Berkeley, California, August 1980.

_____. "An Analysis of the Counting-on Procedure in Addition." In T. Carpenter, T. Romberg, and J. Moser (Eds.), *Addition and Subtraction: A Cognitive Perspective.* Hillsdale, N. J.: Lawrence Erlbaum Associates, 1982, pp. 67–82.

Fuson, Karen, and Hall, James W. "The Acquisition of Early Number Word Meanings: A Conceptual Analysis and Review." In H. Ginsburg (Ed.), *The Development of Mathematical Thinking.* New York: Academic Press, 1982.

Fuson, Karen, and Mierkiewicz, Diane. "A Detailed Analysis of the Act of Counting." Paper presented at the annual meeting of the American Educational Research Association, Boston, April 1980.

Fuson, Karen; Richards, John; and Briars, Diane. "The Acquisition and Elaboration of the Number Word Sequence." In C. Brainerd (Ed.), *Progress in Cognitive Development, Vol. I: Children's Logical and Mathematical Cognition.* New York: Academic Press, 1982.

Fuys, David, and Tischler, Rosamond. *Teaching Mathematics in the Elementary School*, Boston: Little, Brown, 1979.

Gelman, Rochel. "Counting in the Preschooler: What Does and Does Not Develop." In R. Siegler (Ed.), *Children's Thinking—What Develops.* Hillsdale, N. J.: 1978, pp. 213–241.

Gelman, Rochel, and Callistel, C. R. *The Child's Understanding of Number.* Cambridge, Mass.: Harvard University Press, 1978.

Ginsburg, Herbert. *Children's Arithmetic: The Learning Process.* New York: Van Nostrand, 1977.

Ginsburg, Herbert; Baroody, Arthur; and Russell, Robert. "Children's Estimation Ability in Addition and Subtraction." *Focus on Learning Problems in Mathematics*, Spring 1982, pp. 31–46.

Glass, Gene, and Smith, Mary. *Meta-Analysis of Research on Class Size and Achievement.* Boulder, Colo.: Laboratory of Educational Research, University of Colorado, 1978.

Goodman, Kenneth. "A Declaration of Professional Conscience for Teachers." Unpublished paper, University of Arizona, Tucson, 1980.

Hamrick, Kathy. "Oral Language and Readiness for the Written Symbolization of Addition and Subtraction." *Journal for Research in Mathematics Education*, May 1979, pp. 188–94.

Hatano, Giyoo. "Learning to Add and Subtract: A Japanese Perspective." In T. Carpenter, T. Romberg, and J. Moser (Eds.), *Addition and Subtraction: A Cognitive Perspective.* Hillsdale, N.J.: Lawrence Erlbaum Associates, 1982.

Hatfield, Larry. Unpublished paper, 1976.

———. "Young Children's Counting Strategies: Constructing and Learning Alternative Algorithms, a Teaching Experiment with Second Grade Children." Videotape, University of Georgia, Athens, 1978.

———. "Invention in Mathematics Learning." A panel presentation at the research presession of the annual meeting of the National Council of Teachers of Mathematics, St. Louis, 1981.

Hendrickson, Dean. Personal communication, 1980.

———. "Language and Symbolic Form in Children's Mathematics." In T. Post and M. Roberts (Eds.), *Proceedings of the Third Annual Meeting of the North American Chapter of the International Group for the Psychology of Mathematics Education.* Minneapolis: University of Minnesota, September 1981.

———. "A Psychologically Sound Primary School Mathematics Curriculum." *Arithmetic Teacher,* January 1983, pp. 42–47. (a)

———. "Prevention or Cure? Another Look at Mathematics Learning Problems." In *Interdisciplinary Voices in Learning Disabilities and Remedial Education.* Austin, Tex.: Pro-Ed, 1983, pp. 93–106. (b)

———. "Verbal Multiplication and Division Problems: Some Difficulties and Solutions." *Arithmetic Teacher* (in press).

Hendrickson, Dean, and Thompson, Charles. "Verbal Addition and Subtraction Problems: Some Difficulties and Some Solutions." Unpublished paper, University of Minnesota, Duluth, 1982.

Hickerson, J. Allen. *Guiding Children's Arithmetic.* New York: Prentice-Hall, 1952.

Hill, Shirley. "Recommendations for School Mathematics Programs of the 1980s." In M. Montgomery Lindquist (Ed.), *Selected Issues in Mathematics Education.* Berkeley, Calif.: McCutchan, 1980, pp. 258–68.

Hudson, Tom. "Correspondences and Numerical Differences Between Disjoint Sets." Paper presented at the Fourth International Congress on Mathematical Education, Berkeley, California, August 1980.

Hunter, Elizabeth. *Encounter in the Classroom: New Way of Teaching.* New York: Holt, Rinehart & Winston, 1972.

Inhelder, Barbel; Sinclair, Hermine; and Bovet, Magali. *Learning and the Development of Cognition.* Cambridge, Mass.: Harvard University Press, 1974.

Irons, Calvin, and Jones, Graham. *"Correcting and Preventing Learned Disabilities in Primary School Mathematics.* Unpublished monograph, Kelvin Grove College of Advanced Education, Brisbane, Australia, 1979.

Jacobson, Marilyn; Lester, Frank; and Stengel, Arthur. "Making Problem Solving Come Alive in the Intermediate Grades." In S. Krulick and R. Reys (Eds.), *Problem Solving in Mathematics.* Reston, Va.: National Council of Teachers of Mathematics, 1980, pp. 127–35.

Jordan, James. Videotape of classroom interaction, Washington State University, Pullman, 1979.

Johnson, Paul. *From Sticks and Stones: Personal Adventures in Mathematics.* Chicago: Science Research Associates, 1975.

Kamii, Constance. "Pedagogical Principles Derived from Piaget's Theory: Relevance for Educational Practice." In M. Schwebel and J. Raph (Eds.), *Piaget in the Classroom.* New York: Basic Books, 1973, pp. 199–215.

———. "Equations in First Grade Arithmetic: A Problem for the Disadvantaged or for First Graders in General?" Paper presented at the annual meeting of the American Educational Research Association, April 1980. (a)

———. Personal communication, 1980. (b)

———. *Number in Preschool and Kindergarten: Educational Implications of Piaget's Theory.* Washington, D.C.: National Association for the Education of Young Children, 1982. (a)

———. "First Graders Invent Arithmetic: Piaget Theory Used in the Classroom." Paper presented at the annual Psychology of Mathematics Education meeting, October 1982. (b)

Kamii, Constance, and DeVries, Rheta. "Piaget for Early Education." In M. C. Day and R. K. Parker (Eds.), *The Preschool in Action: Exploring Early Childhood Programs.* Boston: Allyn & Bacon, 1977, pp. 365–420.

———. *Group Games in Early Education: Implications of Piaget's Theory.* Washington, D.C.: National Association for the Education of Young Children, 1980.

Kamii, Mieko. "Children's Efforts to Find a Correspondence between Digits and Number of Objects." Paper presented at the annual symposium of the Jean Piaget Society, Philadelphia, May 1980.

———. "Children's Ideas About Written Number." *Topics in Learning and Learning Disabilities,* October 1981, pp. 47–60.

Kilpatrick, Jeremy. "What's a Problem?" Paper presented at the annual meeting of the National Council of Teachers of Mathematics, Toronto, 1982.

Kohlberg, Lawrence and Mayer, Rochelle. "Development as the Aim of Education." *Harvard Educational Review,* Nov. 1972, pp. 449–543.

Krulik, Stephen, and Reys, Robert (Eds.). *Problem Solving in School Mathematics.* Reston, Va.: National Council of Teachers of Mathematics, 1980.

Labinowicz, Ed. *The Piaget Primer: Thinking, Learning, Teaching.* Menlo Park, Calif. Addison-Wesley, 1980.

Labinowicz, Shirley. Videotaped classroom lessons and interview, Los Angeles, 1977.

Lankford, Francis. "What Can a Teacher Learn About a Pupil's Thinking Through Oral Interviews?" *The Arithmetic Teacher,* January 1974, pp. 26–32.

Lerch, Harold. *Active Learning Experiences for Teaching Elementary School Mathematics.* Boston: Houghton Mifflin, 1981.

Lester, Frank. "Problem Solving: Is It a Problem?" In M.

Montgomery Lindquist (Ed.), *Selected Issues in Mathematics Education.* Berkeley, Calif.: McCutchan, 1980, pp. 29–45.

Leutzinger, Larry. "The Effects of Counting-on on the Acquisition of Addition Facts in Grade One." Unpublished dissertation, University of Iowa, 1979.

Levin, James. "Estimation Techniques for Arithmetic: Everyday Math and Mathematics Instruction." *Educational Studies in Mathematics* 12 (1981): pp. 421–34.

Lindvall, C. Mauritz. "Some Implications for Teaching of Arithmetic Story Problems Derived from Research in Cognitive Sciences." Paper presented at the annual meeting of the National Council of Teachers of Mathematics, Toronto, 1982.

Lowery, Lawrence. *Learning About Instruction: Questioning Strategies.* Berkeley: University of California, 1975.

Madell, Robert. "Addition and Subtraction—What Do Children Naturally Do?" Paper presented at the National Council of Teachers of Mathematics annual meeting, San Diego, 1978.

_____. "Children's Natural Arithmetic Processes." Unpublished paper, 1980.

Madell, Robert, and Larkin Stahl, Elizabeth. *Picturing Subtraction—From Models to Symbols.* Palo Alto, Calif.: Creative Publications, 1977.

Martin, David. "Your Praise Can Smother Learning." *Learning,* February 1977, pp. 43–51.

Markman, E. M. "Classes and Collections: Conceptual Organization and Numerical Abilities." *Cognitive Psychology* 1 (1979): 395–411.

Moser, James. "What Is the Evidence that Children Invent Problem Solving Strategies?" In R. Karplus (Ed.), *Proceedings of the Fourth International Conference for the Psychology of Mathematics Education,* University of California, Berkeley, 1980, pp. 312–316.

Musser, Gary. "Let's Teach Mental Algorithms for Addition and Subtraction." *Arithmetic Teacher,* April 1982, pp. 40–42.

National Advisory Committee on Mathematics Education (NACOME). *Overview and Analysis of School Mathematics, Grades K–12.* Reston, Va.: National Council of Teachers of Mathematics, 1975.

Nesher, Pearla; Greeno, James; and Riley, Mary. "Semantics Categories Reconsidered: Developmental Levels" Unpublished paper, Division of Study and Research in Education, Massachusetts Institute of Technology, 1981.

Nichols, Eugene. *First and Second-Grade Children's Interpretation of Action Upon Objects.* PMDC Technical Report #14. Tallahassee: Florida State University, 1976.

Nuttall, Thomas. "Reader's Dialogue." *Arithmetic Teacher,* October 1980, p. 55.

O'Brien, Thomas. "Addition and Subtraction." Unpublished paper, Teachers' Center Project, Southern Illinois University, 1975.

_____. "Forward to Basics." *The Genetic Epistemologist,* July 1977, pp. 1–2. (a)

_____. "More on Basics." *The Genetic Epistemologist,* October 1977, p. 1. (b)

_____. *Puzzle Tables.* New York: Cuisenaire Company of America, 1980.

_____. "Five Anchors." In T. O'Brien (Ed.), *Toward the 21st Century in Mathematics Education.* Edwardsville, Ill.: Southern Illinois University, Teachers' Center Project, 1982, pp. 95–106.

_____. "What Is Basic? A Constructivist View." In *What is Basic? Issues and Choices.* Washington, D.C.: National Institute of Education, 1982 b.

O'Brien, Thomas, and Casey, Shirley. Personal communication, 1982 a.

_____. "Inference and Number—Work in Progress." *Seedbed.* Edwardsville, Ill.: Southern Illinois University Teachers' Center Project, 1982 b.

Omanson, Susan; Peled, Irit; and Resnick, Lauren. "Instruction by Mapping: Its Effects on Understanding and Skill in Subtraction." Paper presented at the annual meeting of the American Educational Research Association, New York, March 1982.

Opper, Sylvia. "Piaget's Clinical Method." *The Journal of Children's Mathematical Behavior,* Spring 1977, pp. 90–107.

Papert, Seymour. *Mindstorms: Children, Computers and Powerful Ideas.* New York: Basic Books, 1980.

Piaget, Jean. "Development and Learning." In R. Ripple and V. Rockcastle (Eds.), *Piaget Rediscovered.* Ithaca: Cornell University, 1964, pp. 7–20.

_____. *The Child's Conception of Number.* New York: W. W. Norton, 1965.

_____. *Six Psychological Studies.* New York: Random House, 1967.

_____. *Science of Education and the Psychology of the Child.* New York: Viking, 1971.

_____. *To Understand Is to Invent: The Future of Education.* New York: Grossman, 1973.

_____. *The Child and Reality.* New York: Viking, 1974.

_____. "Comments on Mathematical Education." In H. Gruber and J. Vonech (Eds.), *The Essential Piaget: An Interpretive Reference and Guide.* New York: Basic Books, 1977, pp. 726–32.

Piaget, Jean, and Duckworth, Eleanor. "Piaget Takes a Teacher's Look." *Learning,* October 1973, pp. 22–27.

Piaget, Jean, and Inhelder, Barbel. *Mental Imagery and the Child.* New York: Basic Books, 1966.

_____. *The Psychology of the Child.* New York: Basic Books, 1969.

Postman, Neil, and Weingartner, Charles. *How to Recognize a Good School.* Bloomington, Ind.: Phi Delta Kappa, 1973, p. 36.

Project Torque. "A New Approach to the Assessment of Children's Mathematical Competence." Project Torque, Education Development Center, Newton, Mass., Spring 1976.

Rathmell, Edward. "Using Thinking Strategies to Teach the Basic Facts." In M. Suydam and R. Reys (Eds.), *Developing Computational Skills*. Reston, Va.: National Council of Teachers of Mathematics, 1978, pp. 13–38.

_____. "What Can I Do for Numeration on Monday?" Paper presented in a work session at the annual meeting of the National Council of Teachers of Mathematics, Seattle, April 1979.

Reid, Kim. "Commentary." *Topics in Learning and Learning Disabilities*, April 1981, pp. 69–70.

Resnick, Lauren. "Syntax and Semantics in Subtraction." In T. Carpenter, T. Romberg, and J. Moser (Eds.), *Addition and Subtraction: A Cognitive Perspective*. Hillsdale, N. J.: Lawrence Erlbaum Associates, 1982, pp. 136–55. (a)

_____. "A Developmental Theory of Number Understanding." In H. Ginsburg (Ed.), *The Development of Mathematical Thinking*. New York: Academic Press, 1982 (b)

Reyes, A. G., and Suarez, J. A. "El Numaro y la Solucion de Problemas Arithmetico." Tesis profesional, Universidad Nacional Autonoma de Mexico, Facultad de Psicologia, 1979.

Riley, Mary. "Effect of Semantic Structure on the Solution of Arithmetic Problems." Paper presented at the annual meeting of the American Educational Research Association, San Francisco, 1979.

Riley, Mary, and Robinson, Mitchell. "An Approach to Assessing the Knowledge Underlying Children's Performance on Arithmetic Word Problems." Paper presented at the annual meeting of the American Educational Research Association, Boston, April 1980.

Riley, Mary; Greeno, James; and Heller, Joan. "Development of Children's Problem-Solving Ability in Arithmetic." In H. Ginsburg (Ed.), *The Development of Mathematical Thinking*. New York: Academic Press, 1982.

Romberg, T.; Harvey, J.; Moser, J.; and Montgomery, M. *Developing Mathematical Processes*. Chicago: Rand McNally, 1974.

Roper, Ann, and Harvey, Linda. *The Pattern Factory: Elementary Problem Solving Through Patterning*. Palo Alto, Calif.: Creative Publications, 1980.

Rowe, Mary Budd. *Teaching Science as Continuous Inquiry*. New York: McGraw-Hill, 1973.

Russell, Robert. "The Cognitive Nature of Learning Disabilities in Mathematics." Unpublished dissertation, Cornell University, 1981.

Scheer, Janet. "The Etiquette of Diagnosis." *Arithmetic Teacher*, May 1980, pp. 18–19.

Secada, Walter; Fuson, Karen; and Hall, James. "The Transition from Counting-All to Counting-On in Addition." *Journal for Research in Mathematics Education*, January 1983, pp. 47–57.

Sharma, Mahesh. "Visual Clustering and Number Concepts." *Math Notebook*, December 1981, pp. 1–4.

Sinclair, A.; Siegrist, F.; and Sinclair, H. "Young Children's Ideas About the Written Number System." Paper presented at the NATO Conference on the Acquisition of Symbolic Skills, University of Keele, July 1982.

Sinclair, Hermine. "Recent Piagetian Research in Learning Studies." In M. Schwebel and J. Raph (Eds.), *Piaget in the Classroom*. New York: Basic Books, 1973, pp. 57–72.

_____. "From Preoperational to Concrete Operational Thinking and Parallel Development of Symbolization." In M. Schwebel and J. Raph (Eds.), *Piaget in the Classroom*. New York: Basic Books, 1973, pp. 40–56.

Sinclair, Hermine. "Epistemology and the Study of Language." In B. Inhelder and H. Chapman (Eds.), *Piaget and His School: A Reader in Developmental Psychology*, New York: Springer-Verlag, 1976, pp. 205–18.

_____. "Intellectual Development, Research and Education." Excerpts from a seminar at the Southern Illinois University Teachers' Center Project, Edwardsville, October 1977.

_____. "The Child as Scientist." Paper presented at the Southern Illinois University Teachers' Center Project, Edwardsville, 1979.

Smith, Mary, and Glass, Gene. *Relationship of Class Size to Classroom Processes, Teacher Satisfaction and Pupil Affect: A Meta-Analysis*. San Francisco: Far West Laboratory, July 1979.

Souviney, Randall; Keyser, Tamara; and Sarver, Allan. *Mathmatters*. Santa Monica, Calif.: Goodyear, 1978.

Stake, Bernadine. "Clinical Studies of Counting Problems with Primary School Children." Unpublished dissertation, University of Illinois, Urbana, 1979.

Steffe, Leslie. Videotape of counting tasks, University of Georgia, Athens, 1977.

_____. Unpublished paper, University of Georgia, Athens, 1978.

_____. Videotape of classroom interaction, University of Georgia, Athens, 1979. (a)

_____. "A Reply to 'Formal Thinking Strategies: A Prerequisite for Learning Basic Facts'." *Journal for Research in Mathematics Education*, November 1979, pp. 370–73. (b)

Steffe, Leslie; Herstein, James; and Spikes, Curtis. "Quantitative Comparisons and Class Inclusion as Readiness Variables for Learning First-Grade Arithmetc Content." PMDC Report #9. Athens: University of Georgia, 1976.

Steffe, Leslie, and Thompson, Patrick. "Children's Counting in Arithmetical Problem Solving." Paper presented at the Wingspread Conference on Initial Learning of Addition and Subtraction, Racine, Wisconsin, November 1979.

Steffe, Leslie; Thompson, Pat; and Richards, John. "Children's Counting and Arithmetical Problem Solving." In T. Romberg, T. Carpenter, and J. Moser (Eds.), *Addition and Subtraction: A Cognitive Perspective*. Hillsdale, N. J.: Lawrence Erlbaum Associates, 1982.

Steiner, Gerhard. "Number Learning as Constructing Coherent Networks by Using Piaget-Derived Operations." Paper presented at the International Congress for Mathematics Education, Berkeley, California, August 1980.

Suydam, Marilyn, and Reys, Robert (Eds.). *Developing Computational Skills*. Reston, Va.: National Council of Teachers of Mathematics, 1978.

Swart, William. *Tricon Mathematics: Objectives and Methods for Addition and Subtraction of Whole Numbers*. Mt. Pleasant, Mich.: Central Michigan University Press, 1980. (a)

————. "Some Findings and Some Conjectures on Conceptual Development of Computational Skills through Manipulative Material." Unpublished manuscript, 1980. (b)

Swenson, Esther J. "Organization and Generalization as Factors in Learning, Transfer and Retroactive Inhibition." In *Learning Theory in School Situations*. University of Minnesota Studies in Education, no. 2. Minneapolis: University of Minnesota Press, 1949.

Taylor, Harold. "Primary School Mathematics." In J. Easley (Ed.), *Pedagogical Dialogs in Primary Mathematics*. Urbana, Ill.: University of Illinois, 1980.

Thiele, C. L. *The Contribution of Generalization to the Learning of Addition Facts*. New York: Bureau of Publications, Teacher's College, Columbia University, 1938.

Thompson, Charles, and Van de Walle, John. "Transition Boards: Moving from Materials to Symbols in Addition." Paper presented at the National Council of Teachers of Mathematics annual meeting, Seattle, 1980.

————. "Transition Boards: Moving from Materials to Symbols in Subtraction." *Arithmetic Teacher*, January 1981, pp. 4–8.

Thompson, Patrick. "Children's Schemas in Solving Problems Involving Whole Number Numeration." Paper presented at the annual meeting of the American Educational Research Association, New York, March 1982. (a)

————. "A Theoretical Framework for Understanding Young Children's Concepts of Whole Number Numeration." Unpublished dissertation, University of Georgia, Athens, 1982. (b)

Thornton, Carol. "Emphasizing Thinking Strategies in Basic Fact Instruction." *Journal for Research in Mathematics Education*, May 1978, pp. 214–27.

Thornton, Carol; Tucker, Benny; Dossey, John; and Bazik, Edna. *Teaching Mathematics to Children with Special Needs*. Menlo Park, Calif.: Addison-Wesley, 1982.

Timnick, Lois. "Electronic Bullies." *Psychology Today*, February 1982, pp. 10–11.

Tougher, Hallie Ephron. "Too Many Blanks! What Workbooks Don't Teach." *Arithmetic Teacher*, February 1981, p. 67.

Trafton, Paul. "Estimation and Mental Arithmetic: Important Components of Computation." In M. Suydam and R. Reys (Eds.), *Developing Computational Skills*. Reston, Va.: National Council of Teachers of Mathematics, 1978.

Underhill, Robert. "Strategies for Diagnosing and Enhancing Learning of Place Value Concepts." Paper presented at the annual meeting of the Research Council for Diagnostic and Prescriptive Mathematics, Bowling Green, Ohio. 1983.

Van Allen, Roach. "A Language Experience Approach to Reading." In M. Douglas (Ed.), *Claremont Reading Conference Twenty-Fifth Yearbook*. Claremont, Calif.: Claremont College, 1961, pp. 59–62.

Van de Walle, John. "An Investigation of the Concepts of Equality and Mathematical Symbolism Held by First-, Second- and Third-Grade Children: An Informal Report." Paper presented at the annual meeting of the National Council of Teachers of Mathematics, Seattle, 1980.

von Glasersfeld. "Subitizing: The Importance of Figural, Non-Numerical Skills." Unpublished paper, University of Georgia, 1980.

Weinzweig, A. I. "Mathematics for the Twenty-First Century." In T. O'Brien (Ed.), *Toward the 21st Century in Mathematics Education*. Edwardsville, Ill.: Southern Illinois University Teachers' Center Project, 1982, pp. 34–53.

West, Tommy. "Looking at Some Hard Spots." *Arithmetic Teacher*, September 1978.

Wheeler, David. "Humanising Mathematical Education." *Mathematics Teaching*, September 1975, pp. 4–9.

————. (Ed.). *Notes on Mathematics for Children*. London: Cambridge University Press, 1977.

Willoughby, Steven; Bereiter, Carl; Hilton, Peter; Rubinstein, Joseph; and Scardamalia, Marlene. *Bargains Galore—Thinking Story Book, Level Three*. Real Math Program. La Salle, Ill.: Open Court, 1981.

Wirtz, Robert. *Drill and Practice at the Problem Solving Level*. Monterey, Calif.: Curriculum Development Associates, 1974.

————. *Banking on Problem Solving*. Monterey, Calif.: Curriculum Development Associates, 1976.

————. *New Beginnings: Guide to Think • Talk • Read Math Centers for Beginners*. Monterey, Calif.: Curriculum Development Associates, 1980.

————. *Evaluation Strategies Focused on Problem Solving Applications, Skills and Attitudes*. Monterey, Calif.: Curriculum Development Associates, 1981.

————. *Think About It Newsletter*. Monterey, Calif.: Curriculum Development Associates, 1982. (a)

————. *Thursday Math Sampler*. Monterey, Calif.: Curriculum Development Associates, 1982. (b)

————. "Counting—The Root of Much Trouble." Presentation made at the annual conference of the Greater San Diego Mathematics Council, February 5, 1982. (c)

Ziajka, Alan. "Reading and Knowing: A Comparison of Ideas of Peter Spencer and Jean Piaget." Unpublished paper, Claremont, College, 1978.

APPENDIX A:
THE PROGRESSION PUZZLE

(De Bono, 1969)

Cut out these pieces to solve the problem presented in
Chapter 1, p. 17.

APPENDIX B:
INTERVIEW HINTS

SOME HINTS ABOUT TAPE-RECORDING INTERVIEWS

Student interviewers are often frustrated at either not getting a recording of their interviews or not getting a good recording. The following suggestions are aimed at helping you avoid such problems.

1. Be sure that you are thoroughly familiar with the operation of your recorder. Know which cord to connect and which button to push. Try it out beforehand and play back the tape to check the sound quality. If your batteries are fresh you will have more flexibility in the location of the interview. Otherwise you will need to use an extension cord.

2. A separate interview room is the ideal situation. In any other location, it is critical that the child's voice be audible on the tape. Placing a microphone closer to the child than to the interviewer, as well as taking precautions to reduce any background noise, may be helpful. If you are using a recorder with a separate microphone, do not rest it on the recorder. Also, use a soft cloth or newspaper to cover the working area where children will be manipulating objects.

SOME HINTS ON USING THE TAPE RECORDER FOR ANALYSIS OF INTERACTION

Most tape analyses will require playing the entire tape, or segments of it, over more than once to capture and check the information. An earplug provides the necessary privacy for concentration on the task. The counter allows you to locate and replay interactions of interest without losing time in a trial-and-error search, provided you jot down the key locations. The pause button allows frequent stopping and starting for verbatim transcription or for interpretation and tally of an interaction. Although you will want to analyze entire interviews to follow the child's thinking, you need only analyze portions of it—five to ten minutes—to check your listening and responding behaviors. Usually, you listen to the tape and focus on the analysis of one type of behavior. However, it is possible to examine your listening and responding simultaneously. The data table on the next page can be used for either separate or simultaneous study—it can be used for a study of your responding behaviors alone or for a combined study of listening and responding.

ANALYZING YOUR LISTENING AND RESPONDING PATTERNS

The purpose of this activity is to gain an awareness of both your listening and responding patterns in your interactions with children.

Listen to a ten minute taped segment containing the most teacher-child interaction and (1) tally each type of response used,[1] (2) measure your wait time to the nearest 0.5 second by using a digital watch or the sweep hand of a regular watch.

---

1. The responding categories are adapted from Curwin and Fuhrmann (1975).

CHILD ANSWERS CORRECTLY

	TALLY	WAIT TIMES
- Praises child:		
"Good"		
"Terrific"		
"Good thinking"		
"Fine"		
Other (indicate) _____		
- Says "OK"		
- Accepts response:		
"Uhuhm"		
"That's an interesting way," etc.		
Repeats answer/rephrases		
Other: _____		
- Thanks the child		
- Adds to answer		
- Asks child for evidence/explanation		
- Uses countersuggestion		
- Asks for another way to show/explain		
- Asks for clarification		
- Uses answer to go on to next task		
- Looks pleased		
- Looks neutral though friendly		
- Other: _____		

CHILD ANSWERS INCORRECTLY

	TALLY	WAIT TIMES
- Rejects response:		
Says "No"		
Says "No, try again."		
- Encourages and asks to try again		
- Praises child for a good effort		
- Corrects response		
- Corrects and asks child to agree		
- Accepts response		
"Uhuh," "OK," "I see"		
Repeats response		
"That's an interesting way," etc.		
- Asks the child for reasoning		
- Repeats question		
- Rephrases question to provide cues		
- Invents a new question/task to check hunch about child's reasoning		
- Uses countersuggestion		
- Ignores response and goes on to next		
- Looks displeased		
- Looks neutral, yet friendly		
- Refers child to materials		

CHILD DOES NOT RESPOND IMMEDIATELY

	TALLY	WAIT TIMES
- Gives encouragement		
- Repeats initial question		
- Rephrases initial question		
- Refers to materials		
- Others: _____		

The purpose of the analysis suggested is to interpret your listening and responding patterns and to identify specific areas for change. Some of the questions also go beyond this initial analysis to follow-up interviews studied for indications of progress. Select and respond to those questions that are appropriate for you or make up your own questions for a deeper analysis.

Listening

1. Look over your data sheet and identify the shortest and longest wait times.

2. Calculate your average wait time $\frac{\text{Total wait time (sec.)}}{\text{\# of potential pauses}}$ showing the figures used.[2]

3. Comment on the effect of your wait time on the quality of the children's responses.

4. If you've done more than one analysis, what similarities and differences exist between the average wait times on your first and second tapes?

5. In what direction do you want to move on the information available here? How will you do so?

Responding

1. Total the tallies for each type of response. Identify the types of responses used most frequently, second-most, and third-most.

2. If you use "OK" often, decide whether it serves as a "gap filler," an acceptance, or a sanction. If it serves varied purposes, can the children detect your shifting intent?

3. Comment on the effect of your responding behaviors on the quality of children's thinking during the interviews.

4. If you've completed two analyses, what similarities and differences exist in your response patterns on the two tapes?

5. Which of these response patterns do you wish to maintain? Which do you wish to change? Devise a plan for achieving this change.

Listening and Responding

1. In looking over your data sheet, do you see that certain responses are accompanied by short wait times? Long wait times? Try to explain any patterns noticed.

ANALYSIS OF CHILDREN'S THINKING

The following table has been used by student interviewers as a convenient method of summarizing children's responses for comparison and analysis of underlying thinking, and communication to others. It is usually done on a large sheet of paper.

---

2. Even if the potential pause is nonexistent, it must be included in a realistic wait time average. For example, the following wait times (1.5 sec., 3.0 sec., 0.0 sec., 2.5 sec., 1.0 sec., 0.0 sec.) would be totaled and divided by 6.

	CHILDREN'S RESPONSES					
TASK DESCRIPTION	Child 1 age — grade —	Child 2 age — grade —	Child 3 age — grade —	Child 4 age — grade —	Child 5 age — grade —	DISCUSSION OF THINKING ON EACH TASK
1. Standard opening questions should be identified. Description may include a diagram of the task.						
2.						
3.						
4.						
5.						
Discussion of each Child's Thinking						

Such an expanded table format facilitates identification of thinking patterns but loses information on spontaneous interactions that arise during the interview. Yet the use of verbatim transcripts for each of the interviews is excessively time consuming. A compromise that fills the gap is to supplement the summary table with transcripts of selected spontaneous interactions in which you rephrased questions and invented new ones in following the child's thinking.

APPENDIX C:
THE TRANSITION FROM COUNTING-ALL TO COUNTING-ON IN ADDITION

(Secada, Fuson, and Hall, 1983)

ABSTRACT

An analysis of the transition from counting-all to counting-on identified three subskills: (1) counting-up from an arbitrary point, (2) shifting from the cardinal to the counting meaning of the first addend, and (3) beginning the count of the second addend with the next counting word. First-grade children were given a test to classify them as counting-all versus counting-on, followed by tests of the subskills. Adequate subskill performance was strongly related to counting-on: All 28 count-on children demonstrated all three subskills or incorporated subskill 2 into subskill 3, whereas 36 of 45 count-all children failed to demonstrate one or more subskills. A posttest indicated that the subskills assessment alone had induced counting-on for 7 of the 9 count-all children who demonstrated all subskills. A random half of the children who initially lacked subskills 2 and 3 were taught them in a single session and retested. Seven of the 8 instructed children counted-on, compared to 1 of 8 not instructed. The conceptual bases for the subskills are discussed.

COUNTING-ON TESTS

Two forms of the test were given, each for three trials. In all cases the first addend was between 12 and 19 and the second was between 6 and 9. A child who counted-on in one or more trials was classified as capable of counting-on.

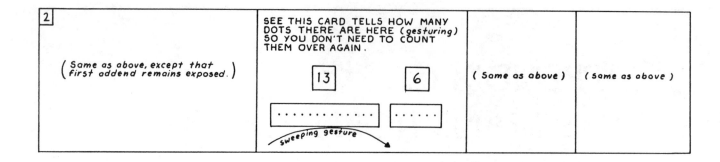

SUBSKILL ASSESSMENT TESTS

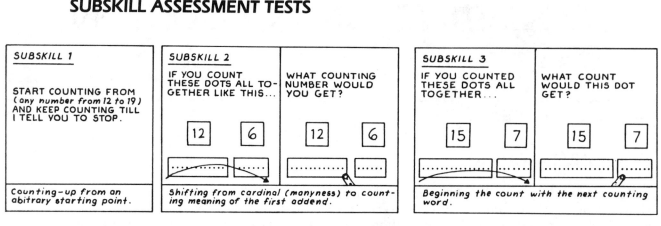

Four trials were given for each subskill. To be considered as demonstrating the subskill, a child had to exhibit it on three consecutive trials. Subskill 3 requires that the child shift from regarding the objects in each addend set as being only in the separate sets to regarding them also as objects within the count of the sum. In 15 + 7, "fifteen, sixteen" serve as a counting-word bridge connecting the two addends within the sum count. The key to counting-on seems to be the ability to consider the addends simultaneously as parts and as composing the whole while counting the second addend.

SUBSKILLS TEACHING PROCEDURE

Success on four consecutive trials was used as the indicator of subskill attainment.

You are invited to read the original report in the *Journal for Research in Mathematics Education* (January, 1983, pp. 45–57) for further details of the study.

APPENDIX D: SYSTEM FOR TEACHING DERIVED STRATEGIES

By knowing the combinations along the inside perimeter of the table and the combinations along the diagonals (the doubles and the combinations totaling 10), the rest of the 121 standard addition combinations can be generated by thinking in terms of "one more or less than" or "two more or less."

+	0	1	2	3	4	5	6	7	8	9	10
0	0	1	2	3	4	5	6	7	8	9	10
1	1	2								10	11
2	2		4						10		12
3	3			6				10			13
4	4				8		10				14
5	5					10					15
6	6				10		12				16
7	7			10				14			17
8	8		10						16		18
9	9	10								18	19
10	10	11	12	13	14	15	16	17	18	19	20

The table is a general representation of a more detailed system for direct teaching of derived strategies developed by Taylor (in Easley, 1980).

The tabular form of Taylor's system is more suited for use by adults than by young children. If children are to be encouraged in their search for patterns and construction of derived strategies, the information must be more accessible. Such an adaptation has been made by Wirtz in his *Thursday Math Sampler* (1982b, p. 30) prior to asking first graders to discuss what they see.

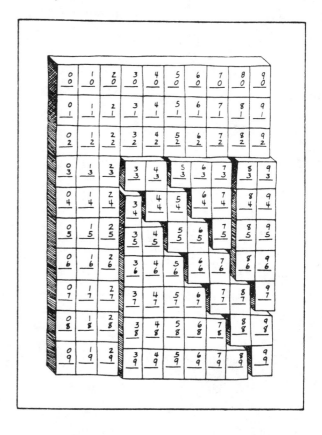

By not using a permanent operational sign on the chart, it can be used at different times to study either addition or multiplication combinations.

APPENDIX E: CATEGORIES OF NUMBER STORY PROBLEMS

(Riley, Greeno, and Heller, 1982)

Change

Result Unknown

1. Joe had three marbles.
   Then Tom gave him five more marbles.
   How many marbles does Joe have now?

2. Joe had eight marbles.
   Then he gave five marbles to Tom.
   How many marbles does Joe have now?

Change Unknown

3. Joe had three marbles.
   Then Tom gave him some more marbles.
   Now Joe has eight marbles.
   How many marbles did Tom give him?

4. Joe had eight marbles.
   Then he gave some marbles to Tom.
   Now Joe has three marbles.
   How many marbles did he give to Tom?

Start Unknown

5. Joe had some marbles.
   Then Tom gave him five more marbles.
   Now Joe has eight marbles.
   How many marbles did Joe have in the beginning?

6. Joe had some marbles.
   Then he gave five marbles to Tom.
   Now Joe has three marbles.
   How many marbles did Joe have in the beginning?

Equalizing

1. Joe has three marbles.
   Tom has eight marbles.
   What could Joe do to have as many marbles as Tom?
   (How many marbles does Joe need to have as many as Tom?)

2. Joe has eight marbles.
   Tom has three marbles.
   What could Joe do to have as many marbles as Tom?

Combine

Combined Value Unknown

1. Joe has three marbles.
   Tom has five marbles.
   How many marbles do they have altogether?

Subset Unknown

2. Joe and Tom have eight marbles altogether.
   Joe has three marbles.
   How many marbles does Tom have?

Compare

Difference Unknown

1. Joe has eight marbles.
   Tom has five marbles.
   How many marbles more than Tom does Joe have?

2. Joe has eight marbles.
   Tom has five marbles.
   How many marbles less than Joe does Tom have?

Compared Quantity Unknown

3. Joe has three marbles.
   Tom has five more marbles than Joe.
   How many marbles does Tom have?

4. Joe has eight marbles.
   Tom has five marbles less than Joe.
   How many marbles does Tom have?

Referent Unknown

5. Joe has eight marbles.
   He has five more marbles than Tom.
   How many marbles does Tom have?

6. Joe has three marbles.
   He has five marbles less than Tom.
   How many marbles does Tom have?

APPENDIX F: STARTING POINTS FOR INTERVIEWS ON OTHER AREAS OF NUMERICAL THINKING

The following collection of tasks is offered as a starting point for continuing your study of children's numerical thinking in areas of multiplication and division of whole numbers, as well as operations on fractions. You are encouraged to select appropriate tasks from the collection, to adapt them as needed, and to devise your own tasks.

MULTIPLICATION

A. The initial tasks test children's intuitive understanding of multiplication. Note that "times" and "multiplication" are not mentioned in the introduction to the tasks.

1. Show a series of dot patterns of the same kind by taping several dice together in a row.

    What do you see?

    How many dots are there altogether? You can use the paper and pencil if you want.

    Is there another (easier/faster) way you could have done it? (if needed)

2. Steven arranged his toy cars on the shelf in this way.

    ```
    o o o o o o o
    o o o o o o o
    o o o o o o o
    o o o o o o o
    ```

    How many toy cars did he have?
    How did you get your answer?
    Could you do it in another way?

What's the shortest way?

How would you write down everything you know about these cars?

3. Suppose you had fifteen cards like this one.

    How would you find out how many dots there were altogether?

    Let's see how you would do it on paper.

4. Let's see you work these out in the easiest way you can. (Paper and pencil are available.)

4444	2509
4444	2509
4444	2509
+ 4444	2509
	+ 2509

    (This task attempts to tap a child's understanding of multiplication as repeated addition.)

Is there a quicker way you could have worked that out? (as needed)

B. Typical language clues, like "times" or the operational sign, are now available.

1. Given a pool of blocks, the child is asked to show "three times four."

2. Given graph paper and a pencil, the child is asked to shade the small squares to show: "five times six" (spoken) or 5 × 6 (written).

3. Suppose you forgot what seven times eight was; show a way to figure it out.

4. Suzie has four boxes with six cookies in each box. How many cookies does Suzie have?

5. Write a number story problem for 8 × 3 (given the notation).

6.
```
  13
 ×12
```
Computation:
Let's see how you work this out on paper.

Explanation:
How would you explain what you just did to a second grader?
When you multiplied 1 × 1, what were you really multiplying?

Demonstration:
How would you show this multiplication with these (Dienes) blocks?

Connection:
When you multiplied 1 × 1 on paper, what were you really multiplying?

7.
```
  89
 ×47
```
Estimation:
Guess how big the answer will be. How did you decide?

Computation:
Let's see how you would work out the exact answer on paper.

Explanation/Probes (as needed):[1]
```
   89
  ×47
  623
  356
 4183
```
How would you explain it to a second grader so that he/she would understand?

When you were multiplying by 7 you said "carry 6."

Tell me what the 6 stands for.

Why didn't you carry the 3 instead of the 6?

When you're multiplying 8 × 7, what two numbers are you really multiplying?

How do you know that the 6 (in 356) belongs in this column under the 8, 4, and 2? Why isn't it lined up under the 9, 7, and 3?

Is there another way to write this (356)?

When you were multiplying 8 × 4, what two numbers were you really multiplying?

How come the 6 and 5 are lined up where they are?

How come we write the 3 (of 356) away over by itself?

Connection:
```
   89
  ×47
   63
  560
  360
 3200
 4183
```
Here's how another boy/girl worked this out.

Do you think that's a good way of doing it? How do you know?

What multiplication does the 63 stand for? 560? 360? 3200?

Show the 623 (original computation) in his/her work.

Show the 356 in his/her work. What does 356 really mean?

---

1. Ideas for probing understanding of long multiplication are taken from Brownell (1945).

DIVISION[2]

1. Jean had 12 cookies. She gave 3 cookies to each of her friends.

   How many of her friends got cookies?

2. Paul had 24 marbles that he gave away to 4 friends. Each friend received the same number of marbles.

   How many marbles did each friend get?

3. Irene has 30 pennies. She has 5 times as many pennies as Pat has.

   How many pennies does Pat have?

4. Donald has 5 marbles.

   Peter has 15 marbles.

   Peter has how many times as many marbles as Donald?

5. Write a number story problem for $9 \div 3$.

6. Estimation

   $13\overline{)317}$

   Guess how big the answer will be. How did you decide?

   Computation:

   Let's see you work it out on paper.

   $$\begin{array}{r} 24 \\ 13\overline{)317} \\ \underline{26} \\ 57 \\ \underline{52} \\ 5 \end{array}$$

   Explanation/Probes (as needed):

   If the child uses the shortcut algorithm shown, you might ask:

   What does the 3 in 317 mean?

   Why is 31 the first number you divided into?

   Do you actually divide 13 into 31 or into 31 of something?

   Why must the 2 in the answer (quotient) be placed where it is?

   How do you explain "bringing down 7"?

An alternative follow-up to the shortcut algorithm is presenting the child with a table of 13-combinations, shown at the bottom of the page.

Finish this table. What do you notice about the numbers in the table? Let's see how you would use this table to solve $13\overline{)317}$. (cover initial work)

What's the largest number of 13s you can take away? (if needed) (Devise your own follow-up.)

If the child uses an expanded algorithm as shown, ask:

   Tell me how you would explain it to a third grader who doesn't understand the method.

   When you take 260 away, what are you really taking away?

$$\begin{array}{r} 24^{r5} \\ 13\overline{)317} \\ 260 \ |\,20 \\ 57 \ | \\ 52 \ |\,4 \\ 5 \ | \end{array} \quad \text{or} \quad \begin{array}{r} 24^{r5} \\ 13\overline{)317} \\ 260 \\ 57 \\ 52 \\ 5 \end{array}$$

7. Here's how a boy/girl answered this division question.

   $$\begin{array}{r} 1\ 7 \\ 4\overline{)428} \\ \underline{-4} \\ 028 \\ \underline{28} \\ 0 \end{array}$$

   Is there anything wrong with the answer? How do you know?

   About how many fours are there in 428? Take a guess. (if needed)

   How would you work this out?

13	13	13	13	13	13	13	13	13
×1	×2	×3	×4	×5	×6	×7	×8	×9
13	26	39	52	65	78	91	104	117
13	13	13	13	13	13	13	13	13
×10	×20	×30	×40	×50	×60	×70	×80	×90

---

2. The number story problems are taken from Hendrickson (1983a), while portions of the long division tasks are found in Brownell (1945) and Wirtz (1974).

FRACTIONS[3]

The following tasks may require one or more kinds of materials:

- Outlines of shapes, like circle squares, rectangles, and triangles, drawn on paper.
- A pool of objects, such as Othello pieces or chips of two colors.
- Fraction kits, like strips of different-colored paper, each cut into a different number of equal-sized pieces.

1. Draw what you think a half (quarter, third, three quarters, and so on) look like.

   How do you know that's a third?

2. Use these chips to show one half, and so on.

   How did you decide?

3. If this piece is one, find the piece that's a half, and so on.

   How do you know?

4. Which of these fractions is bigger? $\frac{1}{3}$ or $\frac{1}{2}$

   (or other pair of fractions presented in written notation)

   How did you decide?

   How could you prove your answer to a second grader using these (selected) materials? (if needed)

5. Who gets the bigger piece of chocolate bar— Tyrone or Jimmy?
   How did you decide?

6. What fractions can you write for this picture?

   $\frac{2}{\Box}$ $\frac{4}{\Box}$ $\frac{8}{\Box}$

7. How would you solve this problem?

   $$\frac{4}{6} = \frac{?}{12}$$

   How would you explain it to a second grader?
   Prove your answer by drawing a picture or using these materials.

8. Is this right or wrong? $\frac{3}{5} = \frac{6}{8}$

   How do you know?

   How could you prove your answer?

9. Let's see how you would solve this problem (presented one at a time).

   $$\frac{2}{3} + \frac{1}{4} \qquad \frac{2}{3} - \frac{1}{6} \qquad \frac{3}{4} \times \frac{1}{2} \qquad \frac{3}{4} \div \frac{1}{2}$$

   How would you explain what you just did to a second grader?

   How could you prove your answer with materials? (if needed)

   How come the answer is smaller in the multiplication problem than in the division problem?

10. Draw a diagram to show how you would solve each of these story problems.

    The hiking trail to Bear Lake is $3\frac{1}{2}$ miles long.

    So far, you've walked $1\frac{3}{4}$ miles.

    How much farther do you need to walk?

    Two fifths of the children in a class of twenty-five are practicing for the concert.

    How many children from this class are in the concert?

    There was half of a pie left uneaten after lunch.
    Sally decided to eat a half of the remaining half.
    Then John ate half of what Sally had left.
    Who ate the biggest piece of pie, Sally, John, or neither one?
    What part of the pie was left after John ate his piece?

    Our class has sixteen boys and twelve girls.
    How many times as many boys as girls are there?

11. Write a number story problem for the following:

    $$\frac{1}{2} \times 2\frac{1}{4} \qquad 3 \div \frac{1}{4} \qquad 1\frac{1}{2} \div 3$$

---

3. Some of the fraction tasks are taken from the unpublished work of both Hendrickson and Kieren.

APPENDIX G:
SOME RESOURCES FOR
INFORMED CHANGE
IN THE CLASSROOM

Easley, Jack (Ed.). *Pedagogical Dialogs in Primary School Mathematics*, University of Illinois, Urbana, Bureau of Educational Research, August 1980.

Studying the teacher's view of the world as a starting point in "helping teachers work out their own evolution" was the goal of a research project conducted within the context of an individualized, on-site, in-service program. Various aspects of this program are discussed in the publication.

Howsam, Robert. "The Workplace: Does it Hamper Professionalism of Pedagogy?" *Phi Delta Kappan*, October 1980.

In this article, the essence of professionalism is defined. Teaching is compared to recognized professions such as medicine, law, and engineering.

Jervis, Kathe. "Teachers Learning from Teachers." *Learning*, August/September 1978, pp. 68–74.

This is the story of how groups of teachers meet voluntarily each week to discuss children's thinking in the context of everyday classroom occurrences.

McKibben, Michael, and Joyce, Bruce. "Psychological States and Staff Development." *Theory Into Practice*, vol. XIX, no. 14, pp. 248–255.

Both professional autonomy and resistance to its development are described in the context of a long-term in-service project offering training in alternative teaching strategies. The relation between teachers' needs/concerns (Maslow's scale) and the extent of their participation and resulting growth is discussed.

O'Brien, Thomas. "Diary of a Teachers' Center." *Mathematics Teaching*, September 1975, pp. 42–45.

A grass-roots operation is described in which a group of interested teachers meet voluntarily to improve what they are doing in the classroom from within a Piagetian framework. The center encourages teachers to be collaborative agents in the process of education rather than the last link in a long chain between publishers or curriculum reform groups and children.

Resek, Diane, and Rupley, William. "Combatting 'Mathophobia' with a Conceptual Approach Towards Mathematics." *Educational Studies in Mathematics* 11 (1980): pp. 423–41.

A description of a course for math-anxious adults in which they reconstruct their view of mathematics, is presented by the authors.

INDEX